337 KILLER VOICE PROCESSING APPLICATIONS

Second Edition

Covering Voice Processing, Audiotex, Call Centers, Fax On Demand, Interactive Voice Response, Outbound Dialing, and Messaging

by Edwin Margulies

A Telecom Library, Inc. Book
Published by Flatiron Publishing, Inc.

ISBN 1-57820-005-9

Manufactured in the United States of America

Second Edition, January 1997
Cover Design by Saul Roldan

Acknowledgments

This book was inspired by people in which I have a great sense of admiration. They are the pioneers of modern voice processing technology. They're the ones that first heard: "You'll never get my customers to call that *thing...*" But just as the early explorers, astronomers, scientists and philosophers of the 17th century were proven right over time, so were these radicals of today's communications industry. For example, Charles Szeto, who built one of the first voice digitization subsystems that gave birth to Voicetek. And there's Nick Zwick, former AMD Microdevices field engineer and now Chairman of the Board of Dialogic Corporation.

I also admire Walt Tetschner, President of Tern Systems, who with a handful of very bright engineers at DEC, fathered the original DEC-Talk text-to-speech system. Graeme Watson, an old neighbor and customer, wire-wrapped and hard-coded custom boards on PDP-based backplanes to launch the original DUNSVoice system. And Chris Bajorek, founder of Telephone Response Technologies, was programming low bit rate encoders fifteen years before Microsoft even dreamed of WAVE files and TAPI. In fact, it was industry notables such as these who were called-in by the software giant to help draft the specifications for TAPI. I don't think they'll be getting any royalties from Microsoft, but they are the ones to thank for the exciting opportunities we have today.

Consider that virtually every major financial institution now offers account information via interactive voice response. At first eschewed like the now prolific ATM, these systems converse with literally thousands of callers worldwide on a 24-hour basis. Before voice mail systems were first installed by the likes of VMX, Sperry, and Rolm, as many as 70% of all business phone calls were not completed.

Now, with voice processing, you can transfer to several extensions, leave messages and page employees all in one phone call. Voice processing and computer-based fax has also transformed the way we distribute collateral. Both big and small companies now provide brochures, data sheets, and technical tips via fax with simple touch-tone commands - transforming the common fax machine into a storehouse of visually based information. And now all these voice processing disciplines are supercharged with the Internet. We now use the Internet to transport voice and fax. We use it for call control. We use it for "mirrored Web-page screen pops." These technologies are having a radical effect on the modern landscape of communications.

The purpose of this book is to inspire, educate, and stimulate. Perhaps you'll see a few "killer apps" you can use in these pages. Maybe you'll create an application you see here to serve customers better. Maybe you're in sales or marketing and you need ideas on how to sell your product. And hopefully, you'll read this book because it's fun.

Lastly, if you've got a great idea and want to promote it, consider making an editorial contribution for the third edition. I'll be happy to take a look.

Edwin Margulies
New York City, February, 1997
ed_margulies@msn.com
www.computertelephony.com

FOREWORD

Who Should Read This Book And Why

by Harry Newton

This non-technical book will help you:

- Increase productivity at your company
- Provide superior customer service
- Let your customers buy things from you without waiting to speak to your people.
- Give customers many more reasons for dealing with your company
- Save your company thousands on long distance telephone charges
- Allow you to develop unique and profitable products for sale

You should read this book if you are the president of your company, if you are the MIS manager or the call center manager, or if you are in marketing, sales or administration. In short, anyone looking to please customers and save precious dollars.

Every one of customers wants information from your company - their account balance, when your next train leaves, where your closest store is, if you have the part they want in stock, etc. Every customer would like to be able to do business with your company 24-hours a day - buy things, put their name in for a catalog, move money around in the bank account, register for a course, etc. All your customers have a telephone. Your customers want to call your company - at any time of the day or night - and touch-tone what they want, fast and easy. They want you to have a simple, usable voice processing system. And they will increasingly seek out companies that do a better job providing those simple, usable systems.

Our studies show that those companies with successful voice processing systems have a larger percentage of their respective marketplaces versus those without this technology. There are powerful business reasons to get into voice processing today. Voice processing is no longer a luxury, nor a gimmick. It's an integral part of the way successful companies are *staying* successful.

The good news is that putting up a customer-useful voice processing system has become much simpler and far cheaper in the last couple of years. Voice processing is now heavily PC-based and increasingly Windows-based. Let's say you have an idea one morning to do a customer voice processing system - say one to read back how many items of a part you have in stock. You could, with today's tools, literally have the system up and running that afternoon, and answering customer calls by the following morning.

What's more, your customers would prefer using your voice processing machine because it would be faster, more accurate and wouldn't take bathroom breaks. Don't think of voice processing as only for customers. Governments use it to give out information to their constituents. (You could think of them as customers. And some government agencies actually do.) Many companies also use voice processing to give their employees information - e.g. their benefit plans. Many companies allow their employees to change plans, directly on the phone.

This book is for modern communications heroes. If you want your company to reap the benefits of voice processing, this book was made for you. Call center managers can gain insights into how other companies have "supercharged" their call centers with voice processing. Information services managers will see how their peers have used computer telephony to combine information stores from disparate computer systems. Marketing managers can use examples to boost sales and launch new products. If you're a software developer, this book can show you how your applications can be applied across many industries to maximize your development efforts. Voice processing technologies and the applications they enable are eclectic. With all the variety in Audiotex, messaging, interactive fax, and call center apps, it's hard to segment the apps so they can be analyzed.

The most intuitive approach is horizontal. Ed organizes this book in broad "horizontal chapters." After an overview of horizontal apps and basic technology, each chapter segments horizontal applications into industry vertical uses of the technology. The vertical applications are arranged in alphabetical order inside the "horizontal chapters."

The best way to use this book is to

1. Pick the horizontal app you are interested in and go to that chapter;
2. Look for the industry vertical inside that chapter that best fits your needs;
3. Jump to other chapters and look at other apps for your industry.

This third step is important because there is significant cross-over in the capabilities of these applications. For example, there are plenty of IVR (interactive voice response) applications that also incorporate messaging and fax. If you're a little hazy on the definition of horizontal applications, or the baseline capabilities of voice processing technologies, you should read chapters two and three for a comprehensive overview.

Make sure you see the list of resources in Appendix A. You'll find great books on other aspects of computer telephony, including Ed's *other* six books. There's also a large glossary of voice processing terms in Appendix B. And in Appendix C, Ed provides a list of the top industry players who provided editorial contributions for this book. If you want to keep up with this exploding field, the easiest way is to subscribe to *Computer Telephony* magazine. Call 1-800-LIBRARY or visit us on the World Wide Web: www.computertelephony.com.

Harry Newton
New York City, February, 1997
harrynewton@mcimail.com

Table of Contents

1

Opportunities In Voice Processing

2

Applications - A Horizontal Overview

3

Voice Processing Capabilities

4

Audiotex

5
Call Center

6

Computer-Based Fax

7 Interactive Voice Response

8

Messaging

9

Outbound Systems

Appendices

1

Opportunities In Voice Processing

The opportunities for end-users, systems integrators, and software developers in the field of voice processing are boundless. This year alone, the global market for voice processing applications will top $4.5 billion according to industry leaders such as Dialogic Corporation, makers of component-level voice processing technology.

We all encounter voice processing on a daily basis as we make telephone calls. The need for more automation and a demand for better customer service worldwide have fueled its popularity. For example, you experience the technology when you make long distance calls and are prompted to enter your calling card number. You encounter voice processing when you leave a message on a "voice mailbox." Voice processing aids you in retrieving bank balance and frequent flyer information as well. Everyone who uses a telephone is likely to encounter some form of automation every day when making service-related telephone calls.

The purpose of this book is to help you to discover how this technology can be put to work for you and your enterprise. If you're a call center manager, you can gain insights into how you can increase the efficiency of your agents and increase traffic handling capability. If you're an information services manager, you will see how you can use voice processing to tie together disparate information stores. Systems integrators and software developers will learn how horizontal applications of the technology can be applied across vertical industries in order to maximize development efforts.

In addition, telephone companies and enhanced service providers will be able to assess a variety of business-related solutions. These solutions can be put to use on a "shared tenant" basis in order to provide special services to increase customer loyalty and usage. With voice processing you can:

- Increase Productivity at your Company
- Provide superior customer service for your clients
- Save thousands of dollars on long distance charges
- Develop unique and profitable products

Capabilities

A variety of technology concepts and capabilities are represented in the discipline of voice processing. Voice processing uses application software control for computer-aided telephony. This definition does not take into account the stand-alone uses for voice processing such as assistive technologies or direct dictation, but rather concentrates on the "over-the-telephone / shared use" model.

This distinction is important inasmuch as this book concentrates on the productivity gains in processing multiple simultaneous transactions over the telephone network. You may wonder just how voice processing technology can help to add value to a telephone call. A comprehensive list of these capabilities is presented later, however, the most important concept is that callers interact with voice processing in order to *route* or *navigate* a call to its final conclusion. If the purpose of the transaction is to complete a long distance call, then that is the conclusion.

If the purpose of the call is to retrieve frequent flyer points from a database - when the point totals are spoken to the caller, then that is the conclusion. If the purpose of the transaction is to automatically switch a caller to a certain agent for a real-time conversation, then the conclusion is when the agent says “hello.”

Voice processing is also used to provide *information* to callers over the phone such as product information, weather reports, or news updates, for example. Voice processing technology can also be used as a “protocol converter” that transforms a caller’s telephone into a remote terminal, so that touch-tone input is converted into host computer *transactions*. These transactions may involve the transfer of cash from a checking account to a savings account, or the billing of a credit card, for example.

Voice processing can also be employed in a store-and-forward sense, allowing the human voice to be played, recorded, and sent to another virtual address. This enables the *distribution* of messages to a variety of recipients. These messages can be represented in a variety of media including facsimile, text, or digitized speech. By reading this book, you will discover the details of how voice processing helps to:

- Route and navigate telephone calls
- Enable information access
- Automate transactions
- Distribute media-independent messages

Applications

There are many ways to segment and explain the application of technology. For example, we could discuss the core algorithms, chip sets, boards, and platforms used for different applications of the technology. Another approach would be to explore the technology from a topology perspective.

For example, desktop-enabled, server-based, switch adjuncts, or large service bureau implementations of the technology. There's also a view from the end-user perspective arranged by vertical market. This describes the technology as it relates to the solving of a specific business problem. Lastly, there's the approach of horizontal application of the technology. Horizontal views take into account a general problem-solving capability that can be applied across different vertical market segments.

Perhaps the most intuitive approach is the horizontal one, so the book has been arranged in broad "horizontal chapters." After an overview of horizontal applications and basic technology, each chapter segments horizontal applications into vertical uses of the technology. The vertical applications are arranged in alphabetical order inside of the "horizontal chapters."

The application examples listed under each industry segment are by no means a comprehensive list. They are provided to stimulate ideas for how you can take advantage of the technology. Each example is presented through a description of a problem that needs to be solved, followed by an example of how voice processing can be a solution to the problem.

Some of the application examples may be familiar to you. These examples are both generic and used widely. In many cases a vendor who developed the application donated editorial material to the book.

Some applications of voice processing technology are patented, so before you dive into a development effort, it would be prudent to contact vendors and conduct other research to determine whether the payment of royalties or other licensing is in order.

Because voice processing technology is "in the middle" and works with other technologies, the last chapter concentrates on communications and computer industry players who supply technologies that are complimentary to voice processing.

These other technologies include PBXs (Private Branch eXchanges), ACDs (Automatic Call Distributors), fax servers, local area networks, application generators and computer telephony components to name a few. In addition, so-called legacy systems such as mainframe and mini-computer connections are often required to put a solution together. If you're stuck on a certain idea or design, most of the players' contact information is in the appendix.

The Market and Your Idea

The overall worldwide market for voice processing systems has grown steadily during the past decade. This year, the factory-shipped revenues representing these systems will top $4.5 Billion. This spells opportunity for those who wish to build and resell systems, because the hardware platform costs associated with each port are less than a third of the selling price. In other words, you can buy speech processing cards and the related computer gear, add your software and then sell the resulting system at a healthy profit. Based on a cumulative average growth rate of over 25%, the voice processing industry promises to be a high-growth market well into the turn of the century.

Add to this the piqued interest of new computer telephony entrants such as Novell and Microsoft, and the opportunities are attractive not only for software developers, but systems integrators as well. Of course, the premium you can demand for software depends on how sophisticated it is and how big of a problem it solves. Then there's the fact that some software, such as voice mail packages for example, are so commoditized, that they can be had for under $1,000 in some cases. You'll certainly sell voice mail packages if you develop them, however, the amount you will get for each copy will be tempered by the fact that "plain old voice mail" can be gotten from literally hundreds of sources. This is why so many developers are beginning to "niche" their packages to fit into certain vertical markets. For example, a voice messaging package that is specifically designed to aid answering service bureaus and doctors with alphanumeric pagers will be sold at a higher price than a generic voice mail package.

Every new technology market goes through initial periods of high cost, shakeout, commoditization, and nicheing. This is certainly true in the voice processing industry.

In the late seventies, most every voice processing system was based on customized mini computers or mainframe computer peripherals. The technology was bulky, inflexible, and an order of magnitude more expensive than it is today. For example, one popular voice mail system of ten years ago was based on a DEC PDP-11 mini computer with customized I/O cards. A 24-port system based on this hardware sold for over $100,000.

Today, you can buy several high-density voice processing cards, a line interface, and a properly configured PC to do the same job for just over $7,000. What with the proliferation of software development tools for voice processing, you can easily put together unique packages within a month or two without "hard coding" in the C programming language. Most voice processing component manufacturers supply DOS, UNIX, OS/2 and Windows drivers for their boards along with C libraries.

This makes the development of custom programs feasible even without tools. For example, Dialogic Corporation provides the source code to an "answering machine" demonstration at no charge with the sale of their popular voice processing cards. Literally hundreds of developers have taken this free demo and used it as the basis for sophisticated voice mail packages over the years.

As a developer, your choice of programming models is diverse. Visual Basic programmers can use products like VBVoice from Pronexus, or Visual Voice from Artisoft. High-level script languages built specifically for voice processing are also available.

There are also a wide variety of forms-based 4GLs (Fourth Generation Languages) that can cut months out of what would have been a C-based coding effort. Whatever model you are accustomed to, you can find just about any tool to fit your budget and the scope of work you have in mind.

But before you rush off to buy an armload of boards and voice processing development tools, please take a moment to think about a few more issues that need to be addressed: Dozens of new and exciting voice processing solutions are launched every month. Many new offerings are really basic packages that can be adapted for specific industry verticals.

In fact, some companies develop software that is fundamentally the same "across the board," yet it is packaged as an entire product line. Even if some of these offerings are mediocre, they may out-sell your own product if they are packaged and promoted well. That's why this book can be an important part of your development effort. Even if your company is well established and well funded, it is a good idea to closely align with your hardware and software vendors. You can take advantage of cooperative marketing efforts such as product inserts, special "package" promotions and lead sharing.

Most hardware component manufacturers receive hundreds of calls each month from end-users that want turnkey software. For example, an insurance company may want to locate an automated system for claims reporting or rate information. A retailer may call asking for an integrated system that works with an ACD for their call center.

Most of these inquiries are handed-off or referred to participants of the aforementioned co-marketing programs. Marketing managers who run these co-marketing programs are reluctant to give leads to participants with generic offerings. They favor vendors with turnkey packages. It may be a good idea to interview the manufacturers you will be doing business with and ask about the popularity of certain offerings.

You may not hear what you want to hear, but you are likely to get some pretty sound advice at no charge. Some of the "hot" applications today include unified messaging, fax broadcasting and Internet-enabled call centers. Examples of these are scattered throughout the book. Messaging applications still dominate the market with over 45% of all the ports installed. Messaging applications are those that provide basic voice store-and-forward functions.

Voice mail, unified messaging, answering services, voice logging and dictation fall into this category. This is followed by call center applications that include operator services, ACD adjuncts, and other call routing applications.

Although interactive fax represents less than 10% of all the ports installed, it is the horizontal segment that is growing the fastest, and it is the least crowded of all other segments from a competitive view. You may very well decide to go into a low-volume, yet high-growth segment without nicheing vertically. This could be profitable, especially if you package and price your products attractively.

There are a variety of voice processing industry research studies and reports that are available to help you with these decisions. For example, Tern Systems of Acton, MA, publishes regular studies on voice processing vendors, applications, and marketplace growth. The Enterprise Integration Group of Danville, CA just published a landmark study on IVR implementation. I wholeheartedly endorse it. These firms specialize in voice processing-related markets and can offer you reports and consulting services saving you months. It's simply not necessary to conduct primary research unless you have the staff and the time to do so.

In addition, there are fine industry journals such as *Computer Telephony* magazine and a number of trade shows with which to augment your research. No matter what your interest is, it's a good idea to spend a little time and money on these before launching a major development effort. Once you've got your idea hammered-out, market research done, and a merchandising strategy roughed-out, please speak with an industry consultant and your patent attorney. You'll be armed with critical information on cross-licensing issues, royalties and packaging strategies.

2

Applications - A Horizontal Overview

Audiotex & Information-On-Demand

Audiotex and Information-On-Demand applications dispense a variety of prerecorded or dynamic information to callers. Users can call into a system and listen to a simple recording as soon as the line is answered (passive Audiotex), or they can interact with the system using automatic speech recognition (ASR) capabilities or touch-tone keypad entry.

With Information-on-Demand systems, the information being heard can also be viewed. This is accomplished by using ancillary equipment along with the voice processing equipment. Information-on-Demand systems use voice, image, and ASCII transmission over the same telephone line.

The equipment dispensing the information has to be able to detect when a caller has the correct equipment to accept the visual data.

This is typified by systems equipped with Radish Communication Systems' VoiceView protocol. If no imaging equipment is sensed, then the call proceeds as a regular Audiotex or Interactive Voice Response transaction. In some cases the caller's phone number can be automatically transmitted to the Audiotex system, thus possibly pre-identifying what the content of the message should be.

In addition, you can use Text-to-Speech technology (Speech Synthesis) to verbalize the contents of large ASCII databases or text files to the listener. There's no end to Audiotex applications that provide for the basic retrieval of information. For example, free information dispensed by corporate sponsors, premium rate information provided by enhanced service providers. In addition, there's information dispensed by local, state and federal governments. Even hospitals and universities have been using Audiotex applications for decades.

At first, Audiotex systems were based on drum announcer technology or endless loop cassette tapes. These older systems were prone to failure and required a lot of maintenance. It wasn't until fifteen years ago that the technology to digitize the human voice was made available on inexpensive microcomputer platforms. With computer-based Audiotex systems, the content of messages or images can be "downloaded" automatically, thus saving much time in administering these systems.

In addition, most modern Audiotex systems allow the administrator to access the system via telephone to change messages or other data. This capability is also available as a standard feature on many voice messaging systems. Figure 2.1 shows how voice processing boards in a microcomputer can be connected directly to the central office switch over regular phone circuits. The Audiotex platform (in this case a PC) can also be connected to a LAN (Local Area Network) so other PCs can access the Audiotex system for administrative purposes. In addition, speech files and other critical information can be stored on the network file server.

Figure 2.1 - Audiotex System

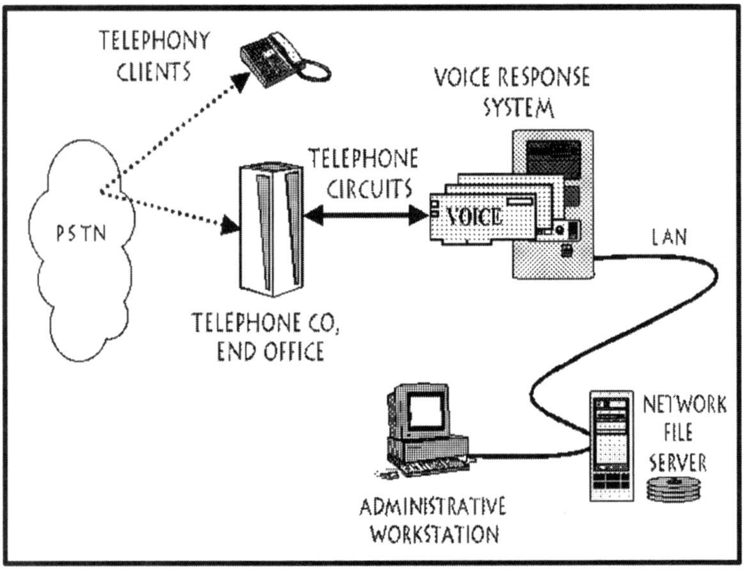

Audiotex systems are closely related to Interactive Voice Response, since host database and outside data links can be used to deliver information to callers. For example, an Audiotex system could dispense local weather information that is funneled into the system via data link from a weather station on the roof in the same building. For that matter, the system could get the same information from a weather service bureau, or by transcribing information from a news wire service.

Audiotex and Information-On-Demand applications can be located on the customer premise (as part of their regular communications gear), or in centralized location. This allows many “information providers” to use the common gear to deliver information to varied constituents. These larger systems are promoted as “gateways” or service bureaus, representing an industry onto itself.

Most simple applications, however, can be operated on-site at your own place of business. For systems requiring only a handful of simultaneous calls, all the software and hardware you need can be purchased for less than $2,000.

Call Center

Classic inbound telemarketing applications include ticket sales, airline reservation systems, and catalog sales. Operator services applications include collect call service, international call-back, and directory assistance, for example. Both inbound telemarketing and operator services applications have one thing in common: The use or augmentation of live operators or agents.

Most telemarketing solutions are found in CPE (Customer Premise Equipment) environments, and are associated with larger PBXs (Private Branch Exchanges) or ACDs (Automatic Call Distributors). It is typical to find the voice response unit (VRU) "sitting behind" the telephone switch, so that station lines attached to the VRU can mimic what live operators are saying and doing. There are a number of ways to integrate the VRUs and switches. For example, out-of-band data links, CTI (Computer-Telephone Integration) technology such as Novell's TSAPI, and proprietary telephone set emulation to name a few. Figure 2.2 shows how a VRU can work alongside a host computer or cluster controller as if it were an IBM 3278 terminal.

In this example, the VRU is using the cluster controller as a proxy for the host computer, so it can play-out information to callers, or get the caller to input information that will be used in a live operator session. The ACD is connected to the host computer either directly or through a data link with a CTI server.

Call center systems assist live operators in helping callers to "navigate" themselves through a telephone call. The call center aids callers in the purchase of a product, repair inquiry, or billing issue. In some cases, the product in question may be the telephone call itself. In this context, a call center can be a telephone company operator center, a computer manufacturer help desk, or an inbound telemarketing outfit such as those typified by airline reservation centers.

Figure 2.2 - Call Center Topology

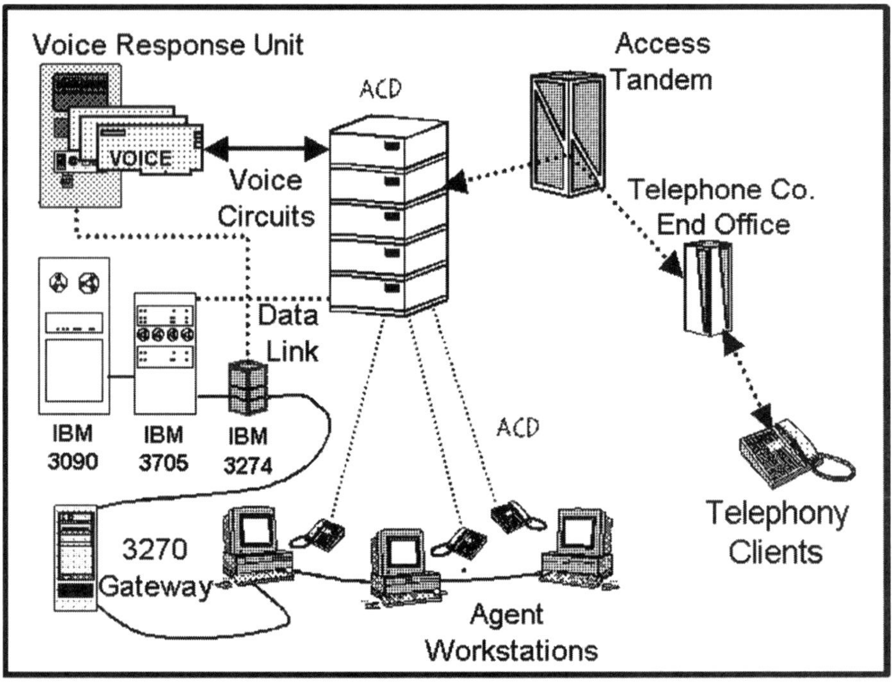

Operator services applications are used to partially automate calls. This is done by retrieving information before the caller is connected to the operator. Collect-call and third-party-billed applications can be completely automated, by recording the collect caller's voice and playing it back to the recipient. The called party may then accept or refuse the charge by pressing "1" or "2" (or any designated digit) or by saying "yes" or "no." This is the technology behind services such as "1-800-COLLECT" and "1-800-CALL-ATT," offered by MCI and AT&T, respectively.

Both carriers use voice processing equipment to greet callers, play instructions, collect called party digits, and to store and forward the recorded name of the collect caller. These capabilities were prototyped by RBOCs (Regional Bell Operating Companies) such as Ameritech in the mid-80's. Ameritech Labs prototyped the "FACCS" (Fully Automated Collect Call Service), a model for dozens of similar applications to this day.

Traditional telephone company call centers are models for smaller telemarketing outfits. These systems support thousands of callers a day, and are connected to a variety of databases and telephone switches. For example, local calls inside of a Regional Bell Operating Company (RBOC) service area are routed to a tandem switch from many lower-order local switching systems.

Automatic Number Identification (ANI) information is sometimes used to pre-identify the caller in order to speed service. These telephone calls are then "funneled" into an ACD, or specialized operator services switch such as a Northern Telecom DMS200, for example.

Here, the operator workstations are directly connected to the ACD with simultaneous access to Directory Services listings, toll information, and intercept announcement equipment. Each RBOC has as many as 3,000 live operators that provide long distance assistance, directory assistance, and emergency services. You can see how there's plenty of incentive to at least partially automate the task of routing and servicing callers.

One way telephone companies automate their call centers is by simply recording the operator's name or number for later play-back when connected to customers. This may sound ridiculous, but if you answered the phone a thousand times a day, you too would get tired of repeating your name. In addition, the actual information about a changed number is also recorded and then played-back by a voice response unit. This shaves as much as ten to fifteen seconds off of the "live" portion of each transaction. It's now pretty standard to hear: "The new number is..." from an automated voice when we call for directory service.

Some telephone companies further automate the process by giving you the option of instantly dialing the new number (referred to as *Call Completion).* Although some regard this as the height of laziness, it is nonetheless a revenue-generating service that cuts down on live operator intervention.

Interactive Fax

Interactive Fax, or Computer-Based Fax (CBF) made its entry in the mid-eighties. Fax machine capabilities were harnessed and put on PC expansion boards so microcomputers could be used to store, display, and transmit faxes. This enabled integrated PC-based fax communication, productivity gains, and a (sometimes) paperless means of document storage. CBF made it feasible to share fax resources on a large scale so it would no longer be necessary to equip hundreds of people with their own fax board. By developing special software that allows these fax resources to be shared, two types of multi-port systems have been developed. One is called the *fax server,* and the other is called interactive fax, which includes *fax-on-demand.*

CBF helps to automate inbound and outbound fax communications with the aid of special software for fax servers and fax-on-demand systems. You can implement the technology together with other productivity software on a local area network (LAN), or with a stand-alone system. CBF software can play an important role in upgrading customer service, increasing the productivity of your sales and customer service staff, or simply to save money on the cost of fulfillment services.

Network-Based Fax Servers

A network-based fax server is a communications device (typically a PC), which houses multiple fax cards or multi-port fax cards and associated fax communication software. Its purpose is to send, receive, and route facsimile communications on behalf of a given enterprise. With a fax server installed, each workstation on the LAN has the ability to create and send documents over the network for ultimate transmission by the fax server.

Most network-based fax servers have the ability to process multiple fax transactions simultaneously, because there are telephone lines associated with each installed fax port. In a sense, fax servers behave much like a shared printer on the LAN. In fact, most workstation-based fax server software makes faxing a document as easy as printing it.

In addition to acting as a shared resource on the LAN for sending faxes, a fax server also has the ability to route faxes that are received. Fax servers can be configured to route incoming faxes to the print queue so they are automatically printed. Alternately, incoming faxes can be electronically routed to administrative workstations, so they can be viewed and then forwarded to the intended recipient, thus avoiding a trip to the printer.

There are also emerging standards for achieving this administrative routing function automatically. This is can be done by instructing the sender to put the recipient's extension number on the cover page. This extension number can be understood by OCR (optical character recognition) software that enables the automatic routing. Another way of routing incoming faxes is the use of DID (Direct Inward Dial) telephone numbers or the new T.30 subaddressing standard. In the case of DID, the last four digits of the telephone number are repeated into the fax card by the telephone network. The last four digits are then used to pre-determine the intended party's extension number.

The sharing of fax resources was not viable until proprietary fax servers were installed seven years ago. Today, however, most network-based fax servers have the ability to interoperate with other devices and processes. There are a number of manufacturers including Alcom and Optus that provide fax server technology.

Fax-On-Demand Systems

Computer-based fax also led to the development of today's fax-on-demand systems. These interactive fax systems were popularized by players in the voice processing industry. Fax-on-demand is a voice response-based capability that allows callers to access documents via simple touch-tone commands. These documents are then automatically transmitted to the caller's fax machine. Fax-on-demand systems are similar to fax servers in that they are typically PC-based, housing multiple fax cards or multi-port fax cards. The primary differences between fax servers and fax-on-demand are:

1) fax-on-demand systems have the additional capability to carry on an interactive dialogue with callers; and 2) the intent of fax-on-demand is to fulfill document requests, rather than to route regular incoming and outgoing faxes.

Fax-on-demand, then, is a means of automating document fulfillment services. For example, a company's data sheets, technical notes, and newsletters can be electronically catalogued and made available to callers 24 hours a day via fax-on-demand. By adding this capability, telephone callers can have documents faxed to them by simply choosing from an audible menu they hear over the phone. The ability to incorporate interactive telephone access to mainstream LAN-based applications is a quantum leap for the communications industry.

Figure 2.3 - Interactive Fax Arrangement

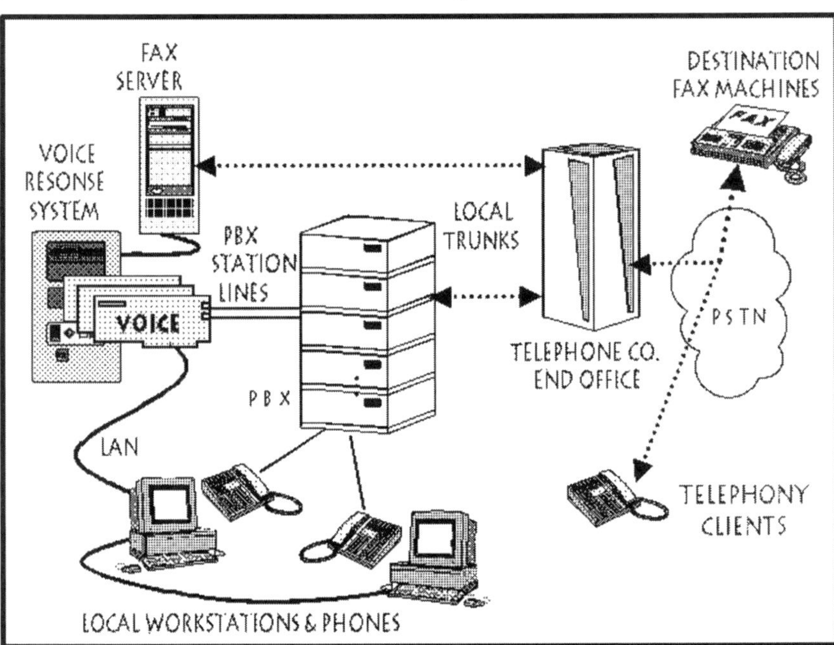

Many voice processing systems are beginning to allow for the transmission of facsimile images. This is true of messaging, IVR, and server-based solutions. Figure 2.3 shows a typical topology for interactive fax wherein a fax server and a voice response unit cooperate over a LAN.

Companies such as V*Channel, CommuniCorp S.A., Telephone Response Technologies, Inc., and Nuntius sell popular packages that support this configuration.

Enhanced Fax Networks

Telephone companies, service bureaus, and value-added network providers offer a variety of enhanced fax services. CBF applications are also used in AIN (Advanced Intelligent Network) environments. There are already a number of network-level fax gateways, including both Sprint and MCI offerings. Other network-based offerings include fax-on-demand, fax mailboxes, and fax-broadcasting to name a few.

A fax broadcast service enables a large number of fax recipients to receive a fax communication from one source. These services use special software to retry busy numbers, confirm receipt, and provide reports on the success of any broadcast. Most fax broadcast applications appeal to high-volume users, but even CPE (Customer Premise Equipment) versions of this broadcast capability are becoming popular.

Interactive Voice Response

Interactive Voice Response (IVR) applications use the telephone keypad as a remote data entry terminal. The voice processing element is used in as a "front end" to some kind of database. IVR systems prompt the callers for touch-tone or spoken input and use the data for host communication. The historical perspective of IVR is that voice processing was used to connect callers to an IBM mainframe computer or an asynchronous host. What with the popularity of computer system downsizing and client-server topologies, IVR has taken on a new meaning in that voice processing technology can be used to access any kind of database.

Figure 2.4 - Interactive Voice Response Topology

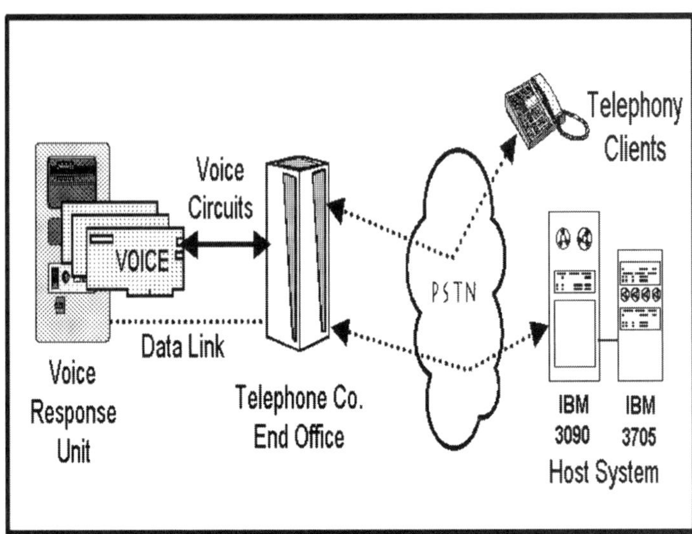

It's the "transaction processing" or "transaction passing" that really qualifies a system to be a true IVR system. This is true of self-hosted systems, ones that deal with RDBMS (Relational Database Management Systems), and of course "legacy systems" (mainframes). These IVR systems can act as front ends, and also batch information that is later uploaded to a mainframe computer or other computer system. Transaction passing systems use voice processing as the primary customer interface, but only "backwards engineer" or emulate what regular data entry terminals are doing for live operators. Figure 2.4 shows how a voice response unit is connected to callers on regular telephone lines and also to a mainframe via data link.

Some IVR systems use terminal emulation cards, while others use cluster controller cards to communicate with the host. Nowadays, the use of common xBASE files on a LAN is popular. Because of the ability to conform to file locking and file sharing conventions, IVR systems can interact with other processes on the LAN in order to "trade" transaction information. This is one means of coordinating "screen pops," so information about a certain caller is transferred to an agent's terminal at the same time the telephone call is transferred to his or her telephone.

Automatic Speech Recognition (ASR) technology is becoming increasingly popular as a means to navigate IVR applications. Dragon Systems and PureSpeech, among others, sell algorithms that enable ASR on PC expansion boards. These same algorithms can be made to run on the host processor, too.

Messaging

This horizontal application is the most popular of all other voice solutions. Messaging systems provide for the store and forward of "non-real time" communication. For example, a recorded voice message can be stored for later play back either locally or remotely, or a fax can be received and stored before it is re-transmitted to the ultimate recipient. Messages, then, can vary in content and media type - the distinction being that they are recorded or stored for pick-up in the future.

The time between original storage and retrieval of a message varies depending on the application. For example, a voice message can be created and stored by a sales manager for later retrieval by multiple (worldwide) sales people. The sales staff can listen to the message at different times over an extended period. This is due to the nature of random retrieval by the recipients in their respective time zones.

Messaging systems are a kind of "shared tenant" answering machine, because messages that were intended for as many as a thousand or more users can be stored and controlled by the same system. If a community of users agree on some basic ground rules, messages can be shared, forwarded, and distributed to multiple recipients in the same fashion as e-mail.

Individual users can be internal or external to an organization. Internal users may be in a meeting or away from their desk only to return to find several external messages and several internal messages waiting in their "voice mailbox." Traveling personnel may use their voice mailbox as a communication tether to the home office. For example, you can pick up and respond to messages regarding work orders, or messages from home.

Voice messages can also be mixed and matched with other media such as e-mail and faxes. This is referred to as "Unified Messaging." Mixed media messaging of this type is mostly dependent on access to a workstation that displays visual indications of the message type and timestamp. If the message is a voice message, the user can select the item with a keystroke or a mouse click.

Once selected, the content of the message is played over a speaker, or the phone may ring to forward the message from a centralized voice messaging system. Retrieval of text or fax messages work similarly, excepting, of course, that you cannot "hear" a fax unless its text is converted using OCR (Optical Character Recognition) and then played-out using speech synthesis technology.

Figure 2.5 - Customer Premise Messaging

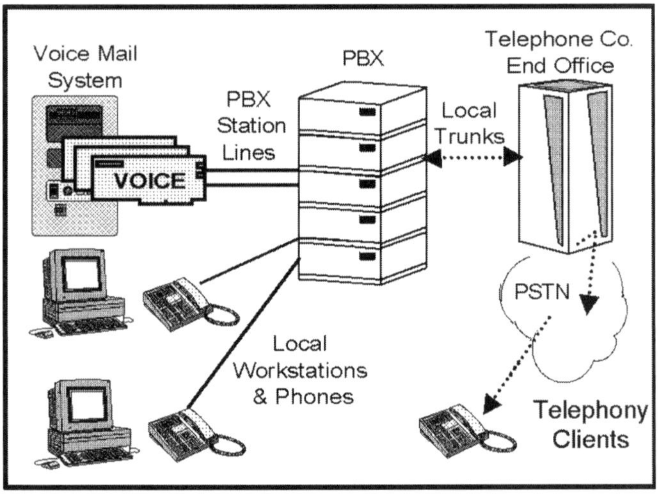

As you can see, the permutations and features of messaging applications can get pretty complex. Bear in mind that the basic job of messaging applications is to deliver information to recipients in a non-real time fashion. In this way, messaging systems can help to overcome time zone constraints, busy numbers, mis-communicated third-party dictation, and security. Figure 2.5 shows how a typical messaging system works in cooperation with a PBX.

The telephone lines attached to the voice processing cards are usually regular PBX extensions. This enables the voice messaging system to answer calls and then transfer callers to other extensions. This transfer capability is called "automated attendant."

Automated attendant capability is standard with most messaging systems. The ability to transfer calls from a messaging system can be used as a "call director," thus transforming a regular key system or PBX into a low-end ACD (Automatic Call Distributor).

Messaging applications are popular not only in CPE environments, but also in service bureau environments. Perhaps the most famous service bureau environments are those provided by the local telephone companies. RBOCs (Regional Bell Operating Companies) and larger independent telephone companies sell messaging services to business and residence subscribers as common fare. Although the actual use of voice messaging is still low (only about 1%), it is gaining in popularity over answering machines.

One explanation for the popularity in service bureau messaging is its growing acceptance amongst consumers. While no-one appreciates poorly executed menus and (longer than necessary) prompts, more and more users of the technology find comfort in being able to leave a personalized message for the called party.

Long distance companies and enhanced service providers use messaging as a stock-in-trade. You'll find messaging services as an option with paging companies, answering services, and even cellular carriers. If you're planning an enhanced service for your customers or prospects, you will probably incorporate some kind of messaging as part of the package.

Fortunately, there are plenty of hardware and software vendors that can help you in your endeavor. You can use turn-key offerings or development tools to put together very unique and productive messaging applications for your enterprise.

Outbound Systems

As the name implies, this horizontal application segment deals with *making* telephone calls. There are many types of outbound systems. For example, there's "personal productivity" dialers, power dialers, predictive dialers, voice & fax broadcasters, and confirmation/notification systems.

The more sophisticated systems (such as predictive dialers), combine elements from IVR, Messaging, and Call Center disciplines. This means that outbound systems can be the most complex and difficult systems to design. On the other hand, some outbound dialer systems use no store-and-forward of voice messages, but rely solely on the call progress detection capabilities of the voice processing subsystem.

Personal Productivity Dialers

Personal Productivity Dialers (PPDs) are commonly found as options with sales contact management systems. For example, Telemagic provides a PPD feature that uses a regular modem in an agent's PC to automatically dial clients with the touch of a button. Although the typical application for PPDs requires a dedicated modem in each PC, many new products are becoming available to enable PPDs on a shared use basis. Microsoft's TAPI, for example, can be used on each PC in tandem with an SPI (Service Provider Interface). The SPI may support a multi-port (shared use) voice response unit. This voice response unit can behave as a "modem pool" so individual agents can pass dialing commands to it simultaneously.

Power Dialers

Power dialers distinguish themselves from PPDs in that human intervention is not required until after the call(s) have been placed. In other words, PPDs are used on a demand basis, meaning that the calling person initiates the transaction.

Power dialers on the other hand, are programmed to initiate the transactions and then "push" the transaction to either a stored message or a live agent. Power dialers are also different from PPDs in that they use sophisticated call progress technology. For example, power dialers need to detect answering machines versus people answering the phone. Power dialers also need to distinguish between answering machines, modems, and fax machines. Figure 2.6 shows how a system of this nature works alongside a PBX and operator workstations.

Figure 2.6 - Outbound Dialing Example

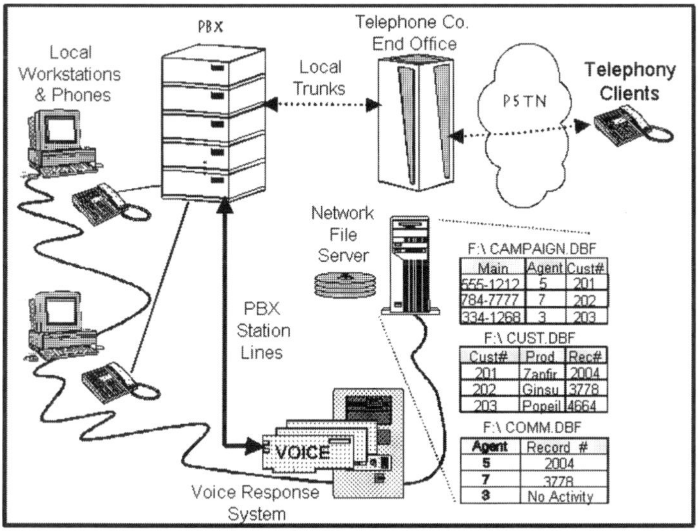

These devices either have their own stored database of telephone numbers to call, or they are linked to a common database on a mainframe computer or network file server. Less sophisticated systems simply dial from a predetermined list of sequential numbers.

Predictive Dialers

Predictive dialers are a represent the most complex of all outbound systems. What really sets a predictive dialer apart from all other outbound systems is the transaction with live agents.

These systems, then are used as productivity tools for call centers in order to reduce the time it takes for each agent to dial a number, wait for the network connection, and then speak to a person. Literally thousands of minutes a day can be saved in some call centers, so the pay-back in productivity is tremendous.

Pacing algorithms that queue-up calls for the agents are at the heart of all predicative dialer systems. These algorithms take into account the number of agents currently available, how many will be busy (based on the number of outbound calls in progress), the history of call completion rates, and the time of day.

As you can see, all of these parameters make for a very sophisticated algorithm, so unless this is going on - you're looking at a power dialer and not a predictive dialer. For this reason, true predictive dialing systems cost five to ten times as much as power dialers. Unfortunately, the names of these systems are used interchangeably so there is much confusion for buyers of these systems.

The other distinction between predictive dialers and less sophisticated power dialers is their ability to connect "live" calls to an agent with virtually no perceptible delay for the dialed party. In other words, if you are called by an automated system that is designed to make you believe you were dialed manually by a live person, and you figured out that it was a machine - you may hang up before the operator comes on the line. These hang-ups are called *call abandonment*. Therefore, the time it takes to detect an answer and then connect the caller to a live operator is critical (less than 1.5 seconds).

Voice & Fax Broadcasters

Voice & fax broadcasters deliver a common message to a predetermined group of recipients. Since no human intervention is required on the calling end, these systems do not have to use pacing algorithms or live operator transfers.

These systems typically deliver a short message to a targeted group of individuals. For example, you may hear the phone ring, pick it up and hear a message from a department store that is following-up on a maintenance contract. In some cases the system may ask you to input digits to indicate your interest in buying additional services or products.

Confirmation and Notification Systems

The intent of confirmation and notification systems is slightly different. For example, you may receive a call after having new phone service installed. The system may ask you what you thought about the installation, or even ask you to type a digit to confirm that the job was done. Notification systems alert a group of people about a certain event or call to action. This is true of emergency notification and dispatch systems - many of which are detailed in the following chapters.

3

Voice Processing Capabilities

There are a core set of functions you should be familiar with in order to fully appreciate what voice processing can do for your business. Depending on your application, you could use dozens of the core capabilities listed. The good news is that very robust software tools are available that allow your application to enlist these capabilities in a straightforward way. The makers of voice processing development tools have put a lot of work into abstracting the developer (and applications) from the underlying complexity.

For example, Dialogic Corporation provides drivers with common application interfaces for answering calls across a variety of disparate telephone network line interfaces. While on the surface, this does not seem like such a great feat, consider that there are over one hundred line interface protocols worldwide. The way a call is answered on the hardware level is different from country to country, so you should take into account the global intent of your application before beginning your design.

These considerations will have a great affect on the overall usability and marketability of your system. This chapter lists the core technological capabilities in alphabetical order. If you need a deeper explanation, check the glossary or contact the hardware vendor you are thinking about working with. Each vendor has a different way of supporting these capabilities, and the semantics can be confusing. Concentrate on the exact function or problem-solving feature when contacting vendors.

CALL ESTABLISHMENT AND DISCONNECTION

Answering Calls

The detection and answering of an incoming call is a science onto itself. When a call is presented to a voice processing system, the type of network interface that is connected determines how the call is answered. The physical interface my be a loop-start line, a DID (Direct Inward Dial) trunk, a proprietary PBX station line, or even a T-1 line. In addition to the physical interface, a specific protocol is used to negotiate the answering of a call. First, the "request for service" is transmitted over the telephone line. In the case of a regular loop start line (POTS - Plain Old Telephone Service), ringing voltage sent from the PBX or the central office is the request for service. In the case of most DID trunks, a battery reversal is the indication; and on T-1 trunks the state of signaling bits represent a request for service.

Once this "ringing" is detected by the voice processing equipment, the system either provides answer supervision by "going off hook," or a series of additional negotiation steps ensue. For example, a DID trunk could be configured for "wink start." This means that the voice processing system causes a momentary *off-hook then on-hook* event to happen. This "wink" event lets the central office know that its O.K. to proceed by sending DID digits. Once these digits are received and verified by the voice processing system, the call is established with another "wink." It is then that the call is considered answered. In the case of T-1, the same basic protocol is used, however, "A" and "B' bit transitions represent the winking.

Take, for example the situation of not wanting to answer a call because the caller is not authorized to use the service. Let's say that the service in question is a premium rate service like a "900" line for entertainment. If it can be determined ahead of time that the caller has not paid his or her current bill, then the service provider does not need to answer the call. The decision on whether or not to answer the call can be made by looking into a database of caller history or a database from a service bureau. As you can see, the matter of answering a call is not always so simple. The same case can be made for other "basic capabilities" of voice processing.

Call Screening

Voice processing can be used to approximate the function of a receptionist screening calls. Call screening takes advantage of either ANI (Automatic Number Identification) or voice store and forward capabilities. For example, if a caller wishes to reach you at your office and you are already on the phone, your voice mail system may be equipped to record the caller's voice and forward the recording so you can decide whether or not you want to take the call.

Once the calling party is identified, you can ignore the call or accept it. If you decide not to accept the call, the caller can leave a recorded message or transfer to another party. Figure 3.1 shows how a voice processing system can be connected to a PBX in the same fashion as a live receptionist. The VRU must emulate the function of the operator or receptionist by answering the call, asking for the caller's name, and then announcing this information to the intended party before transferring the call.

This is the same store-and-forward capability that fully automated collect call services use. The person making the collect call is asked to speak their name. The recording of the caller's name is then played back moments later to the recipient of the collect call. This is done to obtain the called party's permission to accept the call. In effect, the automated collect call service is a networked based call screening application.

Figure 3.1 - Call Screening

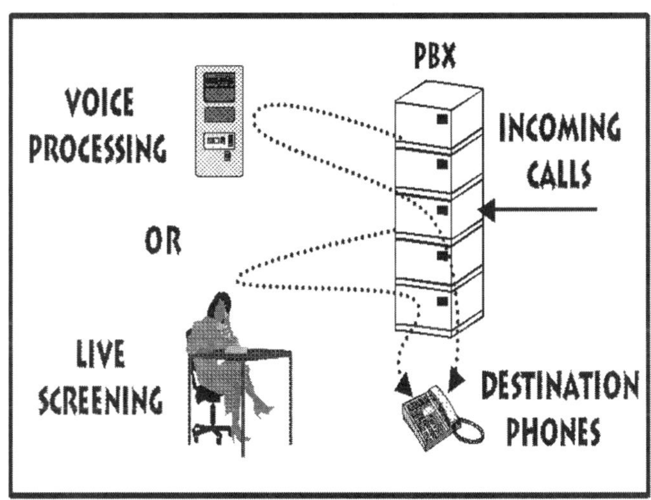

Dial Pacing

Pacing calls is a very special capability. It is important to draw a distinction between the simple placement of the calls (dialing the numbers) versus the logic that goes into when calls are made and how many to make at the same time. These decisions are made automatically by pacing algorithm software. In effect, the voice processing equipment is a slave to the pacing algorithm. The VRU dutifully makes the calls that are requested under the control of this software.

Pacing software is based on huge database records. These records include information on time of day, anticipated odds of the called party answering, the number of live agents that are available to take calls, the work schedule for the call center employees, and a number of other factors. Figure 3.2 shows how a database is connected to a host computer that controls a VRU for this purpose. It is possible for a voice processing system to pace calls based on a canned algorithm. For example, the software may instruct the VRU to dial 25 calls a minute no matter what. This may be suitable if the need for operator transfer is not critical, or if the information to be delivered does not require operator intervention. Keep in mind that the word "pacing" implies that some means are being employed for intelligent dialing.

Figure 3.2 - Automated Dialing

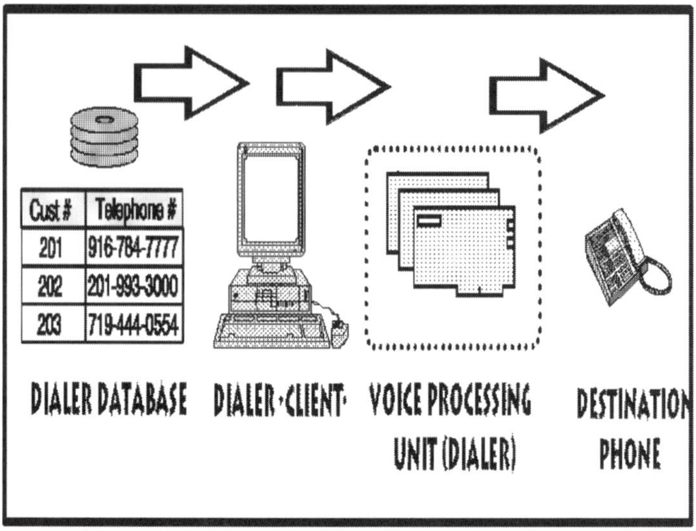

Hang Up

The completion of a call is an important event for voice processing systems. This is true because if a call is not properly "hung up," then the system may not quickly clean-up the last transaction and make ready for a new one. Hanging up the call implies not only the physical disconnection of the line, but also the event reporting of same to the application. If the caller presses a digit that indicates to the software that it's time to hang up, then the voice processing system can do a clean disconnect and get ready for the next call. The situation is not so cut-and-dried when the caller hangs up without "logging off."

For example, if the VRU is behind a PBX, the system may have to wait for a time-out period before the PBX sends a disconnect signal to indicate the end of a call. These disconnect signals can come in the form of battery interrupt, howler tone, or dial tone. In addition, the PBX may provide no signal at all, in which case the VRU has to use it's own time-out software to end the call. On the other hand, hanging up to complete a call can be pretty simple depending on the application.

For example, a voice processing application can be programmed to simply play a message of predetermined length when it is called - and then hang up immediately at the end of that message.

Make Calls

Making telephone calls is a sub-set of pacing the dialing of phone calls. There's and endless variety of examples of how and why a system would make a call. For example, a voice messaging system may need to make a call to alert you of a message just received. If you have a beeper with paging service, the number that the system dials is your pager number. The pager telephone number is stored in a database associated with your subscriber record. The number could alternately be your home phone number or your car phone.

Phone calls can also be made as part of an automated routine. This is the case for broadcast systems and power dialers. A list of phone numbers in a database can be loaded into the voice processing system and then dialed based on a predetermined set of timers.

Making calls is also a part of message waiting light illumination. For example, some PBXs can be accessed and then sent a series of tones to light or extinguish message waiting lights on phones. Calls into PBXs are also made to route and transfer calls. The physical means to make these calls is diverse. For example, the voice processing equipment may cause a phone line to go off-hook and then wait to "hear" dial tone before pulsing the touch-tone digits to make the call. In some cases, there is no dial tone. This is true of proprietary telephone set emulation, digital circuits, and E&M trunks.

We take for granted that a dial tone means that it's O.K. to proceed with dialing, but voice processing systems must also wait long enough to confirm that the sound is indeed dial tone. Some systems are programmed to wait several seconds in order to sample the tone to ensure it's O.K. to dial.

Less sophisticated systems do a "blind dial," by simply pulsing the digits as soon as the line is off hook.

ROUTING

Automatic Transfer

Transfer of calls implies that the call was already answered and that it needs to go somewhere else. To make the call go somewhere else, the voice processing equipment can conference or "patch" the callers to a separate line that is dialed-out on, or execute a hookflash followed by dialed digits (as in what we do on regular phones to transfer someone). In other cases, the voice processing system can send an out-of-band signal over a serial port. This out-of-band method is used in CTI (Computer-Telephone Integration) schemes.

Transfers can be unsupervised or supervised. In the case of an unsupervised transfer, the call is switched somewhere else, and the voice processing system is no longer part of the transaction.

In the case of supervised transfer (conference, patch, or broker call), the voice processing system may have the capability to taking the call back to complete subsequent transactions. If the transfer is supervised, the voice processing system can also keep track of what happens to the call for billing or logging purposes. This billing or logging may still be possible if a data link is established between the voice processing system and the system that the call was transferred to.

Call Forwarding

Forwarding is a transfer that occurs as a result of predetermined criterion. Call forwarding does not typically occur as a result of caller interaction, except in the case of ANI (Automatic Number Identification), DNIS (Dialed Number Identification Service), or DID (Direct Inward Dial).

Even then, it could be argued that the caller did not consciously interact with the system. ANI digits may be collected, matched to a database record, and then be used as a basis for a call forwarding action. In addition, DNIS or DID digits may identify the purpose of the call, and thus pre-identify the destination of the call.

For example, a help desk system may forward calls to paging systems for one type of product, or to a live person for another type of product. The last four digits of the number dialed could represent the type of product. If you dialed 1-800-FOR-GOLD, you could be forwarded to a different number than 1-800-FOR-PLAT. In this example, GOLD could represent "gold" service, and PLAT could represent "platinum" service. The physical telephone circuits that terminate in the voice processing equipment may be the same circuits, the only distinction is the last four digits of the dialed number.

Call Hold

There are times when it may be necessary to put a caller on "hold." Hold status provides for the maintenance of the circuit itself without disconnecting the call. This holding capability could be achieved by transferring the caller into a PBX or ACD call queue, or by simply idling the line. This is done while the voice response system is waiting to find an available agent to transfer the caller to, or to "buy time" while the system checks a database or retrieves information from a host computer. It is important to use this so-called "holding" strategically, so as not to alienate the caller thereby causing them to hang up. For this reason, some voice processing systems will play-out a recorded message, provide music, or transmit "courtesy tones" (short audible blips) to let callers know they have not been hung up on. Since most voice processing systems are designed to avoid call abandonment, the use of holding should be applied sparingly.

Called Number Identification

Incoming calls can be automatically routed to specific individuals or programs based on the intended purpose of the call.

For example, DID (Direct Inward Dial) numbers can be used so that the last four digits of the number dialed can pre-identify a called party's voice mailbox. If the called party is not at his or her desk, the voice processing system can be programmed to automatically greet the caller without the phone ringing for a long period of time. In the case of fax transmission, a DID number could also be used to automatically route the fax to a person's PC or printer.

A long distance network version of DID, called Dialed Number Identification Service (DNIS) can be used in the same fashion. DNIS is used extensively in call center environments to pre-identify products that are advertised in the print or electronic media. For example, 1-800-FLOWERS may route callers to agents who can take orders for roses, whereas 1-800-FOR-CARS might route callers to an agent familiar with car care products. The use of DNIS allows a group of call center employees to answer calls for more than one product in this fashion. In fact, one agent may answer calls for a product in one call, and then take a reservation on the next call.

These "multi-use" agents are typically linked to a host computer and database. The terminal screens are populated with data that coincides with the DNIS number. Used in conjunction with ANI, DNIS can drastically reduce the time it takes to complete a transaction, thus increasing the overall satisfaction of the customer and the performance of the call center agents.

Caller Initiated Transfer

Callers can cause a voice processing system to transfer calls so they can "self navigate" the destination of the call. This is true of automated attendant systems and most voice messaging systems. For example, you can call a messaging system and leave a message for someone who you know is not in, and then ask the system to transfer you to someone else. You can ask the person you are having the live conversation with to then connect you back to the messaging system to leave additional messages or to attempt other transfers.

In addition, long distance carriers provide the same type of transfer capability with "chain calling." Chain calling allows long distance users to press a digit (usually the # sign) to get a "second" dial tone in order to place another call. What is really happening is that the long distance switch is disconnecting the first called party and then reconnecting (transferring) the caller to a second or third destination.

Call Waiting

This is a popular feature offered by most telephone companies to their residential subscribers. With call waiting, a short audible tone alerts the subscriber that a second phone call is coming in during an existing conversation. The subscriber can either ignore the tone or may "flash" to take the second call. The flash is a momentary on-hook state caused by depressing the hookswitch on the phone or by pressing a special button. After the call waiting feature has been used, the first party is put on hold. By alternately "flashing," the subscriber can go back and forth between the phone calls. This feature is mutually exclusive with conferencing, since the same "flash" function would cause all parties to be connected in that case.

With voice processing systems, second calls can also be taken in the same fashion, since the technology has the ability to detect special tones and to change the hook state of the phone line. For some call center applications, the voice processing system itself can generate call waiting tones and then patch the tone into an existing conversation between an agent and a caller. In this case, the voice processing system has control over the agent line, the caller's line, and the line that the tone is being played-out on. By using microprocessor controlled switches, the same software that controls the voice store and forward function can also be used to put the first caller on hold and then connect the second caller to the agent.

Conferencing

Voice processing is used in conference scenarios to the extent that touch-tones or other commands can be used to enter or exit a conference.

The conference is achieved by patching together two or more telephone lines using a matrix switch or by summing all conversations using conferencing chips or DSPs (Digital Signal Processors). Some systems use "tone clamping" to mute the tones entered by callers so they are not heard by other members of the conference. Conferencing gear can be separate from the voice processing gear, so the voice processing system is a "front end" to the conferencing system. In this case, the voice processing system will transfer calls into the conferencing system. In still other scenarios, all of the technologies required to answer the phone, collect digits, and add and delete people from conferences can be housed in the same platform and controlled by the same software.

Queuing

This is a special kind of call hold that puts callers in a sequential or priority order for ultimate answering by a live agent or alternate system. Callers who have an emergency may be "jumped" in the queue, and thus preempt callers who actually logged-on earlier.

Callers who may demand special service, or who pay for premium service may also be "jumped" in the queue. A typical example is the way you are put on hold when you call for bus or airline reservations. Some of the more modern ACDs (Automatic Call Distributors) have the capability of predicting how long a caller will have to wait. In this case, a message stating the length of the waiting time is played out to the caller.

For callers who do not want to stay in the queue for whatever reason, they may be given the option of leaving a message so they can be called back, or attempt to complete their transaction by using an IVR (Interactive Voice Response) application.

Supervised Transfer

As with the example of long distance carrier "chain calling," the idea of a supervised transfer is to provide options to the caller after the disposition of the called party has been determined.

In other words, supervised call transfers are necessary when the application is designed to retain supervision of the call. This means that a caller can be transferred to another system or telephone extension without the VRU dropping out of the call.

This is especially important with some messaging systems that have automated attendant capability. In this case, the VRU will momentarily put the caller on "hold" while determining if the called party is available. This takes into account the call progress tones that are encountered when calling the destination phone. For example, if the voice processing system detects a busy tone, it will take the caller off of hold and indicate that the extension is busy. The caller is typically offered a chance to try again in a few moments, or to leave a message for the called party.

PROVIDING INFORMATION

Broadcast Fax

There are many ways to distribute information to a large audience. There's conference calls, television, radio, print media, voice messaging, and e-mail to name a few. With fax broadcasting, a predetermined list of recipients receives one or more pages of information directly via their fax machine.

For example, special clipping services can design customized faxes appealing to a sub-group of readers. Medical alerts can be faxed to hospitals to advise of epidemics or other hazards. Traffic updates complete with maps can be automatically broadcast to city traffic engineers and public safety officials. Travel clubs can fax trip planners or ads to a large group of customers. There's much controversy over "junk faxing" as a result of this capability. Some people take offense at receiving unsolicited information via fax and view it as an electronic form of junk mail. There will always be abuses of automated technology, but there are hundreds and maybe thousands of legitimate uses for "bulk faxes."

Broadcast Message

Message broadcasting works much the same way as fax broadcasting. The only difference between the two is the media required (a recorded voice message rather than a fax). In the case of a voice messaging system, a message can be created by one individual and then "sent' to a pre-defined list of recipients. These recipients are typically users of the same voice messaging system, so this would be a broadcast to a "closed community."

Most messaging systems have the capability of broadcasting a message to every user simultaneously. The system administrator may record these messages on an ad hoc basis, or they may be canned system messages. Canned system messages include warnings such as: "The system storage capacity is almost full - please pick up and delete your messages," or "the system will be taken out of service this Thursday at 1:00 PM for maintenance and upgrade - please remove all of your messages before that time."

Other broadcasts may be general announcements or even automated solicitations that are received by a large closed community (such as existing customers of a department store). The message could also be a solicitation or survey to an unqualified or random list of phone subscribers.

Forward Forwarding

Fax forwarding is a handy feature for people who travel often and for people who are concerned about confidentiality of messages. With fax forwarding capability, a voice processing system can alert users that they have a fax stored and give them options on how to retrieve the fax. For example, the voice processing system may ask the caller to enter the fax phone number where they would like to have the fax delivered, or ask if they are calling from a fax machine.

Since a number of voice processing manufacturers have combined fax and voice capability, it is possible to retrieve faxes on the same telephone transaction that was initiated as a voice call. This is sometimes referred to as a "one call" capability.

This enables users to wait until they are at a secure fax machine to retrieve their faxes, or allows them to retrieve faxes from any location. This gets around the problem of having to have colleagues intercept faxes at the home office only to have to re-transmit them to the "new" location.

Menu Choices

Voice processing systems are perhaps most known for the sometimes humorous and sometimes frustrating menu choices presented to callers. A menu choice is a list of options explained with a prerecorded prompt or system message.

A typical menu choice for a voice messaging and automated attendant system that is often heard is: "At any time, you can dial the person's extension number or press 1 for sales, 2 for service, or 0 for the operator." A fax-on-demand system menu might say: "Thank you for calling to learn more about our products. Press 1 if you are calling from a fax machine or 2 if you a calling from a regular phone..."

Although these verbal menu choices are simple recorded prompts, they can represent not-so-simple logic steps in the voice processing program itself. For example, if you are asked to enter your PIN number to check your checking account balance, the voice processing system must establish a link with the bank's host computer. The VRU then submits a request for the balance information, retrieves the request, parses the digits from the host response and lastly converts and plays-out those digits to you. What you're hearing sounds real simple, but what is happening underneath these menu choices is very complex.

Herein lies the beauty in well-designed voice processing systems. The goal of your system design should be to abstract the caller from the intricacies of the transaction and make it behave and sound very simple. This is the very basis for why voice processing has become so popular - the fact that your company's information stores can be turned inside-out and made available to users via the most prolific "terminal" in the world - the telephone.

Message Forwarding

The forwarding of messages is a basic voice processing function. In the case of a messaging system, a message can be listened to by one party, and then comments can be attached to it by a second party and then handed-off to yet another party. This is especially helpful if the parties concerned are not in the same physical location. For example, a customer could call a sales person and leave a message about how to use or a product that was purchased. The sales person could listen to that message, add a comment onto the message such as the customer's phone number, and then forward the request to a technical support person. In this fashion, the content, inflection, and general meaning of the message is not altered in any way. In addition, the message is difficult to "lose."

Message forwarding is also used in payphone communications. In this case, you may call someone from a payphone before boarding an airplane to indicate that your flight is going to be late. If the phone is busy, or there is no answer, you may be given the option of recording a message. This recording is then delivered to the same number you dialed. Some special message forwarding services will attempt to deliver the message over several hours.

Verbal greeting cards are another use of message forwarding. With a verbal greeting card, you can record and post a message for future delivery to someone in another time zone. For example, you could record a message you want someone to receive that's 12 hours ahead of you when you leave work. The system will then call the intended party after you are asleep to deliver the message. This has a benefit in delivering important messages when you are not able to do it manually.

Message Notification

There are a number of ways voice processing systems can notify you of a message. Perhaps the easiest and most obvious means is for the VRU to make a telephone call to your current location and deliver the message to you after you type in your password.

This presumes that the system in question is programmed with information regarding your whereabouts. Some systems use a subscriber database that stores alternate phone numbers. These alternate numbers are often based on your schedule and on the time of day and day of week. This is helpful in the dispatch of service personnel.

Message notification can also be achieved by calling into a paging system to activate your beeper. Depending on the system in use, you could get a special code on an alphanumeric beeper indicating the nature of the message, or you could get some of your message played-back on beepers equipped with miniature speakers.

You can also be notified of messages with a message waiting light on your phone, or a "gallop" tone on your phone. Message waiting lights can be turned on by a voice processing machine in a number of ways. For example, the voice processing device can make a call into the switch and send a string of in-band tones to turn on a light. The VRU can also send a message over a proprietary feature phone link, or a serial data link to the PBX. In other cases, the VRU can write message status information into a common database. Other programs can then use this database to alert message recipients on their PCs that they have a message. In this case, message indications can come in the form of a "pop-up" screen or an icon embedded in an e-mail list box.

Play-Back

Although it may seem obvious, the ability for a voice processing system to play back messages is critical. Play-back assumes that a recording has already been made, either by the developers of the system, or by users during run-time operation. For example, the prompts or messages associated with menu choice instructions are usually recorded professionally and then transferred onto the voice processing system in the form of numbered prompts. The prompts are linked to the logic flow of the application so that the "hello" prompt, for example, is automatically played-back upon answering a call. In the same vein, the "good-bye" prompt is played at the end of the call, and so on.

In addition to system prompts, the recordings that callers make during the course of the voice processing transaction can also be played-back to the caller. It is typical to encounter this feature with voice messaging systems.

The purpose of the play-back feature is to allow callers to review their messages. Users can both delete the message and try again or post the message for delivery. Recipients of the message are also users of the play-back capability to the extent that they must retrieve the messages from their mailbox and review them. In addition, the same messages can be forwarded to other system users with comments attached (forwarding a message) for later play-back. Recorded messages are not the only type of play-back, however. It is also common for prompts representing entered digits to be played-back. This is useful in confirming a caller's input before proceeding with certain transactions. For example, a fax-on-demand system may prompt the caller to enter his or her fax phone number and then "repeat" the number back to the caller to make sure it was entered correctly.

Send Fax

Voice processing systems are commonly used to initiate fax communications. There are many examples of this throughout the book. For example, a voice processing system can greet callers by saying: "Please enter your user number and PIN code... Thank you... You have one fax and three voice messages waiting. Please enter the fax number where you want your fax to be sent..." In this case, the messaging system has the capability of forwarding a stored fax for later retrieval by a traveling user. The system will collect the digits input by the caller, command the fax board to go off-hook to make a call, and then identify the location of the fax file and download the file onto the fax card for transmission.

Broadcast faxes are initiated in the same way. In this case, campaigns are initiated by a system administrator from a computer console versus caller input. In the case of a broadcast, the same fax file is being sent, however, the recipient information will change on the cover page as the calls are being made.

For example, a person's name, extension number, or company name will be inserted as a unique item on each fax. The basic content of the fax would typically be the same - just the phone number dialed and the cover page information is different.

A recent advance in this technology is attributed to Copia International, Ltd. of Wheaton, IL. Their fax broadcast system does a mail merge with Microsoft Word documents in order to personalize the letter portion of the fax as well as the cover sheet.

Speech Synthesis

Text-to-speech (TTS) technology allows an alternate spoken method for conveying textual information. As the name implies, text-to-speech converts ASCII text into the spoken word. Instead of a digitized recording, however, a synthesized voice is used in order to speak-out words and phrases on-the-fly.

Applications for this technology include order-entry systems (providing verbal confirmation on hundreds of products), customer name and address systems (access to thousands of addresses and names for locator capabilities), and telephone access to e-mail messages. In addition, TTS can be used to gain access to news information for stock quotes so brokerage firms can give callers access to corporate news headlines along with stock quotes. Newspaper access for the blind is another use for TTS technology.

Services based on TTS technology allow customers and prospects to interact with document databases over the phone. Upon greeting the caller, the system responds to touch-tone or voice input from the caller regarding the data requested. For example, brokerage clients could enter their account number to hear certain stock issue information. This information is dynamically assembled and spoken to the caller based on the stocks in their portfolio.

Speed Control

This capability is useful when playing back recorded messages. Message play-back speed on most voice processing systems is controlled via touch tone digit entry. In the case of a dictation system, a foot pedal can be used. With this function, the user has the ability to adjust the speed, both faster and slower, at any time during a playback sequence. With modern implementations of speed control, the pitch will remain constant throughout any changes in speed. The corrected pitch of a speed controlled message avoids a "Minnie Mouse" effect associated with simply accelerating the play-back. With pitch-corrected playback, extraneous data that would manifest itself in the "Minnie Mouse" effect is sampled and discarded during play-back.

It is important to plan carefully what DTMF (Dual Tone Multifrequency or Touch-Tone) digits will be used to modify the speed during play-back. This is because you can "run out of buttons" if you're not careful while building the application. Figure 3.3 shows how a touch-tone pad can be laid-out for speed and volume control.

Figure 3.3 - Example Speed Control Key Assignments

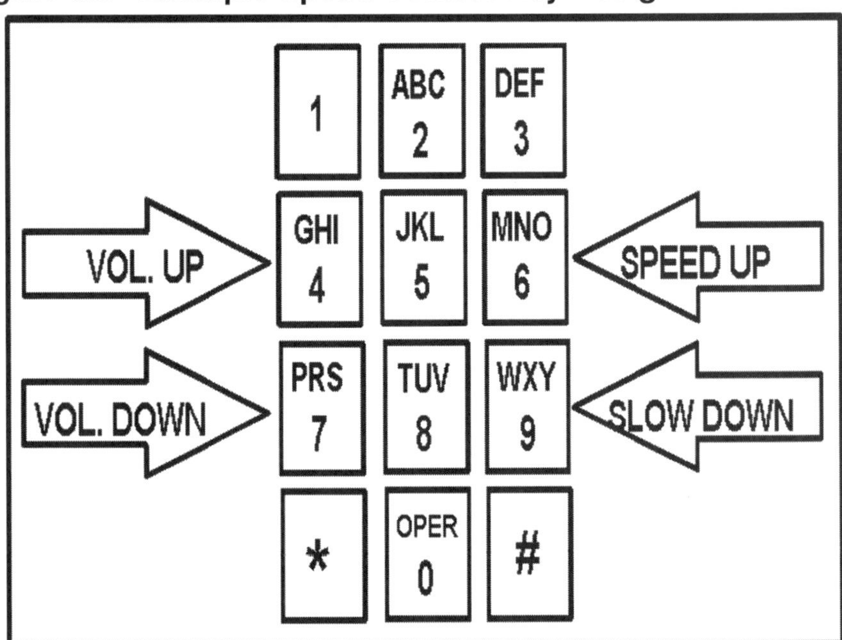

Volume Control

Although most voice processing systems do not use this feature, the control of message play-back volume can be important depending on the user and the location the user is calling from. For users who are hard of hearing, volume control can be a real lifesaver. In addition, callers in busy airports, bus depots, or those calling from noisy payphones find this feature to be quite useful. With this function, the user has the ability to adjust the volume, both louder and softer, by simply using a touch-tone digit during a play-back sequence. Application considerations for this capability include what amount, in dB, to adjust the volume when a certain digit is pressed. As with speed control, it's important to map-out a strategy for what buttons to use, so there are not conflicts with other features.

RECEIVING INFORMATION

ANI

Storage Automatic Number Identification (ANI) or Caller ID services provide the voice processing system with the telephone number of the calling party. This data can be used to route, terminate, or otherwise treat the call. In addition to the treatment of the call during the transaction itself, the ANI data can be "logged" or stored in a database for later use.

Let's say, for example, that you have designed an application to poll a constituency on their opinion of a certain subject. Without ANI storage, it would be ease to invalidate the results of the poll by counting multiple calls from the same individual. If, on the other hand, the application stored ANI, you could discard the additional call counts based on the original call.

There are plenty of other examples of how ANI storage can aid in the automation of telephone-based transactions. Perhaps one of the most popular concepts is the association of ANI with a customer record.

In call center environments, fast ANI lookups can significantly increase the speed and accuracy with which a customer service agent handles transactions. For example, when a customer calls in, the voice processing system can collect the ANI, and then execute a search in the customer database on an index of the ANI digits. If a match is found, the customer record associated with the ANI digits is sent to the workstation or terminal of the next available agent. This saves time in asking the customer for their name and decreases the odds of mis-keying the information.

Collect Digits

Digits can be collected by voice processing systems either by getting them one-by-one with caller touch-tone input, or by collecting them from other transmission equipment in the telephone network. For example, an Audiotex system may greet a caller and ask him or her to touch-tone a "1" or a "2" to hear sports scores or weather. Digits are buffered by the voice processing system based on logic flow instructions.

The application may wait for 4 digits before it processes the input in the case of a four-digit extension number on an automated attendant. For simple "3-choice" menus, the system maybe programmed to process the digits as soon as each digit is collected.

These digits can be detected and collected from DID (Direct Inward Dial) trunks or "TIE" trunks from a central office switch or PBX (Private Branch eXchange). DID digits or dial-repeated digits from TIE trunks can be used to pre-identify the extension numbers to route calls to, or what person's voice greeting to play in the case of a voice mail system.

The collection of digits is also used when the voice processing system is receiving a "networked" message from another voice processing system. For example, one voice mail system may forward a stored message to another (destination) voice mail system as part of an analog networking scheme. In this case, thc digits are sent to the receiving machine to indicate that:

1) the transaction was a "networked message" call; 2) that the message was intended for a certain mailbox; and 3) that the message was done playing. The collection of digits by one voice processing platform from another is also used in testing or "bulk call generator" systems. These special systems are designed to call other systems (the ones being tested) and then to mimic the digit input of regular callers. In this case, the system that is being tested collects the digits in the same fashion as with a regular call.

Cut-Through

Voice cut-through describes a feature wherein a caller's voice can "override" what is being played-out by the voice processing system. This is done in order to interrupt a prompt or message. This is easy enough when a touch-tone digit is pushed by the caller in order to "punch through," but it does not apply if the user is calling from a rotary dial phone, or wishes to use speech recognition instead of touch-tone input. This capability is important for regular users who wish to speed-up the process of a routine call. In other words, for an experienced caller, it is annoying to have to wait for the entire menu or prompt to play before continuing. In addition, voice cut-through can be used to detect speech that matches word spotting criterion.

For example, if you were greeted by a system that asked you to speak the name of the department you want (sales, service, accounting), you could anticipate the correct department name and simply speak it before you are given the prompt for that department.

Logging

Call logging accounts for system and caller activity. For example, the time of day, day of week, month and year a person made a certain telephone call. Other call logging statistics have to do with the kind of activity that occurred during the call itself. This would include what messages were listened to, what fax was requested, or what extension was tried before leaving a voice message.

In the case of operator service applications, call logging is used to bill callers for long distance service. A service bureau might use call logging to bill its IP (Information Provider) clients for the amount of system resources (disk space, trunk time, etc.) that was used during a certain month. A service bureau specializing in fax-on-demand fulfillment may wish to use call logging as a means to report what fax documents were most requested. A voice mail system administrator may use call logging to forecast when more telephone lines need to be installed.

As you can see, there are many uses of call logging, but the data that's being collected must be stored somehow... Most voice processing systems keep a system log, so the time stamp for calls can be spooled-out to a comma delimited or RDBMS file on the system's hard disk. Still other systems allow you to customize the stored data by providing you with database access capability.

Database access modules allow you to read, write, and append database records with whatever information you wish. Other programs format the data into readable reports.

Record Information

Another basic voice processing capability is to simply record a caller's message. These messages can be recorded at different sampling rates in order to economize on storage space. For example, a 24kbps / 6Mhz digitization rate using ADPCM (Adaptive Differential Pulse Code Modulation) encoding will require approximately 10 Megabytes of hard disk storage for 50-60 minutes of recorded speech.

The digitization rate, length of average recordings, busy hour traffic, and number of phone lines used is an important consideration in sizing and engineering any system. Storage capacity is most critical if messages are not retrieved in a timely fashion. Messages that pile-up on occasion may have an effect on the system being able to take new messages.

Messages can be recorded by a recording studio, manipulated for the best quality and then imported onto a voice processing system that will play the messages out to callers. In addition, it is possible to record messages on PC-based sound cards and then transmit the data over a LAN for later play back on a stand-alone voice processing system.

Storing Faxes

With a voice processing system, faxes are either stored on the same platform, or they are retrieved from a host or common directory on a file server. Faxes can be stored in a variety of formats including PCX, DCX, ASCII, and TIFF. PCX files are single-page images that are stored in the PC Paintbrush format. DCX files are actually multi-page PCX files. ASCII files are straight text files, and TIFF (Tagged Image Format) is the native format most faxes are sent and received in by fax processing cards. Scanning documents and then converting them into a faxable format can create fax images. In addition, you can "print" a fax using a print driver utility from a word processor. In this case, the print driver software automatically converts the file into faxable format. Faxable files can also be created with some graphics programs. These programs have a variety of output options including EPS, TIF, BMP, WMF, and PCX to name a few.

Voice processing applications have the ability to locate these stored faxes and take action on them. For example, a fax may have been received and stored that was intended for an individual. That fax can be stored along with a pointer that identifies the intended party's mailbox number. That same party can call into the system at a later time to retrieve the fax by entering in a series of codes. A VRU can also locate a specific fax that was stored for the purpose of sending it to multiple recipients. This is the case with fax-on-demand systems, since the primary purpose of these systems is to automate document fulfillment.

Service bureaus can also use voice processing systems. They can gather new faxable files for re-transmission on behalf of their clients. Let's say, for example, that a fax-on-demand service bureau collects new documents from its service bureau clients on a weekly basis.

A client could call into the system from their fax machine and be prompted to enter the document number of the fax that is about to be added or replaced. Once the document has been identified, the caller can then hit the start button on their fax machine to transmit the new document. The VRU could then capture and store the image (the new document would replace the old one). For example, a travel agency may wish to periodically update a stored document that advertises trip specials, or a department store manager may wish to change his or her weekly coupon document.

Speaker Dependent Recognition

This type of speech recognition allows for individuals to "train" the voice processing system in order to take command from their spoken input. The user identifies a chosen vocabulary of words or digits, and then makes recordings of these words. The "training" program will then use these multiple samples of the chosen words as a "template" to distinguish repeated inputs from that same caller. Most self-training software will prompt the user to speak the words in question many times. The more times the word is sampled, the more accurate the template that will be used for that word.

Speaker dependent systems enjoy a high rate of accuracy because the software does not have to take into account the way in which multiple users speak. This makes speaker dependent systems very desirable for assistive technologies and for those who do not have the use of their hands. Over-the-phone uses of speaker dependent recognition include speaker verification. Speaker verification is used for security purposes (covered below).

Speaker Independent Recognition

This type of speech recognition is the most popular for over-the-phone use. Speaker Independent implies that the technology is capable of identifying utterances from many different people. This is achieved at accuracy rates as high as 95% regardless of the caller's gender or accent.

Most vocabularies for speaker independent recognition are "canned" and come in language sets depending on the geography of the intended callers. Voice processing applications using this technology can detect callers speech when they say, for example: "Yes, No, Help, Cancel, Operator, Oh, Zero." In addition, the numbers one through nine are typically used in a base vocabulary for this type of speech recognition.

Speaker Verification

This is a special type of over-the-phone speaker dependent recognition. The purpose of verification is to confirm the identity of a caller. This is achieved by comparing the caller's spoken word against a pre-stored sample of that same utterance.

For example, a person may call into a special service and ask for access to private information, such as financial records or benefits data. The VRU prompts the person to speak their password after hearing a beep. The password is then recorded by the system and it is compared with the stored file of the same person making that same utterance. If the pattern matching is not consistent with the pre-recorded data, then the transaction will not be allowed to continue.

Figure 3.4 - Pattern Matching Example

Although simplified, Figure 3.4 shows how a sample recording of one person saying the word "password" is different from another person saying "password." With this type of speech recognition, the password itself is the first line of security, and the pattern matching is the second line of security. This is in addition to knowledge of the telephone access number and any other first-order security such as PIN numbers.

Speech Recognition

Speech recognition describes a group of special technologies that allow callers to speak words, phrases, or utterances that are used to control applications. In the case of voice processing, speech recognition is used to replace touch-tone input, make for more intuitive menu structures, and add a level of simplicity and security to some systems.

Speech recognition is often confused with voice recording or voice digitization. Voice digitization and recording explain how the human voice is captured, digitized, and stored for later play-back. Speech recognition, on the other hand, is a technology that uses the spoken word as input that has an effect on the logic flow and execution of the program in question.

Voice Stop

Voice stop is a means for callers to interrupt a menu prompt or other instruction on a voice processing system by merely speaking into the phone. This is a capability that is similar to pressing a digit on the touch-tone pad in order to stop a recording or prompt from continuing. Unlike Voice Cut-Through, voice stop does not actually analyze the word being spoken. Voice Stop technology senses energy on the telephone line and stops execution based on that energy. For example, a train whistle or over head loudspeaker could be transmitted over the phone and have the same effect.

Word Spotting

Word spotting is a special type of speaker independent recognition that has a customized vocabulary for each application. For example, financial applications can use word spotting to find out if a caller wants account *balance*, *checking*, or *savings* information. Systems that support word spotting sound a little different than systems without word spotting. Instead of having to wait for each prompt to be spoken, a caller can simply say "checking," and then be routed to the part of the menu having to do with checking balances and transactions. If the system is a travel-oriented system, word spotting can be used to focus on words like *arrive, depart, flight, reservation, frequent flyer*, and so on.

NETWORK FUNCTIONS

Answering Machine Detection

For outbound dialing systems, the ability to recognize when a call is answered by an answering machine is very important. Take, for example, an application that automatically calls a list of numbers from a collections database for overdue payments. If the called party answers the phone and hears a "machine," they may simply hang up. If, on the other hand, an answering machine answers the line, it would be a waste of time to continue with the call if a live conversation was intended.

There are several ways to detect an answering machine. The first way is to analyze the cadence of the answer. This method is based on the way in which users record most answering machine greetings. For example, most people create an outgoing message that says: "Sorry, but no one is here to take your call, so please leave a message at the beep..."

This type of message is typically one continuous segment of "non-silence." This is different from a live greeting that is typified by a salutation (such as the word hello) followed by the name of a company or a second hello.

Another method for answering machine detection is to analyze the sound that is made when the called party line is answered. In many cases, the outgoing message of tape recorders includes a certain amount of "tape hiss." This hissing, which is imperceptible to the human ear, can be picked-up by voice processing equipment. DSP (Digital Signal Processing) technology is used in this case to check for the frequency range typified by tape hiss.

Call Progress

There are numerous standard tones that are generated by telephone company switching equipment. Ditto for CPE equipment such as PBXs and Key Systems. We take these tones and signals for granted when we make regular telephone calls because we are so used to hearing them. For example, dial tone, busy signals and ringing signals. The ability for voice processing equipment to detect these call progress signals is the basis for many applications including automated attendants, international callback systems, and predictive dialers to name a few.

Busy Tone

A busy signal indicates that the called party's line is engaged, or in use. This goes for regular phone lines, PBX trunks, and dial-up data lines as well. As pictured in Figure 3.5, a standard busy tone is characterized by alternate periods of noise and silence in 500 ms increments. Voice processing equipment can sense this condition by either analyzing the cadence of the noise/silence events, or by detecting the specific frequency used for the "noise" portion of the tone itself.

Figure 3.5 - Standard Busy Tone

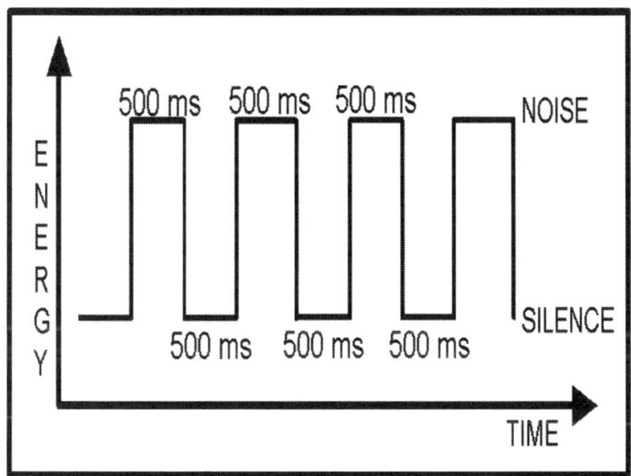

Busy tones sound different depending on the equipment at the end of the telephone call. For example, a busy signal from one PBX extension to the other will sound different than a busy signal generated by a central office when you make a regular local call. In addition, there are "network busy" or "fast busy" signals that indicate that the telephone network does not have enough circuits to complete a call. Network busies are typified by a dual tone with frequencies between 950 to 1,050 Hz and 450 to 550 Hz. These special busy signals usually have an "on" time between 190 and 210 ms and an "off" time between 990 and 1,010 ms.

Connection

There are numerous ways for voice processing equipment to detect a connection. These include "DC signaling," human voice detection, cadences, and tone detection. DC signaling is a form of answer supervision wherein the answering circuit causes a special condition once the call is answered. For example, in the case of a T-1 (digital telephone line), robbed bit signaling is used. Robbed-bit signals are imbedded in the digital stream of communication on each voice circuit. The least significant bit on every sixth frame of the data stream is used to indicate call status.

The bits (A & B bits) are either high or low to indicate whether the phone is ringing, answered, or disconnected. In the case of some DID (Direct Inward Dial) trunks, a battery reversal is used to signal the CPE equipment that a call is coming in from the central office. Depending on the specific protocol used for that trunk, the CPE equipment will “wink” back to the central office to indicate that it is ready to take the call. Once the dial repeated digits are pulsed into the CPE equipment, the system will send an answer supervision signal to the central office to indicate that it has “answered” the call.

Sometimes, simple tones are used to signal to calling equipment that a call has been answered. For example, some paging terminals will answer with a series of beeps to indicate connection. This is useful for outbound dialing functions on voice messaging systems, so a person who has a message waiting in their voice mailbox can be paged automatically by the voice mail system.

Figure 3.6 - Forward and Backward Signaling

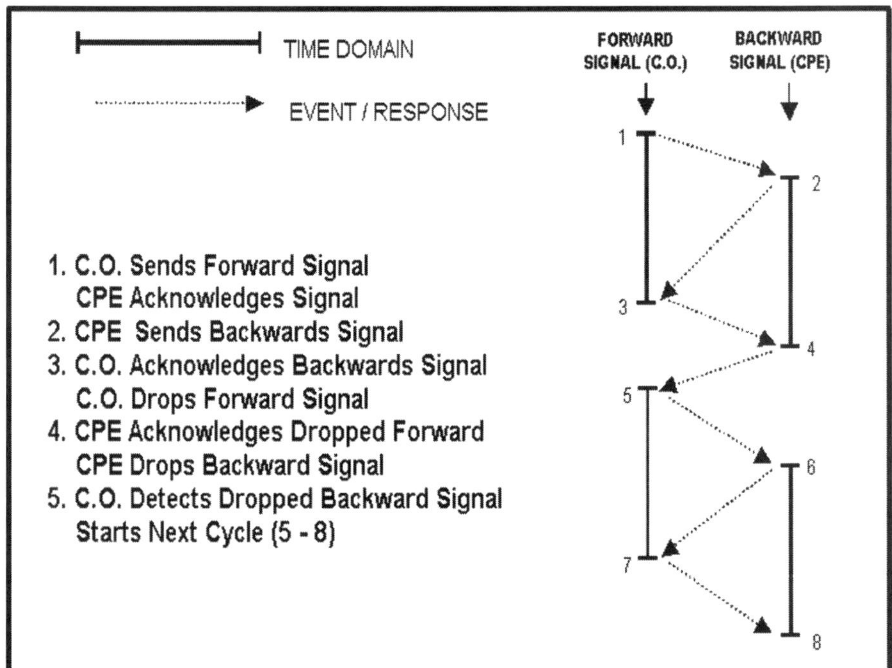

As you can see, the means for detecting answered calls is not as obvious as you might think. In fact, there are even more complicated protocols that are used to "set-up" and "tear down" calls. For example, in many countries, special R2 MF (forward and backward) signaling schemes are used to negotiate and establish calls. Figure 3.6 shows the negotiation process between a central office and CPE equipment using the R2 MF scheme. In some respects, the protocol emulates the kind of call set-up that fax machines use, except there is no carrier signal being generated. This cycle of negotiation is used simply to set-up and connect the call and then to disconnect the call. Voice processing equipment uses special line interfaces and software drivers that take these protocols into account.

Dial Tone

Voice processing systems that make outbound calls usually "look" for dial tone before dialing any digits. Dial tone (usually a continuous pair of tones) is an indication from the central office or PBX system that the circuit is ready to process calls. Applications are usually written so calls can be made after detecting dial tone for about one second. If no dial tone is sensed, the systems will usually hang up, wait a few seconds, and then try again.

Even for systems that do not make outgoing calls, the ability to detect dial tone can be used to test the incoming lines. In other words, a system can be programmed to routinely go off hook, detect dial tone, and then go back on hook.

This can be done sequentially for each telephone line to establish the "ready" condition of the circuits to receive calls. The idea here is that if the circuits are providing dial tone when the system goes "off hook," then the circuits are in good condition.

There are also special types of dial tones called "gallop tones." These sound like dial tone, except that a syncopated cadence is used to indicate a special condition. For example, telephone companies use this "gallop" tone to indicate that a message is waiting in your voice mailbox.

Although the gallop tone is different from a regular dial tone, you can still make telephone calls by simply dialing digits in the regular fashion.

Depending on the type of telephone circuit, there may be no dial tone present at all. The absence of dial tone is inconsequential with some circuit types such as E&M, TIE trunks, and DID, for example. Since DID circuits are incoming only - no dial tone is required. E&M circuits use a series of ground/battery seizures and winks to start calls, as do most TIE trunks.

Fax Machine and Modem Tones

It is important for voice processing systems to detect fax machines and regular data modems. Once a modem tone is detected, the voice processing machine will either hang up, or it will process the transaction as programmed. For example, a fax-on-demand system will store a fax phone number based on caller input and then deliver the fax to that number. When delivering a fax, the voice processing system is expecting to hear a fax machine (or fax board) carrier tone. If the system does not encounter this tone, it will hang up and try the call later. In the case of a "one-call" scenario, the voice processing machine assumes that the person is calling from the phone attached to the fax machine. After the caller is instructed to do so, he or she will press the start button on the fax machine.

This causes the fax machine to generate a fax tone that is sensed by the voice processing system. The VRU software will then patch the circuit over to a fax processing board that detects the tone and negotiates set-up for transmission of the fax.

Voice processing systems can also be programmed to store customized dialing instructions for called parties who have fax/modem switches. These switches are line sharing devices that allow an answering machine, fax machine, and modem to use the same telephone line. The devices are usually set-up to receive touch-tones once the call is established to indicate the purpose of the call. For example, a recorded message may instruct a caller to press "1" to leave a message or "2" to send a fax.

The voice processing machine can be set-up to automatically dial a "1" after the call is established to forward a message to an answering machine. Likewise, the system could be programmed to dial a "2" to automatically send a fax. This capability is useful for outbound dialing systems that routinely deliver recorded messages or faxes to a pre-determined set of users.

Figure 3.7 - Multifrequency Tones

Multifrequency Signals (MF - CCITT R1)		
CODE	Frequency Pairs (Hz)	
1	700	900
2	700	1100
3	900	1100
4	700	1300
5	900	1300
6	1100	1300
7	700	1500
8	900	1500
9	1100	1500
0	1300	1500
KP	1100	1700
ST	1500	1700
STI	900	1700
ST2	1300	1700
ST3	700	1700

MF and DTMF Tones

Voice processing systems routines both generate and detect these tones. Multifrequency (MF) and Dual Tone Multifrequency (DTMF or touch-tone) signals are used for network call set-up and call routing. Both MF and DTMF tones are actually tone pairs that represent both numbers and "begin/end" sequences. Figure 3.7 is a table showing the tone pairs used for MF signaling. These are the in-band tones used on T-1 circuits to transmit the caller's telephone number Automatic Number Identification (ANI), and also the called party telephone number. This is the means by which telephone companies and long distance carriers account for, and then bill long distance calls.

The calling and called party numbers are used not only by the network equipment to route calls, but also voice processing systems to pre-identify callers and the purpose of their call. For example, call centers use ANI in order to automatically route certain customers to a help desk, order entry agent, or customer service representative. The called party number can also be used to route calls to a specific agent, since the number itself can represent a unique product or service.

DTMF tones are the signals that are generated by all touch-tone phones, modems, and fax machines. The same scheme is used with DTMF as it is with MF, except the range of frequencies used is different. Figure 3.8 shows what tones are used for each digit dialed. These tone pairs are programmed into voice processing equipment so that they can be automatically detected. There are several ways to detect these tones in voice processing. The first way is by using computer chips that are designed specifically to understand DTMF or MF tones. This is called "discrete technology" detection. A more modern approach is the use of DSPs (Digital Signal Processors)..

DSP chips are special microprocessors that can be programmed to detect a variety of conditions. For example, they can detect tape hiss, DTMF tones, and certain spoken words. Because DSP chips use downloadable firmware, they are used at an increasing rate in voice processing systems.

The ability to download new versions of DSP algorithms makes system upgrades and maintenance much easier than systems using discrete technology. Discrete chips must be physically extracted and replaced in order for upgrades to occur, whereas DSPs can be remotely programmed.

Figure 3.8 - DTMF Tones

Dual Tone Multifrequency Signals (DTMF)		
CODE	Frequency Pairs (Hz)	
1	697	1209
2	697	1336
3	697	1477
4	770	1209
5	770	1336
6	770	1477
7	852	1209
8	852	1336
9	852	1477
0	941	1336
*	941	1209
#	941	1477
A	697	1633
B	770	1633
C	852	1633
D	941	1633

No Answer

Voice processing systems are usually programmed to look at “no answer” conditions as a unique event.

The detection of a non-answering line is important, because the condition is often logged into a database so the number can be tried again automatically. This is the same type of technology that is used in fax machines and other devices that sport "automatic redial" capabilities.

By logging "no answer" events, more sophisticated outbound dialing systems can characterize the patterns of answering and non-answering history for each telephone number, thus anticipating the best time of day to call the number in question. In order for the voice processing equipment to decide if a number is being answered, the system must be programmed to accept certain thresholds or parameters to characterize the condition.

For example, a system might be programmed to abort the call if the line rings for six "on/off" ringing cycles. Another approach is to simply program the voice processing gear to quit a call that has not been answered after a certain number of seconds since the number was dialed. This, of course, has to take into account the number of seconds typically needed to get the call through the network.

No Ringing

If no ringback tone or associated data message is received by the voice processing equipment, the application can be designed to retry the same number, abort the call, or execute any number of "error condition" steps. For less sophisticated systems, the absence of ringing can be interpreted as a busy signal, network busy, or a faulty equipment condition.

Ringing

The detection of ringback tone is a staple function with virtually all voice processing systems that make telephone calls. The cadence for ringback varies from country to country and from PBX to PBX. Some countries use a "double ring" to indicate that the line called is being requested to answer.

Figure 3.9 - Ring Cadence

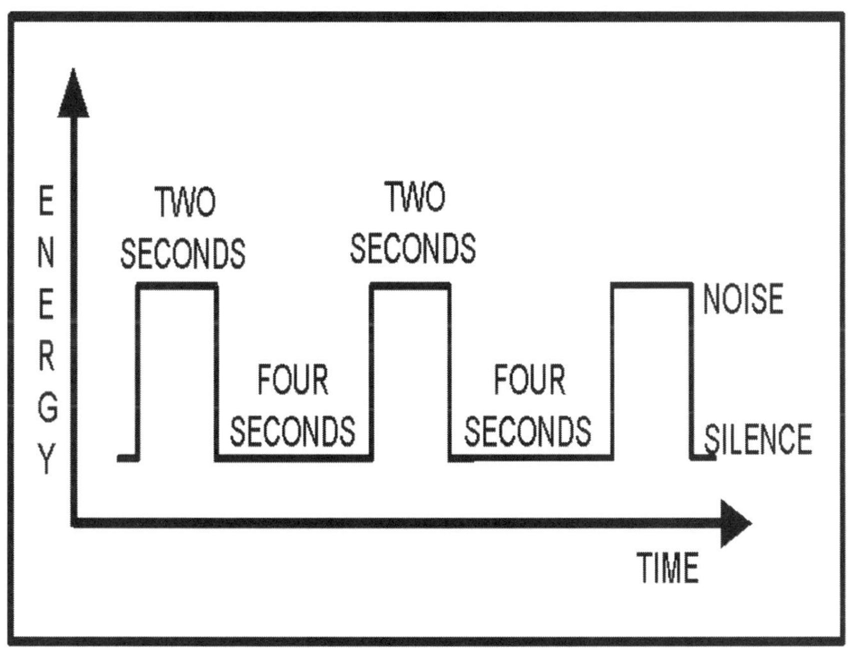

Figure 3.9 shows to basic “two seconds on / four seconds off.” ringing signal typified by the North American telephone network. In addition to standard ringing signals, some voice processing systems need to use an alternate method for detecting a request for answer on the called number. This is the case with out-of-band signaling and proprietary set emulation with certain PBXs and key systems. A voice mail system can use proprietary telephone set emulation cards as the line interface. These cards can send a signal over a data circuit to the PBX to indicate what phone needs to be signaled.

Many PBXs will send a data signal back to the phone that originated the call to confirm the extension number that was dialed and to display that same number on an LCD (Liquid Crystal Display). In the case of a voice processing system, all of the circuitry that provides this data signal is on the voice processing card itself. The card will detect the data signal as if it was a ringing signal, but it does not sense a ringing signal in the standard fashion.

As with proprietary phone emulation, a separate out-of-band data link is simply an alternate means to supply the same call progress information that we are accustomed to hearing over the phone.

Special Information Tones

SIT tones are a series of three sequential tones that are used by telephone companies to indicate a variety of *intercept* conditions. They are used to route invalid numbers, disconnected numbers, out-of-service conditions, and changed numbers, for example. Each condition is assigned a discrete SIT tone in order for network equipment to take special action on the call. For example, some central office equipment will generate a SIT tone if you call an old number that has been disconnected. If a human being hears this tone (usually followed by a spoken recording), the person will simply hang up and call the operator for the new number.

A voice processing system will hear the same tone and perhaps log the condition into a database of "bad" or intercepted call attempts. This data can then be the basis for a report used to investigate the attempts at a later date, or to remove the numbers from an automatic dialing list.

Detect Disconnection

The ability to know when the party on the "far end" has hung up is important for voice processing applications. This is because the system cannot process any more calls on the circuit until the program knows that the line is free. There are many conditions that indicate disconnection. For example, if you make a call to someone and they hang up before you do, you have the choice of hanging up the line or waiting. If you wait, you will sometimes get a recording, followed by a fast busy or other network tone. Many central offices will eventually reset the circuit and provide you with dial tone again.

Voice processing systems can listen for these different conditions and then take action based on the tones the system hears. For example, if a fast busy or “howler tone” is detected, the application can be programmed to assume that the call is disconnected and then hang up the line to make ready for another call.

Central office switches will sometimes send a short “battery interrupt’ signal to indicate that the “far end” has hung up the line. This battery interrupt signal is a short on hook/off-hook event that can be easily detected by most modern voice processing systems. As with answer detection, voice processing systems can be programmed to monitor “A and B bit” transitions on a T-1 circuit to get the same disconnect information.

Many voice processing applications will request the caller to dial a specific digit to terminate the call. Of course, this is the best scenario, because the voice processing system will then clear the line and get ready for the next call. This way, the system does not have to wait for a disconnect signal from the central office or a “time out” to occur.

Tone Generation

Voice processing systems have the capability to generate an endless variety of tones. By using DSP technology, voice processing applications can generate standard DTMF (touch-tones). You can specify the frequency (Hz) and amplitude (dB) of the first tone in Hz; followed by the frequency and amplitude of the second tone.

In addition, voice processing systems are programmed to specify the duration of tones and their repetition if a certain cadence needs to be established. The most common use of this capability is to dial telephone numbers. These numbers can be dialed to place regular telephone calls, automated calls, and to establish calls between two machines.

Human Voice And Grunt Detection

Applications can be programmed to understand when a person is speaking by using human voice and grunt detection. This capability is used to determine when a person versus an answering machine answers a call. This capability can be achieved by characterizing and then detecting a "greeting envelope" as shown in figure 3.10. In this case, there is a pause in energy or noise between the initial salutation followed by some identifying speech. For example, a voice processing system can call a telephone number and sense the cadence associated with "Good Morning... Hydra Corporation."

Figure 3.10 - Greeting Envelope

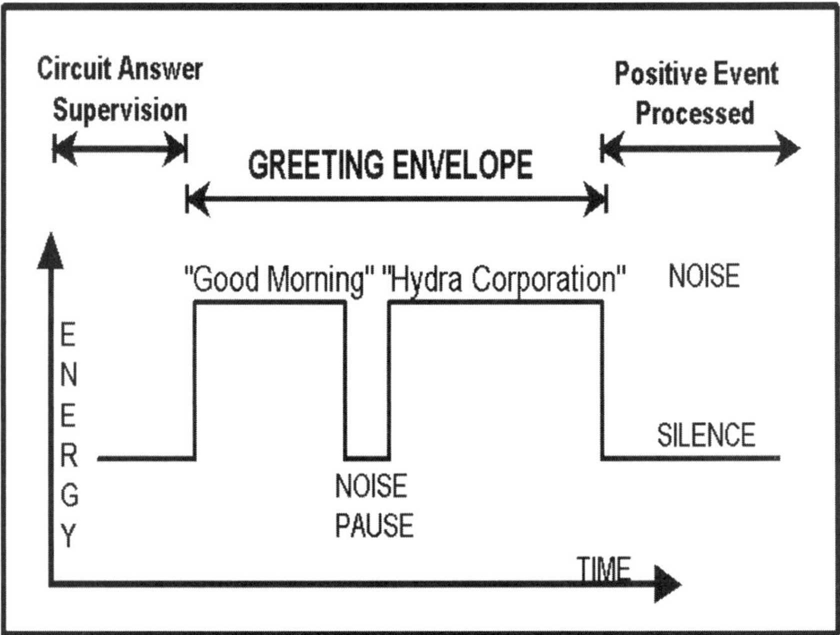

This is different than the typical cadence established by someone simply answering the phone with a single "Hello." By using this same logic, a voice processing application can distinguish live answers from an answering machine, because answering machine greetings are usually recorded as a continuous (and longer) salutation.

The detection of individual "grunts" or utterances can also be used to process calls. This is useful in accepting caller input other than touch-tone digits. For example, an application can be programmed to listen for a "grunt" input, such as a "yes" or "no" spoken by a caller. A grunt is any user utterance, non-specific tone, or other period of non-silence. A grunt may be detected as a single sound of greater than a pre-defined minimum duration and less than a maximum pre-defined duration.

Although grunt detection is not the same as speech recognition (where the actual content of the utterance is analyzed), the capability allows applications to be accessed by callers using rotary phones at a far lower cost than voice recognition. Of course, the use of grunt detection is limited to binary decision logic. That is to say that the application either hears a "grunt" or it doesn't.

Non-Standard Tone And Signal Detection

The ability to detect nonstandard tones is important for voice processing applications that are used to integrate to PBX and special signaling equipment. With this capability, applications can be programmed to recognize a variety of single and dual-frequency tones as well as special cadences.

The specified frequency, amplitude, and duration of tones can be programmed into most modern voice processing systems. These specified tones and signals must first be characterized or "learned" by the system.

There are a number of ways to achieve characterization. The first easiest way is to obtain a published listing of the tones from the manufacturer of the equipment in question. The second way is to record the tones (off-line) and then analyze the tones using test gear.

The third way is to use automated characterization programs that cause the voice processing equipment to make repeated telephone calls based on pre-determined conditions. For example, a special characterization routine can be set up to make calls to extensions behind a PBX that are purposely put in on-hook, off-hook, and out-of-service states.

By making repeated calls to these extensions, the systems can log the results by sorting the sampled frequency, amplitude and cadence of each condition. This data is used to build templates for the tones associated with those events. Once these tones are programmed into the DSPs (Digital Signal Processors) of the voice processing system, they become "standard" tones. Applications that use non-standard tone detection are programmed to consider the frequency and deviation of the first and second tone, and whether the tone will be detected on the leading edge or the trailing edge of the tone event.

Figure 3.11 - Cadence Events

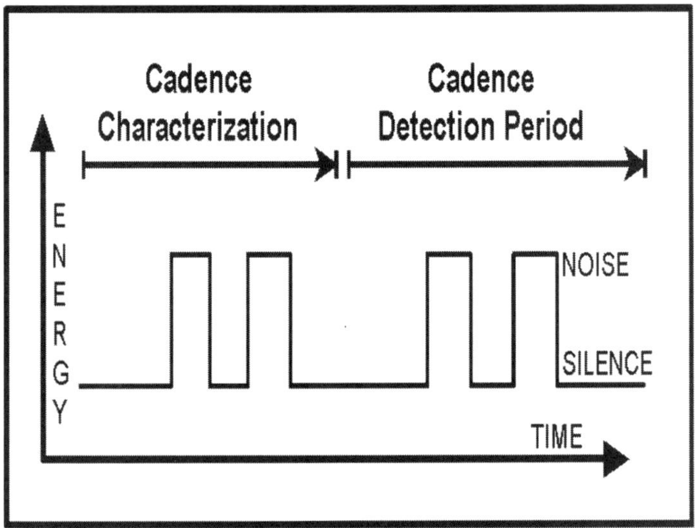

In addition, cadences of detected tones can be taken into consideration. Figure 3.11 shows how cadence events are detected initially and then used to compare against subsequent (similar) events. Application considerations include indicating the cadence *on time* and deviation, and the cadence *off time* and deviation. Programs can also specify the minimum number of cadence repetitions required. For example, to detect a constant tone after a specific duration, such as one second of a non-standard dial tone.

4

Audiotex

Audiotex is the most popular voice processing application in the world. Even though more telephone lines use voice messaging applications, per se, it can easily be argued that most voice messaging systems also have some kind of Audiotex capability.

If your company dispenses information on a repetitive basis, and if this information is already stored in some way - you should seriously consider the use of an Audiotex system. A voice processing solution will reduce the stress on your staff and ensure accurate delivery of information. And most of the time it will significantly extend your "service hours."

ART

Art Locator

There are a number of communication vehicles used to promote the sale of art and collectibles. These include special trade publications, flyers from art galleries, and "word of mouth" in the art community. Prospective buyers are either casual onlookers who might browse through a gallery during an opening, or they could be serious aficionados who frequent galleries and attend every private showing and auction they can.

An Audiotex system can be designed to serve the needs of buyers, artists, and dealers alike. Such a system can greet callers and ask for input on a specific artist, gallery, or type of art. The system can also be used to promote openings, auctions, and other special events.

In fact, to add a little excitement to this Audiotex application, the artists in question could make their own recordings to promote their work. For example, a greeting might say: "Thank you for choosing the watercolor menu. This is Frank Sulley, and I'd like to tell you about the Rainbow of Color series. I've created a line of paintings and reproductions unlike any of my earlier pieces. If you like what you've seen so far, please plan on meeting me at the Rainbow opening this Saturday at the Village Palette. I look forward to seeing you there."

A twist on this application example would be the inclusion of a "private sale" option, where users are required to enter an I.D. code to hear about articles for sale that are limited to a select audience. In addition, the ability to route callers to dealers or galleries could be included. This routing can be achieved by using a zip-code sort, caller input, or ANI (Automatic Number Identification).

Automotive

Auto Dealer Locator

The electronic and print media are filled with solicitations for cars, boats, motorcycles and recreation vehicles. From time to time, you may hear an ad for a certain vehicle and be attracted to the one being advertised. We tend to "tune out" most advertisements, although some latent bit of information may linger.

Audiotex systems can be used to tie print or electronic media advertising to telephone response advertising. For example, an ad might say "Call '1-800-FOR-CARS' to get information on the nearest dealer or other promotions." By listing an 800 number, the advertisement itself can concentrate on content and appeal, and therefore minimize the air time or column inches required to list dealer locations.

By using DNIS (Dial Number Identification Service), a different telephone number can be used for each advertisement, and the Audiotex system will automatically play-out messages having to do with a specific promotion. DNIS is a long distance telephone network version of local DID (Direct Inward Dial) service. DID and DNIS-based services collect the last four digits of the dialed phone number in order to pre-identify the purpose of the call. This is important when many different products or campaigns are being handled by a single system or call center. The benefit of using DNIS or DID in this scenario is the fact that a physical telephone circuit is not required for each discrete telephone number.

In this fashion, a dealer locator system can be set up as a service bureau in order to serve more than one automobile manufacturer, each with many products and campaigns. Of course, it would be less difficult for such a system to be used by a consortium of regional dealers - all of whom sell the same type of car.

For the sake of good public relations, these Audiotex systems can also be used to handle contests, trivia games, and other "fun" activities to promote the use of the system. A typical locator system might say, for example: "Thank you for calling the Tri-State Dealers Association locator system. To enter our car giveaway contest, press 1. To find out about the nearest dealer for the XT-5i, press 2, or you can press 3 to get clues for the trivia game."

Auto Recall

The recall of unsafe, inoperative, or suspect components can be very costly for automobile manufacturers. First, the owners of vehicles have to be located, informed, and scheduled for a maintenance visit. There's also the issue of working closely with dealers and repair shops to replace or fix the part in question.

Recalls are usually communicated by personal letter, newspaper advertisements, and television broadcasts. In most cases, the recall notice will specify the model, year, and type of car. In some instances, a range of vehicle identification numbers is used to identify a "lot" that was bad.

An Audiotex system can be used to greatly simplify the process of getting the correct information out to the consumer. For example, a print media advertisement or radio announcement can alert all owners of a certain type of car to call a single number. Once the caller inputs the car model, year, and type, an audio message can be played to explain the procedure for getting the car serviced. It is also possible for the system to prompt the caller for a vehicle identification number, or other identifying information about the car.

This application can also be used to compile other important information from callers. This may include when they intend to bring the car in for service, and where they wish to bring the car. This information can be used to anticipate parts demand, availability, and scheduling of appointments. In addition, the system could have a "hotline" that automatically connects the caller to the closest dealer in order to schedule a maintenance visit.

Corporate Communications

Employee Directory

Most larger companies maintain lists of departments, functions, contact phone numbers, and locations in one form or another. In the case of companies that have multiple locations, the task is somewhat daunting, because the data is not always unified or updated frequently. It is typical for receptionists to be queried for information about what department a person works in, or when an individual will be back from vacation.

An Audiotex system can be a big help in this area, because the information can be easily stored in a database file and then accessed by many people at the same time. The automation of directory services then, can reduce unwanted traffic from each operator and centralize access to information.

Callers can be prompted to enter the last three digits of the intended party's name to be automatically routed, or to hear information about that person's meeting or travel schedule. In addition, it is possible to construct an Audiotex system that will prompt callers for input regarding alternate choices.

For example, the system might say: "If you would like to locate another person in our directory who has a similar job function, press 1 or press 2 to hear information on another department. In most cases, the automated means for providing this information is quicker than waiting for the receptionist to do it in your behalf.

Crime Prevention

Crime Prevention Hotline

Crime prevention is a hot topic in many communities. To deal with this issue, crime prevention review boards and neighborhood liaison programs have been developed.

These programs seek to provide education and a two-way dialogue between citizens and agencies for the purposes of squelching crime. There are hundreds of pamphlets, booklets, presentations, and videos that have been developed that explain how to spot crime before it happens, and how to steer children away from undesirable activities.

Virtually all of this information can be recorded and conveyed over the telephone with an Audiotex system. A system of this nature can include menus on "crime hot spots," crime prevention tips, and even have a section for guest speakers. For example, radio, television, and sports figures can be enlisted to make regular guest recordings on special subjects. A local football personality can make recordings about avoiding gangs. News team figures can make recordings on crime prevention while you're away on vacation. In addition, local authorities can serialize a number of recordings on statistics, a local crime scorecard, and special meetings and events sponsored by the police.

MacroVoice Corporation of Boca Raton, FL developed such an application for the Toledo Police Department. It's called the Crime Prevention Hotline - a community service that provides crime and safety information. The system freed police officers to take more urgent calls, while dramatically enhancing the ability to distribute much needed information. The system provides the latest data on crime prevention and local Block Watch programs as presented in figure 4.1.

The system is based on MacroVoice's MVX-200 platform that can also double as a voice mail and automated attendant system. By using a special "Auto Scheduling" feature, the Crime Information Hotline is updated automatically to provide current information. The system is also configured to allow Centrex transfer directly to 911 with the touch of one button. According to Lieutenant David Vnuck, the system frees-up the Toledo Police Department so they can concentrate on community policing. Since the system also includes descriptions on suspects, the system does a lot more than dispense statistics. Says Vnuck: "The best way to catch a criminal is to see them in the act. Our MacroVoice MVX system has increased our department's eyes and ears from 700 police officers to those of close to a million citizens."

The eight-port system uses a standard PC with Rhetorex speech processing cards.

Figure 4.1 - Menu Choices - Toledo Crime Prevention Hotline

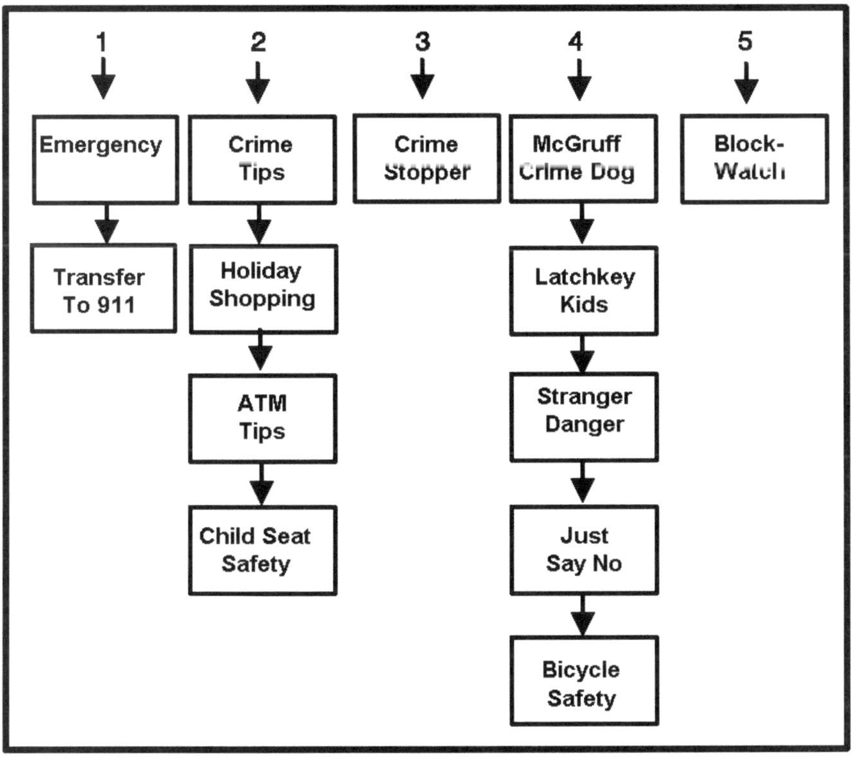

- 1 → Emergency → Transfer To 911
- 2 → Crime Tips → Holiday Shopping → ATM Tips → Child Seat Safety
- 3 → Crime Stopper
- 4 → McGruff Crime Dog → Latchkey Kids → Stranger Danger → Just Say No → Bicycle Safety
- 5 → Block-Watch

Wanted Hotline

Most of us have seen the FBI top ten wanted list and other printed material at the post office. It's typical to find a photograph, list of crimes, and other information regarding the habits, appearance, and disposition of the criminal. In addition, television crime shows have popularized the notion of making telephone calls to report suspicious activity and to identify wanted individuals. Countless apprehensions (and even confessions) have come about as a result of these hotlines.

An Audiotex system can aid in the apprehension of these same criminals. For example, after a radio or television show, the key descriptions, last known whereabouts, and activities of the wanted persons can be recorded. These recordings can be arranged in such a way that they allow callers to browse by state, person, or crime. Anyone who thinks he or she may know the whereabouts of a wanted individual can call into the system and get a more detailed description.

Education

Dining Hall Menu By Phone

Students who live both on and off-campus use university dining halls. Due to the fact that menus are sometimes posted at the last minute, many telephone calls are made to campus information, the operator, and the dining hall regarding meals. For the most part, these phone calls are considered a nuisance. For the students, however, the idea of making a trip to the dining hall for a meal that won't be appreciated is even less appealing. Some students have dietary constraints, and still others do not have full meal plans. For students who pay for meals on an "as-you-go" basis, it is desirable to know what's on the menu. In addition, there are certain plans that allow for only 5 meals a week, versus 21 meals. If a student only has a 5-meal plan, then they are likely to be more critical of the menu. These students will plan their menu so that some meals are taken at the dining hall, some at the concession in the student union, and some off-campus.

An Audiotex system can be updated with recordings the moment the menu is decided upon by the food services staff. The program logic can lead callers through the meal schedule in a variety of ways. For example, the default message could be a recording of the current day's menu. By pressing certain digits, the system could announce menus for any day of the week. Since Audiotex systems can handle multiple telephone lines, the traffic on the campus switchboard will decrease significantly. In fact, the operators can be instructed to always transfer students to the system, or ask them to hang up and redial the correct number.

Homework Helper

Ask any child and you're likely to get a rousing "No," when you ask them if they like homework. Due to the fact that many children avoid or delay doing their homework, even conscientious parents can do little to help when the assignment is overdue. Another problem with homework assignments is that children often forget, or at times ignore the assignments. It is for these reasons that the popularity of homework helpers or homework hotlines is growing. An Audiotex system can be programmed to allow teachers to record the details of homework assignments. For example, the class, subject matter, due date, and reference materials for certain assignments. Teachers can even record helpful hints for parents on how to assist children in completing the assignments. Parents then have the ability to confirm what homework assignments their children have, and also provide guidance.

A special access code and PIN number can be assigned to each teacher, so only authorized recordings can be made. The system can be programmed so no one can change or erase the content. Parents can access these systems after school hours, when most teachers are not available to talk. In addition, they can be equipped with multiple telephone lines, so many parents can confer with the system simultaneously.

Library Reference

Reference material that can be accessed at public and private libraries has increased in sophistication over the years. Now it is commonplace to find microfilm, periodical, on-line databases, multimedia, and video as well as books. It takes a lot of training and expertise to be able to navigate through the multimedia jungle of a modern library. An Audiotex system can be structured to provide queries by subject, author or medium. Systems of this nature can prompt callers to enter the date of a publication, the reference number, and other identifying information to let the caller know if the publication is available (and when it is scheduled for return). Such as system can also be used as a means to announce special events, including book signings, speeches from popular writers, book fairs, and children's storybook readings, for example.

Tutoring Services

Students and tutors alike can access an Audiotex system to listen to lesson tips and record them, respectively. Of course, nothing beats a face-to-face session with a tutor, however, scheduling conflicts and case loads are often too heavy to appropriately address the need for help. An Audiotex system allows the tutor's messages to be "multiplexed," so dozens of students can literally be helped at the same time, in a "virtual classroom." This concept applies to the use of e-mail and Internet conference threads, but has the benefit of providing the same type of information to students who do not have access to a computer.

Lessons can be arranged by skill level, specific classroom assignments, authorized texts, or can be individualized on a per-student basis. Another level of sophistication that can be added to such a system is a messaging capability, so questions posed by students can be replied to in the form of a recorded message. The benefit of this type of Q&A is that the tutor can schedule access to the system, answer perhaps ten questions in a row, and thus satisfy the needs of many students at once.

Electronic Media

Radio Call-In

Radio stations receive hundreds of calls each day as a result of talk show programs and special contest promotions. The call-in rate can easily go to thousands a day based on popular syndicated shows. Listeners ask for schedules, want to hear follow-up information on guests, and have many special requests for viewpoints, contests, and general information. Most calls to popular stations go unanswered or get busy signals. It's also typical for calls to be answered and immediately put on hold for long periods of time. Radio station personnel are ill-equipped to deal with all of these inquiries, so many stations have installed fax machines, e-mail gateways and even voice messaging systems.

Audiotex systems can greatly reduce the congestion by providing information to callers. For example, an opening prompt might say: "Thank you for calling the KXTG hotline. If you want information on how to get a transcript from any of our programs, press 1. To hear bout our guest schedule, press 2. For information on our contest, including clues and rules, press 3. If you would like to be connected to our request line, you can press 0 at any time."

Other uses for an automated system include menu choices for public service announcements, jokes, song parodies, and sponsored advertisements. Due to the fact that voice talent is readily available at radio stations, DJs and other personnel can record their own "air time" messages for their favorite listeners, or do follow-up editorial or comments regarding subjects from their last program. Locus Corporation in Seoul, Korea developed such a solution for the Korean Broadcasting System. The radio network wanted to provide access to broadcasts for listeners who may have missed the news, weather, or other information they tune in for. In addition, like most radio networks, they were concerned about the manpower required to handle two-way (real time) communications with callers.

According to Joongwan Cho, Director of the Korean Broadcasting System, the Locus Audiotex system has literally doubled the "listening rate" of the programs. Says Cho: "The system is dealing with approximately 3,000 calls a day. Now foreign correspondents, reporters, and professors from universities can 'tune in' from any time zone."

Entertainment & Sports

976 Services

Local telephone companies and IPs (Information Providers) jointly provide premium call services. The services, sometimes called "976" (for the telephone number prefix often assigned to them), were once the sole domain of the telephone companies.

Historically, they included public service announcements, time, and weather services. Part of the deregulation of the phone companies called for the separation of information transport from the actual creation and content of the information. This meant that the services having to do with creating and delivering information had to be divested by the phone companies.

This resulted in the birth of the 976/Audiotex industry. The arrangement was that IPs would create and manage the information content, and the telephone companies would handle the transport, billing, and collection. In essence, this amounted to a division of tasks which put the responsibility of purchasing Audiotex equipment (and loading the equipment with commercially viable messages) on the IPs and the management of the network connections and billing on the phone company. At the end of any given month, the phone company collects the charges for each 976 call, subtracts their tariffed billing, transport, taxes and collection charges - and then remits the remainder to the IP.

Depending on the telephone company, the amount IPs could charge could be based on a flat fee per call, a measured usage, or some combination. The actual content of the messages is also restricted from phone company to phone company.

Few telephone companies actually disconnect a subscriber's phone for non-payment of 976 bills. This forced a migration to nationwide 900 services (IPs were getting fewer dollars each month due to "uncollectables").

Both passive and interactive 976 services are available. Some of the most popular programs have been soap opera updates and gossip, weather information, traffic, horoscopes, and sports trivia lines - some of which are detailed below.

Concert & Tour Schedules

There are a number of ways to find out about concert and tour information. There's record and music outlets, ticket counters, arcnas, and concert promoters who can be called.

In addition, the "Living & Entertainment" sections of newspapers are a great source of information. One drawback in calling stores and ticket outlets is the limited hours that they are open. In addition, the print media cannot effectively deal with the immediacy of schedule changes due to disaster, rain, or other delays.

Voice processing applications can do a lot to solve this information distribution problem. First, an Audiotex menu can serve many different concerts, shows, and events. This is not true of answering machines installed on single lines, because there is a limit to the amount of information that can be stored on a tape. In addition, even long tapes can only be sequentially accessed.

Secondly, an Audiotex system has the ability to offer secure access for sponsoring organizations and promoters, thus allowing simultaneous and ongoing updates 24 hours a day. For example, if three big events are scheduled for a certain day, the menu can be changed to deal with the bulk of the callers for that day. If it's raining, the most important message will have to do with whether or not the show is still on, and if not, what the refund instructions are.

In addition, show and concert promoters can use the voices of the performers to help promote the events. Singers can record a few bars of one of their hits, for example. Comedians can toss-out a few sample jokes, and announcers can make recordings on parking, mass transit tips and other helpful information.

One good example of how Audiotex has been put to use in this way is the Creative Loafing Network, sponsored by Eason Publications. The publisher has established a nationwide Audiotex network that is used in conjunction with newspapers and other publications in Alaska, Hawaii, Florida, Massachusetts, and about ten other states so far.

By dialing an 800 number (1-800-LOAFING), callers can access nationwide and city-specific information on nightclub acts, concerts, fairs, and festivals.

Developed by LeasNet of Atlanta, GA, the Audiotex system is packed with high-density speech processing and fax cards including the D240/SC and FAX/120 from Dialogic Corporation. Several LAN-connected workstations and an events database are used to keep the voice processing nodes up-to-date.

According to Debbie Eason, Publisher of Creative Loafing: “Our network saves newspaper Chief Operating Officers thousands of dollars per year and provides their readers with more information than can actually be printed.” Ms. Eason explains that because of the 800 number access, everyone in the country has access to the system. Some of the planned enhancements being developed by LeasNet include the ability to covert newspaper text into automated fax images, so customized publications can be sent to the readership of the network’s client newspapers.

Crossword Puzzles

Countless readers scour their crossword dictionaries every day in an attempt to beat the newspaper crossword puzzle. It’s not only a great challenge, but it’s also fun. What’s surprising is how impatient some of us are in getting the answers to today’s puzzle. Readers will call syndicated writer guilds to get answers, call the newspaper, or call other readers to “crack the puzzle.”

Several years ago, New York Times rolled-out a special service for crossword enthusiasts. The Audiotex system allows callers to input the "down" and "across" numbers of individual puzzles in order to get clues read out to them automatically. The clues are updated daily and old puzzles can be stored and accessed by the date of publication. Readers find these services to be valuable enough to call on a regular basis, and newspaper executives consider services such as this to be good for circulation. In addition to crossword hint lines, there are dating services, want ad messaging services, fax-on-demand systems, and opinion polls that use voice processing technology.

Events Calendar

Audiotex systems can be used as a "talking events calendar" in order to promote tourism and trade with the cooperation of local vendors. Such is the case with an application developed by Automated Business Communications (ABC) of Windsor, Ontario. The company developed an application called "Infoline: Your Key to the City."

The Infoline is a special information service that combines Audiotex information delivery with call routing and locator service. Service bureau providers can install and promote the Infoline system and train local vendors on its use. The system provides a means to promote the products and services of local vendors.

Prospective tourists and even local patrons have a challenge in not only determining where they want to visit, but also what activities would be desirable to pursue on their arrival. This usually results in many phone calls to information bureaus, chambers of commerce, travel agents, and individual vendors. In short, it is a problem to deliver timely, accurate, and needed information about a certain "target' city without making multiple phone calls at great expense.

ABC conducted a study in order to determine the main informational interests of end users (in this case, the callers are primarily tourists). Once these needs were assessed, the unique information delivery needs of local vendors were taken into consideration. All of these factors played a role in determining the most effective solution to the problem.

This needs assessment led ABC to design of a completely automated system that would provide timely information 24 hours a day. Since most tourists inquire about vendors, hotels, attractions, etc. via the telephone, this seemed a natural extension to the concept and design of the system. It was determined that a common number could be published, thus providing easy access to all information, and also centralized use of the system resources for vendors. The concept of providing an "interactive radio broadcast" over the phone was developed.

Of course, a real-time broadcast would require a full-time staff of "announcers," and thus drive the cost of advertising on such a system too high. ABC worked with it's key prospects to discuss the involvement and cooperation of advertising vendors in the city in order to overcome this staffing issue.

In order to ensure a minimum of maintenance on the information by any one individual, a constant update of new and interesting data could be provided to callers by the vendors themselves. Vendors call into the system on their own with special security access codes, allowing them to make new recordings of their "broadcasts."

In addition, it was determined that once a caller navigated his or herself through the menu and information sources on the system, that they should be able to "connect" live to the vendor in question. In this fashion, reservations, tickets, or other pertinent information could be gotten directly from the vendor. Although this adds a level of complexity to the application, it provides a high grade of service to callers, who can not only "browse" through needed information, but also complete transactions in the same phone call.

All call activity is completely documented, and monthly call activity reports can be supplied to advertisers in order to determine the effectiveness of their broadcasts. Service bureau providers have complete access and control of the system and can provide service for up to 5,000 subscribing vendors. "Infoline: Your Key to the City" service bureau providers promote the use of the system to groups of advertisers, chambers of commerce and trade groups by accentuating the benefits of calling ahead, planning ahead, and minimizing paperwork for tourists and local patrons.

According to the system designers, the feature of remote vendor access to recordings has the benefit of instantly changing broadcast messages. This allows for keeping sale, occupancy, and other data current for worldwide callers. The fact that the system allows for transfer of calls increases chance of completing actual transaction with caller due to conveniencc of single phone call, thus turning the phone into a point-of-sale terminal.

Figure 4.2 shows how the system is connected to these businesses by using Centrex lines. According to Rick Buzzeo, President of the City Action Hotline in Windsor, Ontario: "We were able to provide a whole new venue of advertisement to our clients they never experienced until now. We now can provide a feel for what our city has to offer to people calling from all over the world. Callers have the opportunity to be connected to local vendors after getting information without having to make multiple long distance phone calls." Mr. Buzzeo added that "this is a great way for any city to be globally accessible... a truly revolutionary way to communicate and get any city on the information highway."

Figure 4.2 - Infoline: Your Key to the City

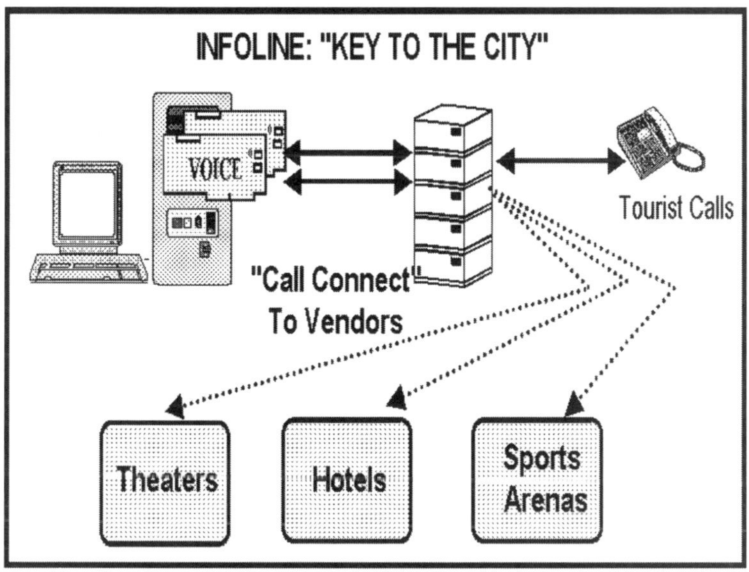

This Audiotex platform was built to support all Dialogic speech cards up to 48 lines per system and has been successfully tested with 4-line Rhetorex boards. PC components include a minimum 386/DX 40 MHz processor; 240 MB hard drive; 8 Mb; keyboard, and monitor. TRT's ProVIDE software was used as a development tool for the application itself. ABC plans on adding host connectivity, fax-on-Demand, and a B.B.S. system to the platform in the near future.

A similar system was developed by BMC Communications of Linburn, GA for National Tele-Sav, Inc. in Myrtle Beach, South Carolina. The system, called "The Travelvision Group Telephone Hotline," is a sponsored advertising vehicle that promotes night life, dining, taxi, limo, dancing, and amusement parks to name a few.

The hotline uses a "rotating ad" feature, so multiple sponsors can promote their products and services on a "round robin" basis. The BMC Group used 4-port Dialogic cards and telephone company Centrex lines to build the final solution. According to Mark Farmer, Director of Applications and Sales, they will be adding speech recognition to the system in order to provide interactive access for rotary dial callers.

Film Producer Music Catalog

BMG Music is the 3rd largest music publisher in the world. Film producers license songs for films and use BMG catalogs to pick the songs they want. It's a fast-paced business that caters to late-working, globe-trotting clientele.

TTM & Associates, Inc. of Fullerton, CA created a special Audiotex application for BMG. It helps automate the process selling film and television rights for their music. BMG music told TTM they needed any easy way for their clients to access songs from music catalog from anywhere. The company also wanted to make sure they could put their entire library of catalogs on line. This represents literally thousands of songs. Film producers would be able to select from these any time, day or night. BMG also wanted TTM to integrate the new system with their "legacy" voice mail system. The integration was needed for messages from persons requesting more information on particular songs.

Now, the BMG Music Catalog system is used all the time. In fact, it was used by the producers of the move *The Fan* starring Robert Dinero and Westly Snipes.

Gaming and Racing

Serious gamblers have tried just about everything to get an edge on "the numbers." There's point spread formulas, historical references, special computer programs, and even e-mail conferences for gamblers. For years, Audiotex systems have provided an alternate means for picking lucky numbers, getting tips on horses, and listening to recorded advice on point spreads.

An Audiotex system for gamblers is usually aimed at a certain sport. For example, a football system will provide menus on upcoming games, information on injuries, weather, and venue. One example of this type of system is the Audiotex application in use at the Hollywood Park horse racing track in California. The system, developed by Alert Communications of Alhambra California, provides up-to-date information to English, Chinese, and Spanish-speaking callers.

Thousands of people from all over the state place wagers both at the park and off-site betting parlors. By calling the 36-port system, you can hear about race results, prizes, post times and scratches. The results are updated every 15 minutes and listened to by approximately 16,000 callers each day.

Horoscopes

There are several types of horoscope-oriented Audiotex systems. The first type is a "passive" system, allowing callers to simply listen with no interaction. Much like horoscopes listed in the newspaper, the caller then hears advice and warnings that are generic, or "canned" for anyone who has the same sign. These passive systems are sometimes updated daily, but still others are fairly unsophisticated and are only updated once a week.

In addition to passive systems, there are more sophisticated interactive ones. Interactive horoscopes ask the caller to input their date of birth, including the year. Based on this input, the Audiotex system calculates a customized horoscope by using algorithms that information providers claim are linked to "star charts."

The benefit of an interactive horoscope system is that callers are more likely to patronize the service on a daily basis, since each day a different recording will be heard. In addition, the fact that the system called for some level of interaction adds an element of fun to each call - therefore the entertainment factor is better than passive systems. Perhaps the most popular example of this type of system is the "Jeanne Dixon Horoscope Line." This program was developed early in 1983 by Dial Info, a preeminent 976 Audiotex provider.

Lottery and Lucky Numbers

"You've got to play to win" is the phrase popularized by a number of state lottery commissions; and for the buyers of lottery tickets, the sooner you can find out that you're a millionaire, the better. This is why the popularity of lottery-based Audiotex systems is a reality. An Audiotex system designed to inform players about winning numbers can provide a variety of services. For example, the opening menu can say: "Thank you for calling the Lotto Hotline. For today's results, please press 1 or press 2 to get information on winning numbers over the past week." Additional menus can be used to allow callers to pick what lotto game they are calling in about, and also information on tips, outlets, and payment options.

An Audiotex system can also be developed to serve callers from multiple states. These systems can be programmed to automatically "download" information directly from lottery officials, or the information can be collected manually and then input into a database that the Audiotex system uses to play-out winning numbers.

In some cases, these systems are sponsored by state agencies, and information providers are compensated directly for maintaining the system for callers. In other cases, 976 lines are used, so the phone company and the Information Provider split the revenue on the cost of the (premium rate) call. Lotto 976 rates range between 25-50 cents. Information Products, of East Granby, CT is an Expert Systems VAR. Their IVR product is called Verify. They developed an IVR system for the Rhode Island Lottery.

The lottery was having a tough time handling all of the calls they were getting for lottery results. The system is handling 10,000 calls a day with 12 lines. Callers touch-tone their ticket date and game selection. Administrators can call in with a secure "back door" code to update the winning numbers.

Movie & Theater Hotline

There are a number of popular Audiotex systems that promote Broadway, off-Broadway, and off-off-Broadway productions. In addition, movie theater hotlines are becoming more popular in major cities. A joint project between Voicetek Corporation of Chelmsford MA, and the New York City Theater Development Fund (TDF) is a good example of the former.

The TDF is an association that was founded to promote Broadway productions. An Audiotex system was prototyped in 1987 by TDF in order to deliver timely and accurate information about curtain times, reviews, and venues for current productions. For example, the system greeted callers with a menu that allows a user to choose "audio samples" from musicals, curtain times, and ticket prices. Due to the large number of theaters, show times, and reviews, a system of this nature can house thousands of prompts and recordings.

It is typical for Audiotex systems like this to provide secure access to many "virtual administrators." For example, a representative from each theater or production company can be given a private access code to update musical recordings, curtain times, and ticket information via telephone. This method significantly reduces the manpower required to administer the application for the system provider.

Olympics

A special challenge for Audiotex providers is presented when the callers speak multiple languages. For example, the Olympics are held in host cities, but people all over the world are watching television, listening to the radio, and scouring the paper to find out about what's happening.

Up-to-the-minute reporting on events is limited to certain radio and television programs, depending on air time and sponsorship rights. A centralized Audiotex system can easily overcome the time zone constraints and language barriers by providing 24-hour access and menu prompts in different languages.

With a system of this nature, a series of prompts can be recorded in separate play-back "bins." Each of these bins can hold the same prompts indicating events, country names, and scores, except that they are recorded in a different language. System administrators can regularly call into the system and type-in the results of each event either by remote terminal or telephone.

Alternately, the Audiotex system can be connected to a computer that automatically sends event results to the Audiotex system over a data link. A system was designed for the Barcelona Spain games that worked in a similar fashion. More than five languages were supported by the system. In addition, speech recognition technology can be used to recognize a discrete phrase for each language to overcome rotary dial telephone problems.

Prayer Hotline

Many churches, religious organizations, and community groups sponsor prayer circles and bible studies for religious followers. Prayer meetings are scheduled either on a weekly or biweekly basis in most cases, and subjects regarding bible interpretation and special community needs are often discussed. An Audiotex system can be used to establish a weekly prayer circle for folks who are hospitalized, homebound, or otherwise unable to travel to organized meetings.

For example, the system can be programmed to say: “Thank you for calling the prayer hotline. Please pray for Mr. And Mrs. Avery Jones, as they where in an accident on Tuesday and are in intensive care. Special prayers also go out to Bobby Miller, who is sick with the flu this week. For bible study subjects, and a special dedication from our pastor, dial 1, or dial 2 to hear about the prayer meeting scheduled for this month.”

Restaurant Guide and Reservations

Applied Language Technologies (ALTech) of Cambridge, MA develops sophisticated speech recognition technology. At Computer Telephony Demo Fall 96 in Orlando, ALTech showed how the power of speech recognition is juicing-up real-world applications in voice processing. The company supplies the technology to a number of enhanced service providers and telephone companies. ALTech chose a scenario wherein a "road warrior" kept in touch with a network-based one number service only to find he's almost forgot his 10th wedding anniversary. A major disaster was diverted when our hero voice-dials an automated florist and restaurant to book a lovely gift and last-minute reservations. Another relationship was saved courtesy voice processing and speech recognition.

All kidding aside, BellSouth now offers a "511" information service featuring a Restaurant Guide. Early prototypes dating back ten years ago were all touch-tone based. Callers can search for restaurants based on type of cuisine, price range, and location. The system connects callers directly to the restaurant of their choice. It's been deployed by the telco in Daytona Beach and Gainesville, FL.

Skiing Reports

Skiers know how difficult it is to get accurate reports on slope and road conditions. In many cases, what's good for the slopes is not good for travel, especially in the height of ski season. For this reason, ski enthusiasts will search out several sources before loading their gear into the car and heading for the mountains. No one wants to get stuck on the side of the road, or find that the skiing is not as good as what a resort reported.

An Audiotex system can provide instant access to comprehensive weather, travel, and promotional information. Such a system can have varied "feeds" in order to avoid bad information or exaggerated claims.

For example, the state highway patrol or local T.V. station can update the weather and travel conditions. "Spotters," who report ski slope information objectively, can recount their personal experiences with slope conditions. In addition, sponsored advertising and reports from individual resorts can be recorded.

Sports League Scores

Like many non-profit organizations, The American Kitefliers Association (AKA) of Rockville, MD is short on resources, yet needs a way to provide competition results to the membership in a timely manner. The association also fields questions on competition rules and AKA policies. Since the organization is staffed by a number of volunteers, it was a burden to dispense information to callers about ranking and regional data.

Darrin Skinner, of Skinner Consulting in Sunnyvale, CA was enlisted by the association to design an Audiotex system to overcome the problems associated with publishing delays, phone traffic, and time zone constraints. The President of the AKA identified the desired information and provided scripts of exactly what to say. Skinner Consulting provided the programming, database expertise, and the voice talent.

Callers to the AKA Automated Information System have access to a host of recorded messages about the association. In addition, callers can request a complete listing of competitive results for over 1,000 fliers competing in 114 different conference-category combinations. For example, by simply selecting Northeast, Intermediate, Individual, Ballet - the caller is presented with a list of 47 fliers, their placement in the Northeast conference and their score. The ability to skip through the list is provided.

The phone system provides simple access to the AKA's competition results database. The database is connected to the VRU by a local area network. The results database is maintained through programs written in Visual Basic running under Microsoft Windows. The ability of the VRU to read database files and act on the information in the database is what makes this system possible.

Information is now provided to the membership 24 hours a day, 7 days a week. By centralizing the AKA system, members can now access expert information from a single source. This has greatly reduced demand on the executive director to answer routine questions. Before the system was installed, competitors had to wait two months for the association's magazine to print outdated information. Now association members can learn about conference standings on demand, instead of waiting.

David Gomberg, president of the AKA, estimates a 10% improvement in staff productivity as a result of installing the system. Says Gomberg: "The system relieves paid staff from taking routine calls such as: 'Who's my director in Region 11? Overall, I can't imagine where we'd be without the AKA Automated Information System. It has helped us enormously."

Figure 4.3 - American Kiteflyers Association - Scoring System

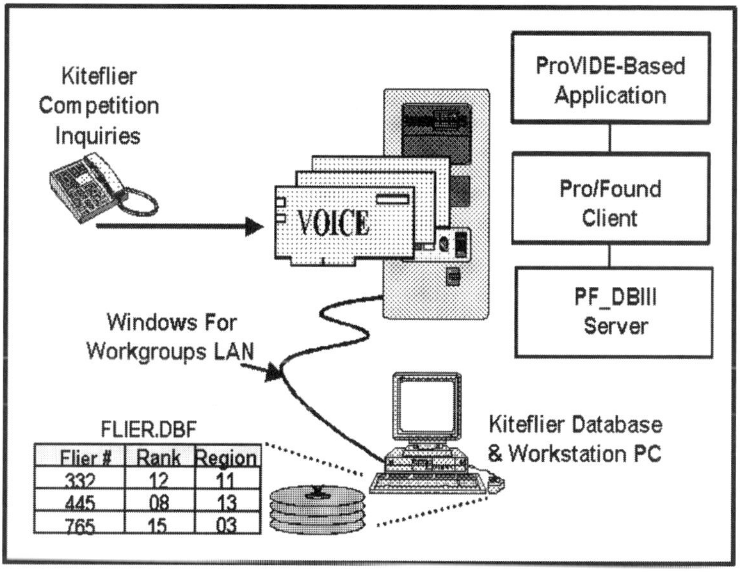

Skinner Consulting developed the AKA system with a 486/40 MHz PC with 4 Mb of RAM memory and a 2l0 Mb hard disk. Mr. Skinner used a Dialogic D/41D speech card and TRT's ProVIDE software with the Pro/Found database connectivity option.

In addition, the system uses an Intel EtherExpress LAN card for connection to the competition results database. The database resides on another PC running Microsoft Windows for Workgroups, as pictured in figure 4.3.

Song Sampler

Audiotex is the perfect medium for wide-area broadcast of musical samples. Take for example, the Unsigned Artists "National Song Network" in Nashville, Tennessee. The application was developed by Designed by Kiman Corporation, also based in Nashville. The song network promotes unsigned artists' work by allowing listeners to choose from a menu of newly recorded songs. Callers can then rate the songs if they wish.

Systems of this type provide artists with instant feedback on their recordings from a nationwide audience. This solves a big problem for musicians who are not yet represented by a recording label or agent, because they do not have access to professional listening groups, test marketing, or large audiences. It is conceivable for an artist to make a recording and get feedback within several days. In addition, the song network provides and automated means for artists to "distribute" their recordings in rapid fashion. For example, radio DJs, reviewers, and prospective agents can access the system to "cruise for talent."

Surfing Reports

Popularized by California 976 Information Providers (IPs), surfing hotlines take full advantage of Audiotex capabilities. Because the coastline covers hundreds of miles, it would be impossible for one person to relay information "down the line." Radio and television stations rarely provide information beyond the service area of their patrons, so Audiotex can cover a much larger area.

With a system of this type, regular "spotters," or part-time employees can call into the Audiotex system on a "back door" telephone number. This back door number is an administrative phone line that can only be accessed by callers with a special access code.

Once the spotter logs-on to the system, he or she can update surfing reports for their stretch of beach. These updates can be made with verbal recordings, or can be further automated with the use of a scoring system.

A scoring system allows administrators and spotters to type certain digits pertaining to the conditions that make surfing favorable. This includes the height of waves, temperature, wind speed, and other conditions. Applications that concentrate on any outdoor sport can use this same method of recording data. In addition, contests, special events, and warnings can also be provided.

Financial

CPA Tax Season Information Delivery

David Tobkin works at a small CPA office. A workload problem emerged due to the company's inability to distribute *one* message to a variety of people in the office. During tax season, distributing needed information to customers is critical, but this puts a strain on the administrator, who was unable to complete other designated tasks.

Voysys Corporation of Fremont, CA designed a solution. The company configured voice mailboxes for all the employees. They programmed the voice mail system with two Audiotex boxes in order to answer repetitive questions during tax season. In addition, the CPA firm was able to offer interesting articles and books to use as tax sources. This gave group partners and administrators an easy way to deliver messages. The result allowed the administrator to capture more time to do other tasks related with the job. This improved customer relations in educating them on tax procedures.

Currency & CD Rates

Audiotex systems can be used to significantly reduce the amount of telephone traffic presented to banks, credit unions, and other financial institutions.

Bank officials have access to up-to-date information on the prime lending rate and other indices that are used to establish interest rates. In addition, the exchange rate on foreign currency can be easily calculated and announced by an Audiotex system.

Users can access prime lending rate information, currency exchange rates, and interest rates on certificates of deposit. For exchange rate information, the system could prompt callers to enter digits representing the currencies they want to compare. The system would then be able to announce the exchange rate expressed by percentage, or prompt the caller to enter a specific amount. Most banks and currency exchange businesses charge a fee to convert cash from one form of currency to another, so the system can also announce the charge associated with exchanging currencies.

Lending Rate Information

Anyone who has borrowed money and filled-out a loan application knows what a time consuming affair it is. First, there's the trip to the bank, a meeting with a loan officer, and then an application is filled-out. Most of the time, the information needed for the loan application is at home, so you have to make a second trip to the bank to deliver the completed form. For consumers who only want lending rate or monthly payment data, this process is a waste of time. Of course, bank employees don't want to go through the entire qualification process only to find out that the consumer just wanted a quote, either.

An Audiotex system can prompt callers for the principal amount they wish to borrow, and the number of months they want to amortize. Based on the type of purchase, the system can then tell the caller what the estimated monthly payment is. The caller may also be prompted to enter an alternate term length for a different monthly payment. An Audiotex system can be enhanced with IVR (Interactive Voice response) capability to automate the filling-out of forms by automatically retrieving stored information that is keyed on the caller's social security number. For example, every lending institution uses credit bureaus to learn the status of your credit history.

These inquiries are typically done via computer, and the data that is transmitted can be stored in a local database. These records include your name, address, employer, social security number, and payment history.

Government

City Ordinance Information

There's plenty of times new residents, visitors, contractors, and homeowners want to know how to apply for a permit, when they can burn trash, and what to do about the disposal of a large amount of reuse. The Chamber of Commerce and city hall also dispense this information.

Audiotex is a way to distribute this information for multiple city departments, the Chamber of Commerce, and community volunteer groups, as well. For example, the opening menu of such a system could say: Thank you for calling the River City Information System. Press 1 for Chamber of Commerce member information, 2 for city ordinance information including trash pick-up, recycling, and permits, or press 3 for special messages from volunteer organizations."

Providing a call transfer capability for callers who access the system during regular business hours can enhance the system. The transfer can be achieved by using Centrex lines. These special circuits can be installed by the local telephone company, or the phone company may alternately suggest call conferencing capability to achieve the same result.

The US Environmental / Recycling Hotline is a program that provides people across the US with information on various environmental topics such as recycling and hazardous waste. It's customized for local areas.

Started in 1991, the Hotline is a public / private partnership, based in Phoenix, which consolidates thousands of federal, state and local recycling telephone hotlines into a single toll-free call (1-800-CLEANUP), available in 22 states.

Winner of President Clinton's 1995 Closing the Circle Award and the President's Environmental Technology Initiative Program, the Hotline saves taxpayers millions of dollars each year and should be available nationwide by the time you read this. A partnership between the US Postal Service, the EPA and several corporate sponsors, including Digital Equipment Corp and Sprint, the Hotline started originally as a "home-grown" voice processing application to manage calls for a limited geographic area.

As the network grew to more states, the Hotline began to experience limitations with its existing telephony application. The government turned to the Show N Tel app gen from Brooktrout (Needham, MA) to build a larger system. When a call comes in, Show N Tel recognizes the state a call originates from using automatic number identification (ANI). The call is transparently transferred to a voice response system linked to that state's databases.

A caller then enters his/her ZIP Code and the Hotline routes the caller to the available local information. If information isn't available for a specific city or town, the Hotline automatically finds information for the nearest city. The Hotline includes multiple IVR clients to provide increased flexibility and fault tolerance. Currently running under Windows NT, the Hot line can handle 72 simultaneous calls on three Digital Alpha systems with Dialogic, cards. Total capacity is 144 simultaneous calls.

Through networking, the Hotline is capable of handling an unlimited number of calls. Currently it receives 40,000 to 60,000 calls each day from the 22 states where it's available. Data access comes from Microsoft SQL Server and the database is replicated daily to enhance its reliability. The Hotline uses Dialogic cards to connect directly with multiple T-1 phone networks. Once expanded nationally, the system will be able to handle at least 400 lines. The government continues to work with states to integrate individual databases into a nationwide system. They expect to have 150,000 recycling sites in the system by the time you read this. Within nine months of going national, The Hotline anticipates 750,000 to one million recycling sites online.

The Hotline recently partnered with the Direct Marketing Association, which will put the 800 number on its 69 billion annual direct mail pieces. After integrating every state's information into its database, the Hotline plans to expand yet again-internationally-again by using Show N Tel. The organization continues to add new information categories to include facts on energy conservation, water quality, and air quality, composting and special sections geared to children and small business.

Congressional Statistics

Voter leagues can use Audiotex systems to provide telephone access to congressional statistics. This would allow voters to access the system at any time in order to conduct research on the history of politicians and their voting record during the last term. For example, the system could prompt callers to enter digits having to do with certain platform issues. Congressional voting history on each of these issues could be linked to the inquiry. Alternately, the system can be programmed to accept caller input on bill and proposition numbers. The system can play-out a message that describes the intent of certain legislation, and then prompt the caller for further input.

By using digit translation, callers can input the last three letters of the legislator's name in order to get a list of names with which to narrow any search. This is similar to automated attendants and voice messaging systems. The system can be programmed to say: "Enter the last three digits of the person's name. There are three matches. Press any digit during the playback of that person's name to receive their I.D. number. Use this I.D. number to make inquiries about that legislator's voting record or other statistics."

Election Results

Audiotex can provide voters with access to multi-district election results via telephone. Although it's hard to beat live television coverage, it takes a long time to hear about your own district, especially in large metropolitan areas.

Such a system can be programmed to solicit district identification from each caller via touch-tone input. As an alternate, the system can receive Caller ID information from the telephone circuit in order to pre-identify the district that the caller is accessing the system from. This can significantly reduce the amount of time spent on the phone with each caller in order to "clear the line" for additional callers. If the caller wishes to hear information on another district, they can be prompted to dial a digit after the automatic play-back of their own district information.

Election systems of this type can be updated via secure telephone access, or by a secure data link. In the case of telephone access, officials can enter a series of codes associated with a hand-held menu. These codes can act as a second level of security to curb unauthorized access to the system.

Highway Patrol Traffic Reports

State Highway Patrols can use Audiotex to deliver regional or statewide information on all types of road conditions. These systems allow callers to select roads or highways to receive information on traffic, flooding, or other conditions.

Traffic systems of this type can be of great help to travelers as well as commuters. Advice on snow chains, slower-than-posted speeds, or construction areas can help callers to decide on alternate routes suggested by the system. Commuters in Long Island, New York have a similar capability by accessing a special number from their cellular telephones. In this fashion, thousands of commuters can be re-routed in order to avoid traffic delays and dangerous conditions.

National Park Advisory

National and state contact rangers and other officials can easily an Audiotex system in order to advise travelers of park closings, special events, and parking alternatives.

Most state and national parks get drive-through as well as reservation-based visitors, sometimes making it hard for park officials to do an adequate job. This results in closing the parks early, or limiting the use of certain facilities.

Much of the overcrowding associated with these problems can be avoided with the use of an Audiotex system. For example, a system like this could greet callers with an important broadcast message before providing other options: "The Skyline Drive is experiencing heavy traffic this afternoon. If you are driving a camper or other large vehicle, you may wish to camp for the night and attempt your entry tomorrow morning..."

Postal Services

Both the United States Post Office and alternate carriers have used Audiotex systems to augment telephone inquiries and face-to-face transactions. For example, the USPS has tested an Audiotex system that uses alphabet-based speech recognition. Callers are asked to spell the street number and name, after which the system announces the correct zip code.

Audiotex technology can also be used to allow callers to enter the zip code, weight, and class of certain parcels in order to determine the basic cost, or price of insurance for an item. In addition, postal customers can access an Audiotex system to hear about postal services and regulations. By making modifications to such a system, payment for services, package tracking, and pick-up services can be arranged with an Interactive Voice Response (IVR) capability.

Services Hotline

Audiotex systems are used in many government applications, including the Department of Motor Vehicles, Immigration, Internal Revenue Service, and Unemployment offices. These system provide information on business hours, special requirements, fees, and instructions on how to get required forms for permits, applications, and tax returns. Some of these systems allow for touch-tone input, and still others are passive systems that simply hang-up after a canned message has been played.

Due to language barriers, it is sometimes difficult for federal, state and local governments to provide up-to-date information. With Audiotex, translators can be hired to record a variety of messages in any number of languages. For example, there can be French, Italian, Spanish, Korean, Vietnamese, and Japanese recordings made for several agencies at the same time. These recording sessions can be scheduled regularly in order to maintain current information for these agencies.

Tax Assistance

State and federal tax offices are deluged with questions on extensions, tax laws, and how to get certain forms. In addition to government employees, there are also volunteers who assist taxpayers on these matters. Before Audiotex systems were put into use, hundreds of tax office employees were required to answer these questions on a full-time basis.

Audiotex systems have no problem in establishing and maintaining the "party line" on any subject. Because they are machines, the same information given to one caller is also given to the next one; and of course the same message is given in the same fashion to people who access the system more than once.

An Audiotex system can also be used to ask certain questions of callers in order to determine what forms the caller needs to submit. These forms may include certain schedules and attachments that would be in addition to the basic tax return form. In addition, there can be an audio menu to lead callers through a "top 10 questions asked" session. This can reduce the number of manual calls made into the tax service by as much as 20 to 30 percent.

Travel Advisory

The Department of State provides citizens with information on virtually every country to which they may travel. The government has established a system of warnings and advisories that help to quantify the amount of danger associated with travel outside of the country.

Advisorys include updates on terrorist activity, political unrest, natural disasters or health hazards. An Audiotex system can provide information on immunization, special visas required, embassy information, and weather. A system like this can also be programmed to play recordings in many languages.

Voter Information

Citizens can be encouraged to vote with the help of an Audiotex system. Popular sports figures, movie stars, and singers can be enlisted to make a variety of "public service announcements" that can greet callers with information on how to register, where to register, and also where to vote.

Voting places are usually set-up according to your street address or school district. An Audiotex system can help new voters by providing them with up-to-date information on the venue, and also with information on how to cast an absentee ballot. In addition, verbal instructions on what to expect in the balloting box can also go a long way in helping new voters.

VISA Information

Most embassies are swamped with more phone calls on VISA requirements than any other type of phone call. Callers speak a variety of languages, and despite the multilingual talents of embassy personnel, it becomes tedious to dole-out the same information all day long. An Audiotex system can be programmed to guide callers through a series of menus on VISA rules, expiration dates, fees, and required documents.

A typical menu for such a system would read: Thank you for calling the UAE embassy. For information on VISA rules and guidelines, press 1. For information on access to forms and application instructions, press 2. All other callers can stay on the line to speak with our receptionist."

Health Care

Disease Center

Audiotex systems can do much to help concerned citizens with timely information. For example, a voice processing system can immediately greet all callers by squelching rumors of outbreaks and health disasters. For example: "You have reached the city disease center hotline. Rumors of the so-called "east side plague" are a hoax. There is no plague confirmed in the city. If you are calling for any other reason, please stay on the line, or press '1' to access our automated disease education service."

The automated service can advise citizens on how to avoid certain diseases and health hazards. For example, there can be a question and answer section on venereal diseases, colds and flu, rabies, and other topics. The system can also provide information on how to get pamphlets, counseling, or other material on the subjects covered by the Audiotex system.

Medical Inquiry

Hospitals and universities have long sponsored medical information systems that allow the community access information on health topics over the phone. Before microprocessor controlled systems were developed, the Audiotex systems were tape cartridge based. In fact, the cartridge on many of these systems had to be manually installed by a hospital employee before the caller was transferred to the system. Due to the sensitive nature of some of the taped material, callers used a coded system of numbers read from a menu so they did not have to ask for the specific subject. Although this helped certain callers to be less shy in educating themselves on certain medical subjects, it was cumbersome and labor intensive. Now, Audiotex systems can be programmed to play recordings based on a variety of (private) inputs from the caller. Subjects can be arranged alphabetically, so callers can use their touch-tone pads to spell the subject in question. By using an indexed database, the Audiotex system can automatically match the spelling with a list of related subjects.

The caller will then be prompted to choose from a list. The Audiotex system can also refer callers to hospital and community health workers or volunteers. AdVoTec, Ltd. of Huron, OH is an Expert Systems VAR. They sell the Audio Tele-Health system. It's an automated marketing and public relations system for long term care providers. Callers are the general public looking for a Medicaid application, care provider or information on illnesses. The first sale was to a small, 50-bed provider in a rural area. They bought a four line Audio TeleHealth system. Their former advertising was getting no response. With voice processing, call volume increased by 20 times. They were instantly impressed and happy.

Patient Information

The most important role a hospital can play is that of life-saver. There are dozens of medical emergencies, operations, and routine tests that are facilitated by hospitals each week. One undeniable aspect of running a hospital, however, is the fact that patients have loved ones wanting to know how they are doing. In addition, friends and relatives visit patients. It's inevitable that the telephone system at the hospital will be overloaded with telephone calls about patient status.

As is the case with an Automated attendant, an Audiotex system can be programmed to react to caller input based on the spelling of the patient's last name. Once the caller identifies the patient, the system can check a database maintained by the nurse's station or administrative staff that advises callers about visiting hours, visiting rules, and directions to the hospital. The system can also provide the patient's phone number, or let the caller know if the patient has checked-out.

Travel Immunization

Medical offices can take advantage of Audiotex by providing instant access to travel-related information. For example, and Audiotex system can be programmed with recommended immunization, health tips, and safety information for each country. Callers can input the country they are thinking about traveling to and input date ranges for their travel.

The system will then play-out a message such as: “You should get the ‘Flu-Pak II’ before traveling this summer. In addition, you may wish to consider a water filter and our pamphlet on food preparation.” A more sophisticated system may prompt callers for the age of family members, so information about infants and children can be provided.

Hospitality & Travel

Travel Auto Club Travel Condition Delivery

Philips Dialogue Systems of Germany recently developed an automated road condition Audiotex system for ADAC, the German automobile club. The platform dispenses critical travel condition information to callers. This includes snow heights in winter sport areas and Alps road conditions. Callers can get prerecorded speech or a fax to determine whether or not they wish to brave the elements. The system uses advanced speech recognition technology developed by Philips to guide callers through the transaction:

System: Good morning, this is the ADAC. Do you want to have information on road conditions?

Caller: No

System: On snow heights in France, Italy, ... For which country?

Caller: Italy

System .Which region? Say yes to the chosen region. Trentino?

Caller: No

System: Belluno?

Caller: Yes

System: Which valley? Say yes to the chosen valley. Livigno?

Caller: Yes

System: In the Livigno valley from Belluno there is .....

Reservation Network

There's plenty of ways Audiotex and IVR can help to promote tourism and resort reservation business. One good example is the Lake Tahoe Hotline (916) 546-LAKE. The Hotline provides Tahoe's numerous mountain towns and worldwide visitors with a centralized information and reservation network. Designed by Rubicon Technology, Inc. Of Tahoe City, CA, the Audiotex system provides callers with information on:

1) Entertainment. This section includes nightclubs, theaters, concerts, casino shows, special events, and attractions. This section is updated weekly.

2) Recreation. This section has sub-categories that change with the seasons. For example, the summer months include water sports, mountain biking, popular hikes, fishing reports, "just for kids" section, etc. Winter categories include Alpine and Nordic ski reports for all Tahoe resorts (updated each morning), snow mobiling, ice skating, sleigh rides, etc.

3) Restaurants. An audio listing of specials, menus, and reservation information guides callers through a variety of choices. Restaurateurs are able to update the menus and specials over the phone for up-to-date listings each day.

4) Transportation. This section covers taxi services, public transit, towing, tours, ski shuttles, etc.

5) Weather. This section contains re-broadcasts of the National Weather Service forecast and is updated every two hours.

6) Road conditions. All callers to this section are transferred to the CalTrans Highway information network.

Figure 4.4 - Lake Tahoe Hotline

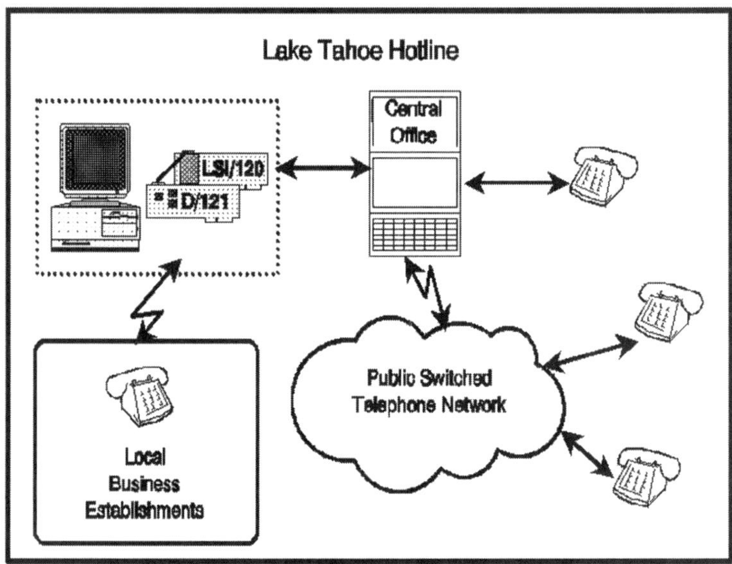

Callers access the Lake Tahoe Hotline by calling a local number from either the South Shore or North Shore of the resort, so callers hear information specific to their area. Each business is assigned a personal access code and is therefore able to administer messages about their own business. After listening to an announcement, callers are given the “direct connect” option. Callers can get transferred to the business in question. Prizes are awarded weekly to random callers, and monthly to the businesses that post the most creative announcements on the system. Tahoe’s economy is chiefly sustained by tourism, which keeps the local hospitality, entertainment, and skiing concerns on the lookout for ways to promote their products and services.

Since Tahoe's weather, ski conditions, road conditions, and entertainment change daily and often hourly, the issue of providing visitors with accurate information while remaining attentive to other duties is always acute. In addition, most visitors find it difficult to navigate through the myriad phone numbers, services, and agencies that provide them with the same information.

Over 250 lodging owner/managers participated in the research survey and development of the hotline. All those questioned agreed that they wanted to provide current information to their guests, since most did not find the time to "stay on top of it all" themselves. The participants of the survey agreed to consolidate the vast information into a general clearinghouse. The concept of a "hotline" was established, so that "one number convenience" would limit the frustration of multiple calls to hotels, concierge desks, restaurants, and resorts for visitors to the area. Each participant was asked, "What could be offered on the hotline to make your life easier?"

Survey participants agreed to promote the use of the hotline with a simple placard accompanied by a map of the area. These placards could be strategically placed in hotel rooms and information desks all over town. Thus, the "do it yourself" display card was created. These cards sit by the phone in 12,000 rooms lakewide. This provides an estimated 6.5 million exposures to Tahoe visitors per year. The Lake Tahoe Hotline and its display card serves as a powerful, promotional tool for sponsors and a valuable, time saving concierge service for the Tahoe and Reno lodging industries.

Although the Lake Tahoe Hotline was devised primarily for visitors to the area, a side benefit of the system is that it provides locals with access to the same information. This is especially useful for weather updates, since so many of the residents work at local resorts and therefore must brave the winter hazards on a daily basis. Says Bob Wayman, D.D.S.; "I just found out about the hotline a couple of weeks ago. We've already used it several times and have grown to appreciate this terrific technology. I wish we had the same capability 20 years ago." An IBM compatible 386 PC acts as the platform for the system, which is connected to a UPS (uninterupptable power supply) that maintains the system during frequent power outages.

As pictured in figure 4.4, the voice processing hardware consists of a Dialogic D/121B and LSI/120 voice processing and line interface card. This provides access to as many as 12 callers simultaneously. The system was developed using Telephone Response Technologies' ProVIDE application generator and run-time environment.

The hotline is scheduled for a fax-on-demand upgrade, so callers can have maps, directions, menus, and special bulletins sent to them automatically. The fax-on-demand capability will allow callers to enter their fax machine number with the touch-tone keypad.

Tourist Day Trip Hotline

Audiotex can be used effectively to help day trip enthusiasts stay current on new and exciting things to do over the weekend. Often sponsored by local newspapers, systems of this nature can include information on nature walks, garden tours, canoeing trips, rafting, and special museum presentations to name a few. The main menu on this kind of system can prompt callers for areas of interest, the date they want to travel, and the age of the parties in their group. A more sophisticated approach would be to ask for the caller's zip code and the number of miles they are willing to travel. This would help to focus on activities that are within the desired range of travel.

Travel Agency Discount Travel

MOSCOM Corporation of Pittsford, NY designed a voice-controlled IVR and Audiotex system for a German-based discount travel agency business called First Travel Group. There are 200 First Travel agency branches, 70 First Travel business centers and 30 offices so far in the new rapidly growing First Travel discount agencies. The company fills vacant airline seats and hotel rooms by booking travelers at the last minute - often several days or even within hours of their scheduled departure.

This is done through agreements with airlines and hotels. The vendors provide First Travel with on-line computerized information about travel openings, and then offering those openings to customers as they call in. Because First Travel Group is a service-oriented business, they are always looking for ways to increase productivity while keeping customers satisfied. That need became more important when First Travel opened discount travel offices. To be profitable and still provide customers with low-cost travel, the discount offices had to find ways to keep the quality of service high while keeping overhead low.

Customers call the agency because they like the convenience and immediacy of making travel arrangements over the phone. For those customers, the key is assuring they can get through to the travel agency, receive concise travel information and then book their trip as quickly as possible. For First Travel to supply the latest information, they have to update their offerings as soon as airlines and hotels alert them to last minute savings. Once a travel agency has established itself and has a good reputation, it only needs to staff its offices with adequate numbers of qualified travel agents and telephone lines in order to do well. But installing additional telephone lines and hiring more agents to run those lines is expensive. The company found that without keeping overhead low, they couldn't pass travel savings to their customers.

First Travel originally considered using an information system that would put callers on hold listening to music until an agent was available. But they realized this was a waste of time. This would not leverage the dead time while the customer waited for the agent to answer. Why not find a system that could offer information while the caller waited, or better yet, a system that would answer the callers' initial questions, before an agent even took their calls? It had been suggested they consider an automated attendant and Audiotex system to answer the phone and allow callers to ask questions and receive responses. However, many of these systems work on the premise of touch-tone phones, which are not common in Germany. For the concept to work, First Travel needed a system that callers could control and interact with using their voice.

After researching various alternatives, we were introduced to a voice-activated, automated telephone service called Infovoice, produced by Moscom Corp., and sold through Siemens AG internationally. Moscom sells the same system in the U.S. under the name TeleVoice.

To date, First Travel has implemented 20 Infovoice systems. Rather than requiring travel agents on every one of their telephone lines, it is offered on a separate phone line and allows customers to listen to a variety of travel options before they speak to an agent.

As agents receive updated travel information, they use a microphone to record a new announcement about the available packages into the system. This not only improves customer service but it increases employee productivity because our agents do not waste valuable time answering preliminary questions about new offers.

Not only can customers answer many of their own questions without speaking to an agent; they can also call and receive updated travel information at any time of the day or night. The system operates 24 hours a day and is updated twice a week with 12 new offers in each travel category. The agency initially directed callers to the system only if agents were busy. However, because customers can call any time of day, they often call the special line first rather than calling the regular phone line, and then being transferred to an agent.

According to Stephanie Scholz, travel associate for First Travel: “Savvy customers use the system and understand they might be offered new travel opportunities if they call back at various times. They even call in the middle of the night, choose a travel package, and leave a message for an agent to call back in the morning to book the trip. This way they receive the true benefits of travel-last-minute savings that might not have been available had they waited.” Customers call the audiotext and listen to current sale announcements, usually presented in three categories of trips, which we call Last Minute, Flight Only, and Travel Planner (for trips booked weeks in advance). A call might go like this:

Infovoice: "We've got some great last-minute travel bargains this week. Make your selection by saying Last Minute, Travel Planner, Flight Only."

Caller: "Last Minute."

Infovoice: "You've selected Last Minute. When you hear an item you want, say agent to speak to an agent for booking."

Infovoice: "First, seven days in Zermatt at the St. Julien Hotel for DM 1585 (DM=deutsch marks, about 1.59 to the dollar), double occupancy...Next, three days in Zurich and four days in St. Moritz, air fares included, DM 1285."

Caller: Agent! (Caller is transferred to agent.)

Because the system plays an excerpt from the chosen announcement to the agent prior to connecting the caller, the agent knows the caller's interest and responds quickly and professionally to the call:

Agent: "Hello, I understand you're interested in reservations for our Zurich/St.Moritz special...shall I book it for you?

Callers may ask to repeat an announcement, hear the next announcement, or listen to announcements under a different topic. Callers can even interrupt message playback using the voice interrupt feature. When no agents are available, the system takes a message and lights the agent's message waiting light.

Marine

Storm Advisory

Boaters, vacationers, and travel advisors can use an Audiotex system to help complete their travel and boating plans. Such a system can be programmed to prompt callers for the type of craft they intend to use, the area they will be boating, and the time of day.

By using a combination of live weather feeds, and a database of craft sizes and skill level, the Audiotex system can provide a detailed list of advice, alternate plans, and warnings.

For example, after a caller selects small craft and novice skill level, the system could say: "The Cape May area is in for some choppy waters due to the trailing edge of tropical storm Edward. We advise waiting until after 2 PM today or postponing your outing until tomorrow based on the craft and skill level indicated."

Human Resources

Benefits Information

The management of benefits and other personnel-related plans is no small feat, even for small companies. There are retirement benefits, vacation time, savings plans, 401K distributions and a variety of other complex issues that are managed by human resource managers. Professionals in human resources estimate that as many as 25% of all calls made to human resources departments have to do with benefits. Here's a perfect scenario for the application of an Audiotex or IVR (Interactive Voice Response) system.

Since most information on benefits is stored in a database, an Audiotex system can interface with the database in order to relay certain data to employees. For example, an employee could call into the system to ask about the future value of their 401K plan. The caller can be asked to input the salary figure and then the percentage of 401K contribution currently applied.

They can further be asked to input the number of years they have worked for the company followed by the date they want the calculation to be based on for maturity of the plan. The system can then announce the estimated dollar value of the plan based on certain assumptions.

Since no "real world" data is being entered into the system, but rather a simple calculation is being made - this function would be regarded as an Audiotex capability. If, on the other hand, a caller were to be prompted to enter his or her Social Security number and personal identification code, the application would likely interact with the "real" personnel files. In this case, the function would be regarded as IVR application. An Audiotex system can also be used to automatically guide the callers through a verbal menu that explains benefits and other plans for the company. For example, a caller could hear about stock option plans, retirement plans, and vacation benefits.

If sensitive and private information needs to be relayed to the caller, then the IVR function discussed earlier would be appropriate. Cascade Technologies, Inc. of New York City has installed many of these systems. The company refers to these solutions as "Capital Accumulation Systems" since they are used in dispensing information about financial plans. Cascade has installed systems that automatically create special reports for the caller. These reports are then either mailed or faxed directly to the employee.

Job Postings

There are a number of ways to communicate job openings, department expansions, and relocation issues to employees. Perhaps the most popular methods are bulletin boards, company newsletters, and e-mail messages. Some companies have adopted policies to "hire and promote from within." In this case, human resources managers have to make a concerted effort in posting job openings in a conspicuous and timely fashion. This often presents a challenge, since many employees travel, or work off-site.

Job postings can be arranged on a departmental basis. In addition, they include information on how many other employees have inquired or interviewed for the position. The actual job description can be recorded by either the human resources manager, or by the supervising manager. By providing telephone access to this information, every employee has 24-hour access and can retrieve descriptions from any phone in the world.

For example, the opening menu prompt might say: "Thank you for calling the company-wide job posting system. For information about jobs in our European office, press 1. You can get information on our west coast postings by pressing 2. For information about openings at the headquarters office, press 3. Press 0 to speak with human resources at any time." Callers can be given a description about working hours, office locations, or information about relocation packages.

Resume Routing and Locator

Voice Processing Plus, Inc. of West Bloomfield, MI is a Parity Software VAR, Nortel Business Affiliate and WilTel Business Partner. They sell custom solutions with a focus on "giving a voice to data." The company does LAN/Voice Integrations, mainframe emulation and GammaLink fax stuff.

Before automating with voice processing, Upjohn's Human Resources Department had only one person dedicated to answer calls for job-hunting students. They want to know where to send resumes. Resumes can go to any one of 17 locations.

The 16-port system prompts callers for their zip code. The system matches the input against a National Zip database, and then speaks out the closest contact info for one of the 17 locations. The system also gives callers the opportunity to enter a secondary zip in case they're moving to a new location.

Military

Duty Roster

Company clerks and military officials can use an Audiotex system to automate access to duty roster information. It is typical for this information to be posted in a common area for military personnel to "read 'em and weep." Due to the fact that the military augments its forces with civilian personnel who live off-base, it is reasonable to provide this information in a more universal way.

Duty rosters are segmented so routine and special tasks are categorized. These lists are assigned depending on rank, seniority, and sometimes a merit/demerit system.

By maintaining a computerized duty schedule that includes personnel profiles, an Audiotex system can automatically dispense this information 24 hours a day. If the Audiotex system is linked to a common database, the prompts used to announce duty assignments can be dynamically spoken the moment new duties are assigned.

Power Utilities

Service Outage

Utilities encourage their customers to promptly report power outages and other problems having to do with service interruption. The sooner the utility hears about a problem, the sooner they can fix it. When power outages occur that effect a large area, the switchboard is often inundated with well-meaning callers. An Audiotex capability can process virtually all of these outage calls automatically. This can be arranged by associating the telephone prefix with the geographic location of the outage. By using Caller ID or ANI (Automatic Number Identification), the system can determine if the caller in question is likely to have been affected by the outage.

In this case, a message is played indicating that the power company has located and is working on the problem.

Audiotex systems that do not have ANI capability can ask callers to input their phone number or zip code in order to find out if the outage in question is already being addressed. In case the caller is contacting the utility for another reason, the option to speak to an operator can be presented. During most large outage situations, some 50 to 65% of all calls are about the outages, so it makes sense to automate these calls "up front."

Print Media

Classifieds

Newspaper classified advertisements and Audiotex make for a perfect marriage. This is because most classifieds customers are concerned about the cost of describing their item, while the readers are more concerned about finding out details.

Readers are often tantalized with listings, but may loose interest if there is not enough information. In addition, not everyone listing an item for sale wishes to advertise their home phone number. In these cases, newspapers can offer and box number or messaging service in order to ensure privacy.

A "Talking Classifieds" Audiotex system can augment the use of print media listings with a verbal description. For example, a reader can jot down an Audiotex code number and then call the paper's Audiotex system. After entering the code, the caller can hear a detailed piece of information on the item. In fact, if the system is administered well, a message can be played to the caller indicating that the item has already been sold. This is a great time saver if the customer who listed the item does not wish to receive phone calls after the sale. This is also a savings of time and money for the reader if the number listed is a long distance call.

Yellow Pages

Businesses pay a premium to have color, bolding, good page position, and large spaces dedicated to their Yellow Page advertisement. Due to the fact that these listings are static snapshots of the business, the Yellow Pages are not a good vehicle for advertising sales, specials, or other time-sensitive information. This is not the case, however, if the Yellow Pages are linked to an Audiotex system. The Donnelly Company pioneered this concept in the late 1980s.

An audio version of the Yellow Pages can provide a host of information that can be updated and accessed easily. A special telephone access number can accompany advertisements in the printed medium. Readers who call this special number can be greeted by a custom message. These custom messages can be created and updated by the advertiser in question. For example, a pet store owner can update his or her recording every week to indicate a new sale item, or to advertise a special on grooming or a dog show.

Product Distribution

Locator Service

Consumers are often attracted to a certain product or service, but don't know where to find the item. In most cases, if the consumer calls the manufacturer, they end up with a service representative who can scan a database for the closest dealer. This usually occurs after the consumer contacts several directory services to locate the manufacturer.

A locator system can use Audiotex as the basis for delivering this same information to callers. The system can either be hooked-up to a database that resides on a host computer, or the database on dealer locations can be downloaded onto the Audiotex system from time to time.

Callers can be automatically identified from their ANI (Automatic Number Identification) so the system does not have to ask for their location. Less sophisticated systems will prompt the caller for their telephone number or area code in order to use the input as the search criterion for the database look-up.

Most locator systems have the ability to announce several choices to the caller based on their geographical preference. If the operation of the system is paid for by sponsorship, then it is common to have references played-out on a "round robin" basis, so that each dealer gets an equal number of references.

Recall of Products

Recalls can be either a nuisance, a danger, or both. Sometimes, manufacturers voluntarily recall products that are defective or outdated. In the case of health risk, manufacturers are sometimes compelled by court order to recall items.

Tainted food is a typical reason for a recall. In most cases, the lot numbers, date ranges, or distribution centers can be used to identify whether or not the items need to be recalled. For example, if baby food is being recalled, the manufacturer may specify that all strained carrots purchased after July 5th with a lot number greater than 88789 are being recalled.

An Audiotex system can be programmed with date ranges and lot numbers for each item. Callers can hear about a toll-free number on television, so they can call to enter their lot number. If the number is a match with the database, then the caller is given instruction to either return the product for a refund, or to go ahead and use it.

Real Estate

Agent Finder

Audiotex systems are beginning to gain popularity with home hunters and real estate agents alike. This is due to the nature of house hunting behavior and the schedules of most real estate agents. Take, for example, a couple who spies a "For Sale" sign in a neighborhood.

They may cruise past the home several times and try to decide whether they are interested enough to call the agent, or whether they should just let the opportunity pass. Since most listing agents and homeowners require some kind of appointment to tour the home in question, it is difficult to get informal data on a property without engaging the sales process. In addition, agents are not always immediately available, so interested parties and agents do not always get together.

Some "For Sale" signs now include the phone number of a special Audiotex system that dispenses details on every home listed by the agent. The details include school and fire house proximity, number of bedrooms, square feet, acreage, and all of the data you would expect to find out by personally visiting the agent.

The use of these systems help to qualify buyer interest with the home, and helps to prepare the agent for setting-up a meeting. Some systems transfer the caller to the agent's car phone or home phone number. These systems can also collect statistics that reveal the number of times callers have inquired about a certain property.

The number of inquiries can be used to gauge the interest for the seller and prospective buyers. By adding a messaging capability, the system can be used to page agents and then forward messages to them from prospective buyers.

Apartment Locator

Hunting for an apartment can be a daunting task, even with a handful of brochures and guides that you pick-up at the supermarket. Perhaps what makes the task so difficult is the volume of data that has to be poured through in order to find a suitable apartment in the correct area, price range, and with all of the amenities desired.

An Audiotex system can be programmed to receive inputs including geographic zone, price range, size, and extras like washer and dryer access, gym, and pools, for example. Callers can be prompted to input their search criterion in order to allow the system to automatically pick from a database of narrowed choices. This can replace what would have been several hours of reading newspapers and apartment guides.

After entering the request, callers can hear a message like: "Congratulations, we have found three apartments that meet your requirements. They are in Shadowbridge, KingsGate, and Hunter Pass. Press any key to hear a description of these apartments and to get contact information."

Retail Services

Audiotex Coupon Book

AltiGen Communications of Fremont, CA has a customer who sells advertising to local merchants and advertises a toll free number to call for discounts. It's and "auto-attendant coupon book." When the people call the 800 number, the AltiServ auto-attendant provides choices depending on their area of interest. This routes them to their advertisers recordings, at the end of which they are give a number to present for a discount or gift when visiting the merchant.

Store Information

Retailers answer a lot of questions over the phone for their customers. Shoppers want to know when a sale starts and ends, rain check usage, where the branch store is, and what the hours are, for example.

By programming an Audiotex system to provide the "Top 20" answers to the most frequently asked questions, retailers can significantly reduce telephone traffic and increase customer satisfaction. This is due to the fact that many times a shopper will call a store only to be put on hold for a long time. Perhaps they only wanted to ask a simple (most frequently asked) question.

By the time the call is serviced, the shopper may be angry and not in the mood to buy anything. The Audiotex system can be made to answer all calls first, thus giving customers an option of staying on the line to talk to the machine, or to be put on hold for the next available operator.

Product Demonstration

There are a number of products that can be described over the phone effectively, but not too many that can be demonstrated effectively. One notable exception is music. Tapes, CDs, and the audio tracks to certain videos can be sampled and categorized for play-back on an Audiotex system. Music can be categorized by type (pop, country, rock), or by artist or label.

Callers can thus be prompted to choose a "clip" from a new or previous recording they are interested in purchasing. The system can also provide information to callers regarding concert schedules, special television appearances, and albums to choose from. If the Audiotex system is sponsored by a record store, the prices for the products can be announced, and an option to be transferred to a sales agent can be programmed.

Social Services

Community Bulletin Board

Local newspapers do a good job of covering community interest stories, activities, and announcements. As a special service to its readers, these same newspapers can record this information so that it can be accessed over the telephone.

Audiotex systems of this nature are the basis for "newspapers for the blind" services that are sponsored by community organizations and staffed by volunteers. In this example, volunteers will call into the Audiotex system and read aloud the passages from a specific section of the newspaper. This section is then assigned a special access code so that callers can access it with touch-tone entries. A technology twist on this method would be the use of speech synthesis technology (text-to-speech), wherein ASCII text is automatically converted into the spoken word.

Community Score Card

Church organizations, social workers, and neighborhood watch societies can use Audiotex systems to post information about crime, fires, community service projects, graffiti, or any other issue of concern. The score card concept comes from the "let's clean it up" mentality of many neighborhood programs. Interested parties can call into an Audiotex scorecard system to hear about improvements made in certain areas on a weekly basis. For example, a general greeting might say:

"The Green Street Pride League is happy to announce that the graffiti clean-up block party was a huge success last Sunday. To date, only several graffiti incidents have occurred since that time. Also, the meeting on security measures was put to good use by one of our members last night when a burglary attempt was thwarted. Our thanks to Sergeant Struthers for his speech on Thursday night."

Disaster Hotline

There's much to coordinate when disaster strikes. In the case of flood, earthquake, and hurricane, for example, the damage done sometimes turns a city's communication infrastructure up-side-down. It is difficult to coordinate the efforts of relief crews and to communicate to outsiders what supplies are needed. When these disasters strike, an Audiotex system can be used in conjunction with radio and television broadcasts. For example, an announcer can say: "To hear more about how you can help with the disaster, call the flood hotline on 555-8878. You will hear about what relief supplies, donations, and work you can offer to victims."

A disaster hotline can also advise travelers on what areas to stay away from and also inform contractors and suppliers of what jobs need to be done and who to contact for work authority. In addition, an Audiotex system can list the addresses for drop-off points for food, blankets, clothing, and sandbags, for example.

Job Lines

Audiotex technology can be used to promote job listings for employers, and prospective employees who are searching for employment. Such as system can be programmed to accept both touch tone and recorded input from employers. Authorized users can use touch tones and recordings to indicate the job description, salary, and other data that would interest job seekers.

Job-seekers are then able to select the type of job they're interested in, and then receive verbal descriptions that were recorded by employers. Companies that are posting jobs internally can use this type of system on a private scale. On the local level, job seekers can record their qualifications and job experiences for potential employers to hear when they call in to post job description messages. In some cases, the employer may erase a posted message based on the hiring of someone he or she heard about on the Audiotex system.

A good example of how Audiotex has been put to use is the JobsLine system developed by Career VoiceLink of Orinda, CA. More than 20 colleges and Universities, including San Francisco State University and University of British Columbia use the system. School officials sponsor the JobsLine and distribute pamphlets that explain how students can find jobs within minutes of using the system. Students are given choices for full-time, part time and seasonal work with menus on each category. In addition, career fairs, counseling advice, and internships are promoted with the system. Says Dr. David Small, Assistant Vice President of Student Services at the University of Houston: "Career VoiceLink provides convenience while giving our students and alumni a competitive advantage in a tough job market."

The staff and students at San Francisco State University helped the company to develop the JobsLine. Based on the enthusiastic response from users, Pacific Bell nominated the JobsLine for the National Golden Phone Audiotex Award in several categories including "Most Innovative Programming" and "Best Quality Information Delivered to the General Public."

Missing Persons

In addition to bulk mail flyers, milk cartons, and television news shows, Audiotex can be applied to the search for missing persons. Since name, age, hair color, build, sex, color of eyes, etc. can identify each missing person - a database can be used to do automatic indexing.

If someone sees a person who they suspect is a missing or kidnapped person, they can call into a centralized missing person's hotline. Since volunteers man many of these hotlines, it is often the case that the caller is put on hold. An Audiotex system can help the caller to remember what they saw by prompting them. For example, a caller can be asked to dial certain digits for hair and eye color. Once an agent is available to speak to the caller, the agent can access the data that was stored by the Audiotex system and read it back to thc caller for confirmation.

Transportation

Rail & Bus Schedules

Amtrak uses an Audiotex announcement system to help automate its Northeast Corridor scheduling pamphlets. Since most calls into bus or railway reservation centers are for schedules, it makes sense to automate as much of this information as possible. Many riders don't need a reservation for the lines they are traveling, so it is a strain on the reservation call center to handle simple information calls.

With this Audiotex capability, callers listen to a menu of "corridor" before pressing a digit associated with the line they want to travel on. Train numbers, departure, and arrival times are then announced. Transfer to an agent is a welcome option on systems of this nature, since the caller has been "pre-screened" by the Audiotex system.

SBB (Schweizerische Bundesbahnen / Swiss Railways) is also using an Automatic Timetable Information System. Philips Dialogue Systems of Germany developed it.

Says Jean-Claude Peng of SBB: "We started in February of 1996 with a pilot. We wanted to test the ability for working in the Swiss environment (different pronunciation of German, special cases of the railway network structure and names; for example the system recognizes Neuenburg as well as Neuchatel) and the acceptance by the customers." All Goals were reached. The success rate reached 90 % and only 5 % of the people asked said the system is bad. So SBB decided to go on.

The system does natural language (German) recognition and supports speaker independent telephone access. 1,500 station names are supported with assembled speech fragments for the output. SBB uses a DEC Alpha Computer running Windows NT for the Automatic Inquiry System (AIS) software from Philips. The initial system handled 55,000 calls with peaks of 1,000 each day.

SBB is working on a new version, with support for 3,000 station names and new features like leafing (asking for an earlier or later connection). The new version will also support "Via"-connections (intermediate stations).

For relevant calls, SBB enjoys a 93% dialogue success rate. Here's a sample dialogue:

System: Good morning. This is the Automatic Railway Timetable. Can I help you?

Caller: Good morning. I would like to go from Zurich to Geneva.

System: When would you like to go from Zurich to Geneva?

I have to leave tomorrow at 4 o'clock.

So you would like to go tomorrow at 4 p.m.?

No, no. In the morning.

So you would like to go tomorrow at 4 a.m.?

Yes, please.

There is the following connection: ...

5

Call Center

Call centers come in all sizes and serve as many purposes as there are businesses. For a ten-person company with two sales people, two PCs equipped with autodialers makes for a small, yet functional call center. These groups of people and machines perform both inbound and outbound functions. Call centers evolve as companies grow and customers demand more sophistication, better service and quicker transactions. These transactions can increase in efficiency when voice processing elements are applied. But getting the phones and computers to "talk" to one another is no simple matter. The beginning of this chapter will give you an overview of some integration schemes:

- Inband Integration
- Out-Of-Band Integration
- Proprietary Phone Integration
- Drop-and-Insert Integration
- Integrated Call Center (No PBX)

Inband Integration

Inband Integration uses analog signaling. These signals are transmitted over the telephone lines connecting the VRU (Voice Response Unit) and the telephone system (PBX or ACD) together. In some cases, an adjunct or secondary device is used to perform the inband functions. In most cases, inband integration uses DTMF or MF tone pairs to emulate a person using a regular telephone. Automated attendants and voice messaging systems connected to the station side of PBXs typify this. On some PBX's, the message desk or secretarial phones are programmed to light message waiting lights on other phones. Message desk employees enter touch-tone digits representing the called party's extension number to turn on message waiting lights. In some cases, the extension number is followed by a code. This same function can be performed by a VRU, since voice processing cards can be controlled by software providing the same instructions (go off-hook, dial the extension number followed by code number, etc.).

This method of integration is usually inexpensive, and can be used on dozens of phone systems without special equipment. Sometimes, the PBX manufacturer will charge a licensing fee to instruct systems integrators and VRU makers on how to achieve integration. In other cases, the development team can "backwards engineer" the interface by using trial and error methods. Perhaps the biggest drawback of analog integration is its fairly limited feature set. In addition, the actual code sequence execution takes much longer versus the more sophisticated techniques explained below.

Out-Of-Band Integration

This type of interface between voice processing gear and PBXs is also referred to as CTI (Computer-Telephone Integration). CTI links are serial communication links between the PBX and a computer. The computer is typically a PC that acts as a gateway to the application processor or VRU.

For example: HP ACT, Lucent ASAI, NT SCAI, IBM CallPath, and ECMA CSTA are all serial link protocols.

With CTI, call progress status or commands are sent back and forth over the serial link in order to manage call processing on the PBX with outboard state control. For example, a person at a workstation can press a button on the keyboard to make a telephone call. The command travels from the PC over to a VRU, CTI server, or some other adjunct processor. These intermediary devices then transform the command into a message that can be interpreted by the PBX over the serial link. The PBX then places the call and automatically connects the call to the phone associated with the requesting workstation.

The same goes for incoming calls, except the process is in reverse. For example, a PBX will identify ANI (Automatic Number Identification), and this data is sent over the serial link. An application associated with the call center then checks a database to see what agent (workstation) is associated with that phone number (customer) record. The VRU or CTI server then instructs the PBX to transfer the call to the correct agent's phone. More sophisticated systems also transfer customer data to the workstation coincident with the call transfer. This is known as a "screen pop."

Figure 5.1 shows how this is done with Novell TSAPI CTI server. The CTI unit receives commands from two directions: 1) from the PBX via a proprietary emulation card or serial link; and 2) over the LAN from the workstation. This arrangement is typified by offerings from Novell and AT&T, Northern Telecom, and NEC, for example. Similar arrangements are achieved with IBM CallPath and Dialogic CT-Connect.

PBX and computer makers are beginning to abandon their former proprietary schemes for the ECMA CSTA standard. Now, virtually all major switch manufacturers have CSTA-compliant CTI prototcols. It is important to note that Novell's TSAPI CTI server and Microsoft's TAPI are *not* CSTA-compliant. It is left to the switch vendors and makers of call center software to write the *service providers* (drivers) to talk directly to the switch.

Figure 5.1 - Novell View of CTI Integration

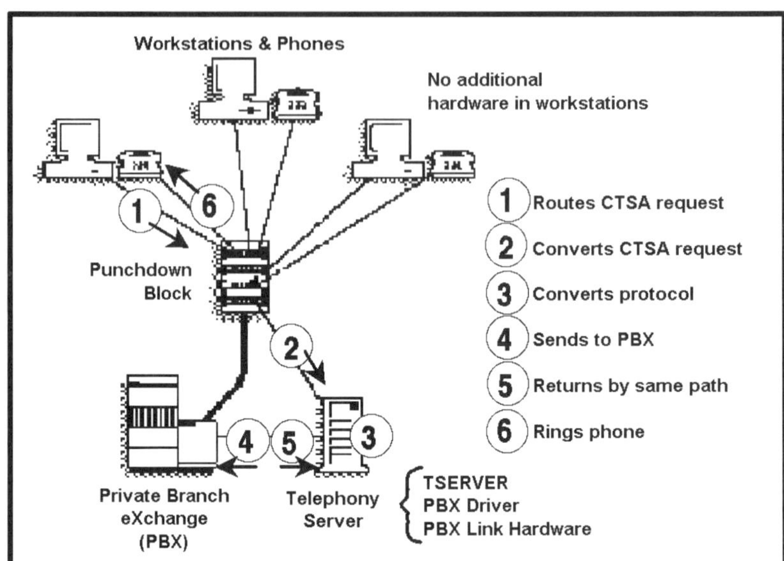

For example, Dialogic's CT-Connect has a direct CSTA link to about one dozen switches. TSAPI and TAPI-based applications can access this link using CT-Connect as a proxy. The benefit of out-of-band integration is its tight coupling between computers and telephone systems. Access to advanced features, such as ANI, extension transfer, ASCII names, screen pops, is also a great benefit. Unfortunately, the special software, servers, and licensing arrangements sometimes make for an expensive proposition. For example, to truly implement a TSAPI interconnection, you're bound to spend at least $20,000 by the time you purchase the platform, APIs, PBX interface card, and associated software. This cost does not take into account the application-level programs.

Figure 5.2 shows how similar interconnection can be achieved using the Microsoft TAPI standard. TAPI 2.0 supports "remote service providers," thus allowing client workstations to initiate third-party calls. So, with the right *hardware*, *application software* and *service providers,* users can call-up both local and shared services with TAPI. For example, a sales contact management system could use a TAPI interface to command an inboard modem to go off-hook and dial a client's number automatically.

Figure 5.2 – Microsoft's View of CTI Integration

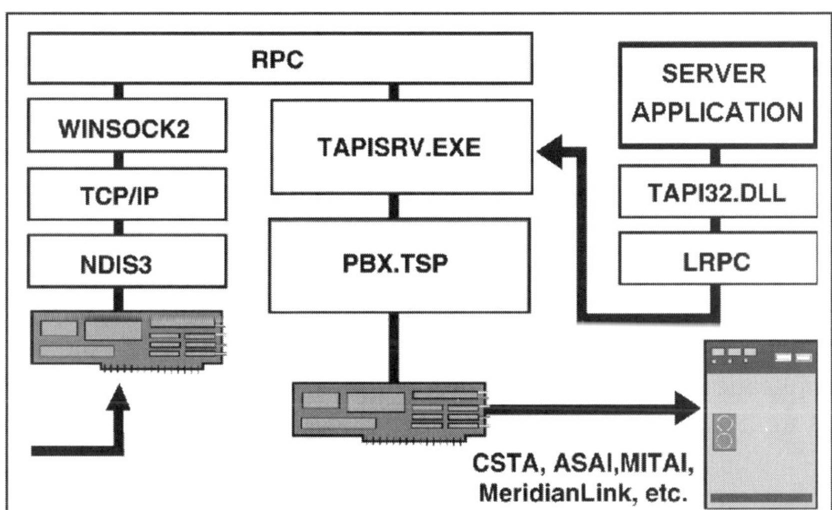

Such is the case with COM2001's TransCom application. Now Microsft promises applications like these will work in boht first-party and third-party call ceontrol scenarios. This is true if the requisite local *and* remote service providers are implemented.

In some cases, you can use TAPI-compliant applications that sit on top of a Novell TSAPI CTI server. Such is the case with Comdial's wide.open.office. And Lucent Technologies recently announced its port of their TSAPI-based PassageWay product to Windows NT. Amidst all these options, Sun Microsystems has launched their Java Telephony API (JTAPI). Sun says their API will sit on top of all the others to provide applications written in Java the ultimate portability.

Proprietary Phone Integration

Proprietary phone integration emulates the electronics of proprietary telephone instruments. This is done by putting the same microprocessors and telephone interface circuitry found in proprietary phones onto a PC expansion board. Examples of this include products from Voice Technology Group and PBX integration cards made by Dialogic Corporation.

Proprietary phone integration allows special PBX features to be controlled by the VRU or CTI server. Proprietary phones use C.O.V. (Carrier Over Voice), FSK (Frequency Shift Keying), B.R.I. (Basic Rate ISDN) and other methods to establish a communication link to the PBX. This is in addition to the regular voice circuit associated with telephone calls. In a sense, these methods are a station-oriented means of out-of-band integration.

This technology enables time, date, trunk number, extension number, names, and other data to appear on the LCD displays of feature phones while you're making a telephone call. The same data can be presented to the application running in the VRU or CTI server in order to automate access to the features.

Feature access is limited to the capabilities of the feature phone that is being emulated. Since these phones are based on proprietary standards, users of the technology are held hostage by the switch maker's whim on revision control and feature access. Some third-party CTI applications are co-marketed or sold directly by the PBX manufacturer. This is true of the PassageWay Direct Connection and FastCall products sold by Lucent Technoloiges and Aurora Systems, respectively.

Figure 5.3 - Passageway Direct Connection

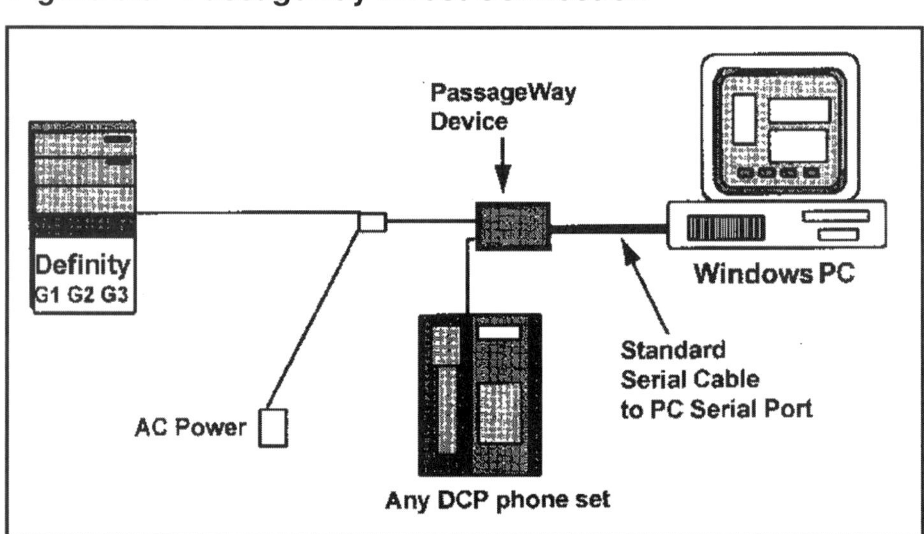

Figure 5.4 - Phonetastic Incoming Call Rules

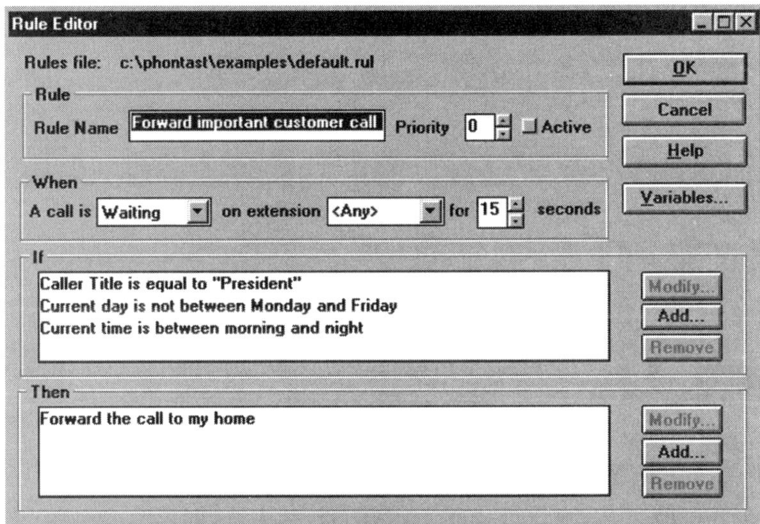

Figure 5.5 - Phonetastic Incoming Call Status Screen

Figure 5.3 shows how a special serial device that plugs into a PC can be connected in-line with a proprietary phone. This is a "twist" on the emulation method, because the device is more or less a "tap" on the proprietary phone. Northern Telecom and Rolm pioneered this in the late seventies with their add-on data modules. What we're seeing now is the "second generation" of these earlier devices.

But now, you can link these serial devices into standard Windows programs. Some programs let you use TAPI, macros or DLLs (Dynamic Link Libraries) to spawn programs when call processing events occur. You can define these events in a set-up or "rules" screen. One good example of a rules-based screen is illustrated in figure 5.4. Here, CallWare's Phonetastic is used to force calls from "presidents" to the user.

By integrating your application with an address book or PIM database, you can trigger visual call screening events. Figure 5.5 shows how the Phonetastic incoming call status screen provides a caller name pop-up when a call arrives.

Drop-and-Insert Integration

Sometimes, there is just no way to elegantly integrate to a PBX. For example, if the PBX does not have an ANI (Automatic Number Identification) interface, you cannot automatically route callers to specific agent extensions. This can be overcome by putting in place certain voice processing gear as an adjunct to the switch. In this case, calls are taken first by the VRU in "front of" the switch.

In effect, the VRU acts as a protocol converter between the Central Office and the PBX. This allows the VRU to control database lookups based on the caller's ANI or an incoming DID (Direct Inward Dial) number. After the call handling for that transaction is determined, the VRU can insert the calls into the front-end (trunk side) of the PBX as if the call were a DID call. Calls can also be taken by the VRU and then inserted into a group of station-sided interfaces like an automated attendant.

But the VRUs must be configured for a one-to-one correspondence between each incoming VRU port and PBX circuit. Since the switch's circuits are not being shared by the VRU, you're buying twice the telephone circuit. In some cases, the ports associated with a certain call center function may only require a small percentage of the overall circuits feeding into the switch.

The primary benefit of this drop and insert approach, however, is that the VRU (and thus your software) is in total control of the call processing. This transforms older, feature-poor switches into intelligent "telephone resource servers."

Integrated Call Center (UN-PBX)

The new age of computer telephony has given birth the concept of combining the functions of the voice processing, CTI links, and switching all into the same platform. We calls this an UN-PBX. This requires the developer to create ACD and other switching logic and algorithms. This takes years of experience to do well, but it is perhaps the most "integrated" environment possible. This approach is typified by PC-based systems Altigen's AltiServ platform. Altigen's systems use Win NT-based PCs with cards that handle voice processing, telephone network interfaces, matrix switching, tone generation, and LAN interfaces. Their product also provides messaging and routing functions.

The benefit of this approach is that there is one main program that is in control over every aspect of the telephone-based transactions. By tightly coupling the telephone functions to host and database activities, no other telephone-oriented resources are needed except for the telephone instruments themselves. Shrink-wrapped PBX and ACD software packages for PCs are just becoming commercially viable. These integration schemes also apply to other environments. Now that these have been covered in overview fashion, here are a number of examples of how call centers use voice processing in vertical industries.

Automotive

Dealer Locator

There's many cases when a customer has to call a car manufacturer to ask about where a certain model can be purchased.

Although this can be achieved with total automation using an IVR system, more often than not, you speak with an agent first. A call center application for dealer locator will use voice processing to partially automate the call. This idea is similar to the way directory service operators help callers with changed numbers and new listings.

For example, a caller dials an 800 number to find out where the closest dealer is for a new sports car. A customer service agent greets the caller and then asks for his or her zip code or telephone number. The agent can then type this information into a database program and then transfer the caller over to a VRU that has access to the same database. If the call center is set-up with a CTI link, then the switch can inform the VRU that the call is associated with the database entries made seconds earlier by the agent. In this fashion, the call is partially automated so that the caller finishes-up the transaction with the VRU. In this case the VRU may say: "There are three choices for Z-5000 dealers in your area. Press any key when you wish to receive more information on each one or wish to get directions to the dealership."

Rental & Reservations

Call centers that deal with reservations employ hundreds of operators to handle thousands of calls each day. Customers typically dial a toll-free number and are connected to an ACD (Automatic Call Distributor). In some cases, there are no agents available, so the call is put on "hold." Due to the fact that car rental is very competitive, it is not desirable to put people on hold for fear that they will hang-up the phone. There are several ways to deal with this.

First, callers who are put on hold can be transferred to a VRU in order to describe the type of car they want, where they wish to pick the car up and drop it off, and the dates of their travel. Since this information has to be collected in order to complete the transaction, the caller is actually saving him or herself the time that would have been spent on the phone with the agent.

When an agent is free to take the call, he or she can pick-up where the VRU left off. The caller would hear a message that says: "Thank you for your help so far, an agent is now available, so I will connect you along with all of the information you just gave us. One moment please while I transfer the call." Secondly, if the system is equipped to handle ANI, the caller's phone number can be captured and used to make a return call. Of course, if the caller became impatient and called another rental company, the fact that the number is stored is a moot point. If the call-back is made quickly, however, a sale can be "saved" in this way. Another approach to automating reservations centers is to play-out an "on-hold" recording that describes different specials, car upgrade polices, and other information the customer may be interested in hearing about. This can be done by either playing a revolving message, or by providing an Audiotex type of capability.

Corporate Communications

Conferencing and Meeting Coordination

Latitude Communications of Santa Clara, CA has developed a special voice processing and conferencing system called the MeetingPlace Conference Server. You can now set up your telephone voice conferences via the Internet / Intranet. You can hear meeting comments and / or complete recordings of meetings with streaming RealAudio or Microsoft ActiveMovie. You can also see meeting "attachments" -- PowerPoint slides, Word notes, Excel sheets, etc.

Latitude's products make for more productive business meetings amongst people working in different locations. Latitude developed the industry's first conference server, enabling increasingly dispersed employees to streamline the entire meeting process. The company has received key patents on the MeetingPlace technology and are widely regarded as delivering the first real-world, collaborative work tools. The company is doing well, having sold over 40 systems -- from $40K to $250K (not cheap).

One of their customers is Microsoft, which has created what they call the "Expert Roundtable". These are conferences held as part of their "Premier" support offerings covering topics such as Windows NT, SQL Server and System Management Server. With MeetingPlace, they can create a "library" of meeting topics and post them to their Web page.

Microsoft also uses MeetingPlace for their Technical Account Manager (TAM) Escalation Calls. Microsoft's TAMs work with customers to resolve problems which frequently require multiple vendors to be pulled together at a moment's notice. In responding to customer problems, Microsoft uses MeetingPlace to initiate immediate meetings to quickly gather the appropriate internal and customer personnel on a conference call. As the problem is better understood, other vendors can be added via the outdial feature. MeetingNotes recording capability provides an accurate history of the issue as well as a record of vendor commitments. Other Latitude customers include Credit Suisse, ADP Brokerage Services, Best Buy Corporation and 3Com.

The MeetingPlace conference server addresses all aspects of the "meeting process," not just time participants are actually in the meeting. MeetingPlace creates virtual meeting rooms that are more like face to face meetings.

The system enables users to "see" who is in the call, provides for breakout sessions (private side conferences), ensures a secure meeting environment, facilitates document distribution and allows participants to record all or part of the meeting. These tightly integrated, face to face like features make meetings more productive.

The system enables users to schedule conferences from any location. Figure 5.6 shows a Web browser's screen for inputting a meeting request over the Internet. MeetingPlace will then notify all meeting participants of their conference call via e-mail or fax. At the start of the meeting, the system will page or dial out to users to bring them into the meeting. Documents may be distributed with the meeting notification, pulled down from the server in advance of the meeting or retrieved during the meeting.

Finally, MeetingPlace makes the meeting available to non-participants by storing all meeting materials (documents and recordings) for those who were not able to attend.

Figure 5.6 – Latitude MeetingPlace Scheduler

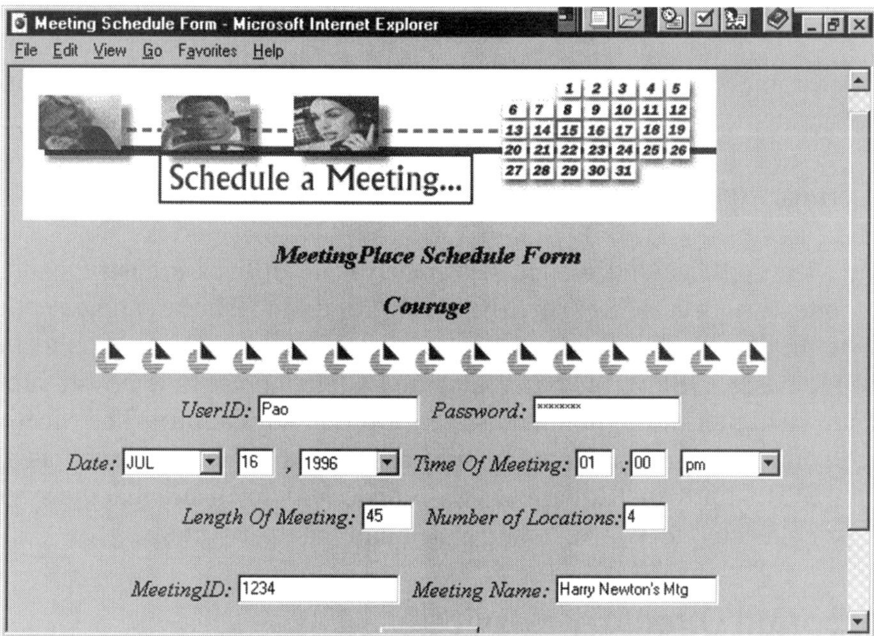

A meeting is much more that just the time participants spend in the actual meeting. A meeting impacts its organizers and participants before, during and after the actual discussion. MeetingPlace addresses five distinct stages of every meeting:

Schedule. Ensure meeting resources are available

Notify. Inform meeting participants

Distribute. Disseminate meeting materials/documents

Attend. Round up participants to start meeting

Follow-up. Provide meeting recording/documents to non participants

As a conference server, MeetingPlace establishes and coordinates multiple voice connections for a conference call. The conference server can support up to 120 ports in any combination of simultaneous conference calls.

At the same time, MeetingPlace's client/server architecture allows users to schedule meetings, access in-conference features, review previous meetings and attach and retrieve documents from their desktops. The server has two primary components: server hardware and server software.

The MeetingPlace conference server hardware utilizes industry standard components such as SCSI disk drives, EISA and MVIP buses, and a Pentium system processor. The system utilizes a distributed processing architecture, multiple buses, DSP/RISC processing and a telephony grade power supply to achieve higher throughput and superior product reliability. The MeetingPlace conference server provides industry standard interfaces such as T-1, analog POTS lines and Ethernet LAN connectivity to ensure that the system can be connected to your existing voice and data networks.

Corporate Directory Service

CCOM Information Systems of Iselin, NJ sells the PhoneLine is a family of directory management products. It lets PC users access comprehensive corporate directory information. It works with TSAPI and TAPI. CCOM has users in Fortune 1000 corporations, international and national government agencies, public utilities, financial and healthcare institutions.

Their first sale was to Johnson & Johnson. At many Johnson & Johnson facilities, you'll find PhoneLine when you walk through the front door, right at the reception desk. And once you're past reception, you'll discover it on J&J desktops throughout the organization -- in fact, PhoneLine can be found on J&J desktops throughout the world.

Corporate Internet-Enabled Voice Conference

Compro Technologies, Inc., of Barnegat, NJ sells solutions to the enhanced services sector of the telecommunications industry. Compro is also a Summa Four reseller and a full service solution provider. Compro offers a wide variety of programmable switching solutions as well as consulting, installation and support services.

The company's PeopleLink ChatCall is a telephone conference calling system designed for chatters on the Internet. PeopleLink demonstrated ChatCall publicly for the first time at the Agenda '97 conference. ChatCall gives people a safe, convenient way to move their conversations from typing to talking. After setting up a ChatCall account from a World Wide Web site (http://www.chatcall.com), people chatting online who want to talk, share an anonymous code then pick up the telephone and dial a toll free number. Upon entering the codes, the chatters are instantly connected on a private line. No telephone numbers are revealed or traceable, even with Caller ID technology.

Behind the scenes, the Internet and telephone systems interact using sophisticated proprietary and secure communications methods. The telephone portion of the system consists of specially configured open computers, provided by Compro Technologies, executing a customized, exclusively-licensed version of Compro's telephone conferencing technology.

As a participant in the ChatCall system development project led by PeopleLink, Compro took its conferencing technology with its standard feature set and modified it to meet PeopleLink's unique requirements. "Our systems are highly data driven," said Mike Gilbert, Senior Engineer at Compro Technologies. "This provides a tremendous amount of flexibility. Typically when people think of a standard application they think of identical off-the-shelf functionality, but we've found that every customer is different. We worked with PeopleLink to integrate our conferencing technology and their proprietary technology creating an innovative, brand new Internet/telephony service."

For businesses offering conventional telephone conferencing service, Compro's standard Meet-Me Conferencing application is an exceptional product. It supports eight talking participants and an unlimited number of listening participants. Meet-Me-Conferencing controls a Summa Four switch using Summa Four's command language. A database is loaded with the local and toll-free telephone numbers designated as conference calling numbers and a set of passcodes. Conferencers typically call a common number with a pre-assigned passcode. When the caller enters the passcode on a touch tone keypad, the application verifies the code, than connects the call to the conference (or rejects invalid passcodes). A variety of options are available for playing prompts that provide help instructions or announcements.

Corporate On-Site Paging

Priority Call Management of Wilmington, MA sells an enhanced service platform called ORYX. It handles one number "FOLLOW ME" calling. One notable success story was their sale to Progress Software. PCM replaced their noisy overhead paging system with the Oryx platform and an in-building paging system. Now people aren't bothered all day by the noise. People are notified by pager and then they pick up a call from any phone in the building. It's easy to sell because it's something that makes office life easier. People are no longer deskbound.

Telemarketing Call Centers

Almost any customer service team can be called a telemarketing center. It doesn't matter if there are 500 agents, or just two people answering the phone. No mater how many people there are, voice processing can increase their efficiency, extend service hours, and make communicating with customers a whole lot easier. Take, for example a simple automated attendant function. With automated attendants, customers can call the system, type-in the extension number or department number they wish to speak to, and then allow the VRU to route the call.

In most cases, callers are prompted to leave a message if the agents are busy. More sophisticated systems will screen the call, so agents can decide whether or not to take a more urgent call and put the first caller on hold. Automated attendants can transform small call centers by providing features sometimes associated with larger ACDs.

Although it's not typical to have supervisory control, management reports, and sophisticated queuing with a simple automated attendant, they still provide call routing and call handling capability. The use of automated attendants or call sequencers is very popular nowadays since a small call center can take on the appearance of a much larger operation.

In addition, many call center mangers are adding fax servers, music on hold systems, fax-on-demand systems, and Internet links in an attempt to augment the services of their sales staff. Every little bit helps, but no one single action can make more of an impact than helping to automate the telephone calls themselves, and that's where voice processing comes in.

Voice processing systems can also be used to ask the caller questions in anticipation of a call transfer. For example, a method called "whispering" (coined by Granada Systems Design of New York City) can be used to partially automate calls. In this example, callers are asked to speak their name or to enter their account number as soon as they call in. The VRU then calls the agent's telephone and "whispers' the record locator number into the phone.

The agent then pulls-up the record and makes ready to receive the call. This avoids the expense associated with CTI links. With a few keystrokes, the agent is able to quickly scan the customer record and take the call after being so prepared. This can shave ten to fifteen seconds from each call as well as "keep the customer busy" during the holding time. The net result is that the overall efficiency of the call center is increased.

Correctional Institutions

Collect Call Service

The penal system faces a special challenge in providing outside communications for the inmate population. The most common problems have to do privileges associated with making phone calls, accounting for the number of calls, and the use of coins to make payphone calls. Prison officials do not encourage inmates to carry money, because it creates a temptation for theft, violence, and corruption. Collect calls are a way around this problem, and many elaborate schemes have been devised to facilitate these calls. For example, the use of special call accounting software, debit cards, and pre-arranged phone calls are all in use.

Since the cost involved in making collect calls is typically higher than direct distance dialing, a number of companies have developed systems to reduce the cost. Notable examples include Telequip Labs and Executone Information Systems. Executone sells, installs, and maintains thousands of hospital, campus environment, and prison-based PBXs, nurse call, and paging systems. Voice processing goes a long way to solving the collect call problem by providing an interface for collecting dialed digits, account codes, and time-stamps for each inmate telephone call.

Among the approaches for automating the calls is the prompting of inmates for their name in order to forward the recording of their name to the called party. The voice processing system "takes" a call for the inmate, "makes" a call to the friend or relative, and then "patches" the calls together to complete the transaction. All of this is achieved without the use of an operator.The voice processing system produces logs used to debit an established account, or to prepare an invoice that is sent to the called party for collection. Another method is to use the voice processing system as an adjunct to an existing "payphone concentrator." The billing and collection are performed by a long distance carrier or reseller, and the voice processing system is used to "multiplex" a shared group of telephone lines that are connected to the long distance telephone system.

Fugitive Finder

Both volunteers and professional call center staff answer special "fugitive hotlines" to follow-up on tips from citizens. These call centers have been popularized by "wanted" shows on television. Because the information being sought is about criminals, there's no telling whether or not the caller is in danger. Voice processing systems can work alongside telephone switches to gather information about the caller to aid in the processing of these calls.

For example, a discrete telephone number can be established for each fugitive, so the last four digits of the phone number are unique. Using DNIS (Dialed Number Identification Service) enables this.DNIS service signals the voice processing or telephone system with the last four digits of the number dialed so the call can be given special treatment. For example, the voice processing system can check a database for the fugitive's name that is associated with the phone number and announce it to the caller: "Thank you for calling about John Doe. A special agent will be with you in a moment - please stay on the line."

The caller gets instant confirmation that they dialed the correct number because the fugitive's name is spoken. This cuts-down on pre-mature hang-ups and encourages the caller to stay on the line.By using ANI (Automatic Number Identification), the voice processing system can log the phone number of the calling party so that they can be called-back in case they do hang-up. In the case of an emergency, it is even possible to match the phone number with a name and address database, so local officials can be dispatched immediately.

The voice processing system can also continually log the ANI/DNIS information so it can be analyzed over a period of time. This has the effect of "tracking" the whereabouts or at least the sightings of the fugitive. Several calls on Tuesday, for example, may indicate that the fugitive was seen in a truck stop in Alabama - calls the next day may indicate a sighting in Tennessee, suggesting that the fugitive is making his or her way for the North. This could be helpful in guessing the fugitive's next move.

Education

Telephone Activation

Students often discover that getting their phone hooked-up in campus dormitories is a tedious prospect. Service activation means waiting in long lines at the telephone company or at an administrative office on campus. Voice processing gear can help to alleviate the problem by partially automating the procedure. For example, phones can be tested and activated ahead of schedule so that the provisioning of the service is limited to billing issues. Students can be instructed to call the voice response system can ask the student a series of questions. For example, they can enter a student I.D. number or social security number, the telephone number they are calling from, and the type of service they would like to sign-up for.

This information can then be made available to an operator if a live transaction is required. Operator intervention my be required if the social security number or student I.D. triggers an alarm indicating that the student has an outstanding bill from the previous semester. A voice processing system can also be used to collect credit card information in order to automatically take payments for overdue bills or establish a deposit for new service. A system of this type can significantly reduce the time it takes to set-up phone service for the entire student population.

Financial

Financial Brokerage Voice Logging

In addition to a back up resource for settling disputes, many countries require trading floor transactions to be recorded. Financial institutions such as brokerage and trading firms, which conduct most of their business over the telephone, need to rely upon their ability to record, store and retrieve voice data of transactions in a timely, reliable and efficient manner.

Financial users record and store recordings of transactions to provide back up and verification of such transactions and to guard against risks posed by lost or misinterpreted voice communications.

Nice Systems, Inc. of New York, NY estimates there are a total of 150,000-200,000 brokers worldwide. Each broker typically uses four lines and this equipment is replaced every five to seven years. All of the telephone lines need to be recorded. Based on an average cost of US $700-US $800 per line, sales to financial institutions amount to US $15-US $32 million a year, including analog and digital equipment.

Financial Call Indexing

Dragon Systems of Newton, MA says institutions large and small that handle large call volumes often have need to record them. Financial institutions to comply with SEC regulations, law firms for liability reasons, and customer support call centers of all sorts to monitor quality and keep records of customer problems. At the end of each day, for example, large institutions may have recorded over 100,000 conversations and the resulting data are not generally indexed or categorized.

Knowledge of the content of those calls has significant potential value to the institution being called. They can use that knowledge to find specific topics of interest, track references to competitors, build databases of problems, and many other facts that can be "mined" from the data contained in those calls. Some of the value to the institution is in cost savings, where relevant, but the biggest value is in accessing data that is currently unanalyzed.

For example, financial institutions would love to have a daily summary of all competitors' "instruments" mentioned by their callers.

However, audio (and video, for that matter) are difficult and time consuming to access and analyzing them has unique constraints:

- Audio takes time to experience—one must listen to it in roughly real time before deciding if it is of interest. That means that there is no good way for a person to rapidly browse large amounts of audio data.

- Transcriptions lose information—important data such as tone of voice, emphasis, and irony are lost when the original data are converted to text. Therefore, an indexed database of the original data is preferable to a reduced text form.

- Human-created transcriptions lose data—Call center representatives find dictating summaries to be a hassle and do not necessarily do it conscientiously.

Also, only some calls are summarized. It is preferable, therefore, to create automatically an index of the original audio data. Recent advances in speech recognition developed by Dragon Systems can solve those problems. Topic identification, speaker identification, language identification, and voice-driven information retrieval can identify topic, speaker, tone of voice, language, and other speech attributes to provide automatically the markers with which to make audio databases *content addressable*.

Credit Dispute Hotline

Consumers usually only call credit bureaus and collection agencies when they are in financial trouble, or there is some dispute about an overdue bill or unpaid balance. Most callers are frustrated by the time an agent comes on the line. For this reason, voice processing can be put to good use.Each credit report has the consumer's social security number on the first page (usually in the upper right-hand corner). A voice processing system can prompt the caller to enter this information so that a credit bureau agent can automatically bring up the record and provide information to the caller as soon as they are connected. In addition, if no agents are available, it is reasonable for the voice processing system to dispense information about the credit report.

For example, TRW, CBA Associates, and Equifax all have some level of automation that allows this. The voice response system can say: "According to our records, the inquiry you have made is still under review. We will be prepared to provide information to you on the issue in 5 days."An enhancement to the call center can also be made by allowing callers to leave a message indicating that they want to order special forms. These forms can either be mailed to the consumer or faxed. In addition, the voice processing system can be programmed to give an explanation of the codes that appear on credit reports. These codes are associated with the aging of accounts, credit rating, type of account, and payment status. Most of the codes are obscure and require explanation. A voice processing system can prompt callers for the code numbers for a full explanation.

Institutional Bank Annuity Inquiries

The Institutional Products Division of American Express Financial Advisors (AEFA) approached Venturian Software of Hopkins, MN in the spring of 1995 with a problem. Based in Minneapolis, MN, the Institutional Products Group was fielding an increasing number of inquires from their customers (insurance department of client banks, client bank's agents and marketing companies) regarding their investments. They needed a faster way to disseminate information.

In less than two months, a solution was delivered by Venturian Software. The system was developed in MultiCall, VSI's computer telephony application development product. MultiCall runs under Windows 3.X, uses Rhetorex, Dialogic or Aculab cards and is scaleable up to the largest size allowed by MVIP switching.

Customers now use touch-tones to navigate through a series of audio menus, extracting account and rate information on fixed annuities from the LAN based administration database. In addition, when needed, the system automatically routes callers to the appropriate local sales office for additional assistance.

MultiCall's TSAPI-based modules support call-center applications such as intelligent call routing, intelligent call distribution, call pacing, voice pops, screen pops and real-time call management. It also supports IVR, fax-on-demand, Audiotex, voice mail and auto attendant.

MultiCall software has two main components: The MultiCall Generator, a menu-driven app gen that lets you develop and application; and the MultiCall Runtime Engine, which is the program that actually runs the "Voice Box" hardware. It's a multitasking program that supports from four to 32 ports per "Voice Box."

Financial Mutual Fund Web-Enabled Call Center

Ellen Muraskin, reporter for Computer Telephony Magazine, researched a killer application at Scudder, Stevens & Clark, Boston. Scudder is one of the biggest U.S. mutual fund brokerage houses. Scudder wanted to duplicate all the functions of its IVR-based trading and inquiry system on the Web-and then some, taking advantage of the added possibilities offered by a visual interface. It also wanted to make it more truly interactive, by tying investors into their call center operation if they needed assistance.

Scudder's mutual fund trading IVR system won high praise when it was implemented about a year ago. According to Mick Chamberlain, VP of automated services for Scudder, independent market researcher Dalbar gave it the highest marks among mutual fund companies for ease of access and use. One of its special charms, says Chamberlain, was that it could be individually customized. You don't want to hear all those options? Fine. For your account, you could banish them from the IVR script or reinsert them. That was a very popular feature. That system, running on NMS cards and 240 software "agents" of Edify Corporation's (Santa Clara) Electronic Workforce, took nine months to develop.

When it came time duplicate those IVR functions on the Web, Scudder again wanted to add a twist.

They started with a secure Web site that would give investors the familiar options to purchase, redeem, and exchange mutual fund shares. The company started to add some bells and whistles that only a graphical environment can provide; displaying, for example, transaction histories and instant pie charts of investor's assets spread across different Scudder funds. The new Web piece, using 48 more Edify agents on a separate Web server, accesses and updates the same mainframe database retrieved by the IVR.

They also wanted to add the same "dial 0" escape to human contact that all good IVR systems offer. They did this by sticking on their Web site a graphical "call me" button that links the whole Web process to their call center.

When a befuddled user clicks that button, the screen asks him or her to enter a phone number he can be reached at. Using Edify software agents, Scudder's system grabs the number and calls the user back immediately. As soon as it reaches its live customer, it throws the call into the call center's ACD queue.

When a live Scudder agent picks up, the system pops the customer record on the agent's screen and bridges the call. Chamberlain says they'll probably use another Edify agent to do "page shadowing." This is also called a "mirrored Web page screen pop." It shows the agent exactly what the caller is seeing by sending the Web page along with the call.

Most of Scudder's Web-enabled transaction processing parallels its IVR functionality; you log in through the Web site, enter your password, and can purchase, redeem, and exchange shares. You can view transaction history, open new funds and view the check-writing, electronic-funds-transfer, and other options that come with your account, as well as other options offered. According to Edify's Senior Marketing Manager Francisco Kattan, Scudder's IVR server now accommodates up to 240 concurrent IVR sessions. And according to Chamber- lain, another Edify server is on order to bolster Web access, adding to the 48 simultaneous sessions it can now support.

Figure 5.7 – Edify's Electronic Workforce at Scudder

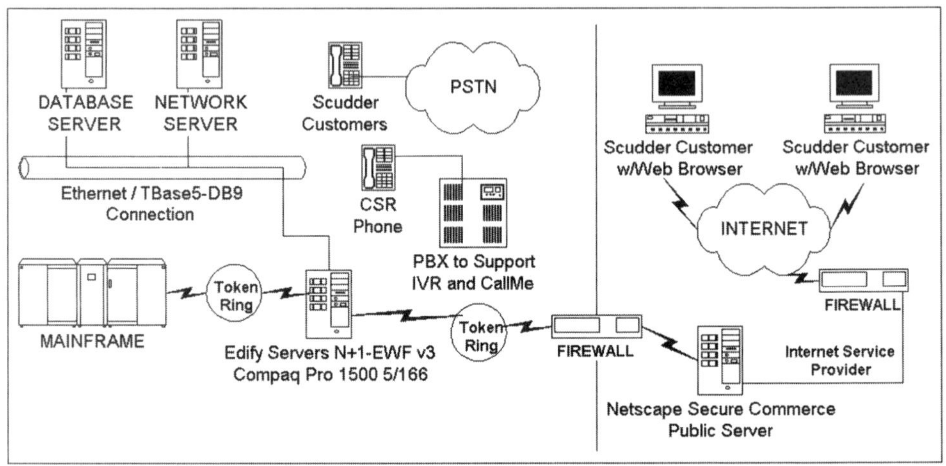

The new server runs on a Compaq machine, inside the firewall behind its off-site Web server. The Edify server fetches customer records from (and makes real-time updates to) the same PC-based database as the IVR system; the mainframe provides mutual fund performance data. The Edify server also works a T-1 connection to Scudder's customer call center, shipping Web customers' "call-me" calls to the center's ACD.

Look for this trend: While Web servers access the same customer records that CT servers use, they can deliver far more information at a time. We're starting to see HTML displays, composed on the fly, that go on for columns: the kinds of individual account information that bank customers are only used to seeing in a monthly envelope.

Scudder's preexisting IVR system used 240 Edify Electronic Workforce "software agents" (software objects that answer calls, retrieve records, speak data), distributed across multiple sites. The database of customer information uses popular PC-based DBMS software. A mainframe host with mutual fund performance data augments this. The voice processing part included Natural MicroSystems voice boards serving 240 ports.

By adding Web access to the system, Scudder used the same PC-based database and mainframe host but added 48 more Edify software agents. They also installed an internal (inside the firewall) Edify server on a Compaq platform, running OS/2 and added a T-1 connection between the Edify server and the call center. In addition, Scudder used an off-site Web host, using a Netscape Secure Commerce Public Server and a Secure Sockets Layer and encryption to ensure secure transactions. All of the real-time telephone calls are tied in to the call center's switch and ACD system as illustrated in figure 5.7.

Financial Real-Time Turret Trading

Technical Telephone Systems, Inc. of Somerset, NJ is a systems integrator specializing in ISDN and computer telephony. TTSI is also reseller of Sun Microsystems and Microsoft products. Their big distinction is the proprietary X-Turret trading turret platform and Integral 33x ISDN switching system. They are exclusive partners with Bosch Telecom (division of R. Bosch GmbH in Germany). Together, they've installed large, specialty call centers for the financial/trading industry.

ABN AMRO Bank's new trading floor in Frankfurt, Germany and Limited Credit Services' Ohio-based call centers are both users of the system. The platform uses a network of integrated ISDN workstations with programmable keyboards. These workstations let traders or call center agents perform traditional trading functions without traditional turret hardware.

The heart of the system is TTSI's proprietary TANK keyboard and X-Turret software. With these pieces, you can create your own customized trading turret in an Open Windows or Motif environment right on the Sun SparcStation. In effect, you get a "turret-on-screen." Add to this video conferencing, access to market data services and other on-line interfaces. This works with Bosch Telecom's ISDN switching technology. Each X-Turret station works on a minimum of one BRI ISDN connection and interfaces to the ISDN Network controller.

TTSI's TANK is the Integral 33x System's integrated keyboard. TANK supports everything you can do with a phone and it drives the workstations and PCs. TANK also handles the emulation of market data services like Reuters and DTB. All this with a single keyboard.

TTSI's X-Turret screen is the front end of each position. It can be redesigned as needed for each trader. You also get "free seating." This allows a trader to access an integrated workstation anywhere in the world, log on, and instantly access his own, customized, front-end.

The ISDN part of the solution gives traders real-time access to vital information. Historically, this data resides on separate networks for voice, data, text and images. The integrated turret displays data on-screen with a programmable keyboard interface. This lets traders execute transactions with instant access to market data and client info simultaneously. The alternate (multiple disparate terminals) is slow and painful.

According to TTSI, you can run as many as 500 lines into each station. This due to the ISDN switching built-in to the Integral 33x platform. In effect, the traders' workstation becomes their telephone. It greatly accelerates the traders' ability to gather, send and react to real-time data.

Financial Remote Positive Verification

Genesys recently demonstrated North America's first functional video call center. The demonstration showed Genesys' ability to route incoming H.320 compliant video calls to the best skilled agent in an operational call center. The video calls were carried over ISDN lines, passing ANI and DNIS information to be used by the skills-based routing engine. As video calls can be routed to any multimedia agent in the call center, the routing engine queried their status to prevent routing a video call to an agent already handling a voice or Internet call. As you may well imagine, the possibilities are endless.

Applications range from customer loyalty schemes, transactions that require face-to-face interaction such as in the financial sector help desk scenarios, etc. Certain transactions require not only initial (positive) identification, but human interaction where body language and facial expression communicate crucial data.

Dan Plashkes, CEO of S&P Data, called his Genesys enabled video call center in Toronto from the stage of Computer Telephony Demo Fall '96 in Orlando On stage was a single PC with an H.320 video system (Intel Proshare in this case). Mr. Plashkes placed the call by dialing a 1-800 number. Two live feeds were presented on the stage screen. One half of the screen had the Video Softphone GUI showing the progress of the video call. The other half of the screen showed a live data feed from Toronto. This detailed the progress of the skills based routing (text lines scrolling).

Figure 5.8 - Video Softphone GUI

Figure 5.8 depicts the desktop Video Softphone GUI, which was projected on the left half of the stage screen in Orlando. Here is a transcript of Dan Plashkes' "Killer Video ACD" phone call:

Dan: I will call 1-800-TRAVELS with this H.320 compliant video system, in this case an Intel Proshare, but it could just as easily have been a Nortel VisitVideo, etc. When my call arrives at the S&P Data call center, split second routing decisions will be made against the ANI and DNIS information related to my call. In particular, because I am calling from Florida, I will be connected to an agent who specializes in meeting the travel needs of residents of the Southern United States.

I direct your attention to the large stage screen. On the left you see the Video Softphone GUI. The agent will fill the large dark box as soon the agent is selected. The lower left box is a feedback image of myself. On the right, you see the text echo of the routing strategy that is being followed by VideoACD.

Agent: *(Southern accent)* Hi, this is Suzie. You all have reached S&P's Travel Service. How is it in Florida, today?

Dan: Hi, Suzie. This is Dan Plashkes and I am enjoying Orlando's sunny weather, but it looks like I'll have to head back to Toronto to take all the orders for this wonderful video service. Please book a flight for me from Orlando to Toronto.

Agent: Certainly, Dan. I have your customer record and preferences in front of me, and will fly you all first class on your favorite airline. Will there be anything else?

Dan: Yes, would you pan your camera to your left to show me your fellow agents handling voice, Internet, or video calls. *(pause)* Thank you, Suzie. I'll be going now. *(hangs up)*.

Well, there you have it. North America's first video call center. I could have called on any H.320 compliant video system using ISDN lines. Furthermore, if greater bandwidth is desired, Genesys VideoACD can aggregate any number of ISDN channels.

Government

Intelligence Agency Voice Logging

Nice Systems, Inc. of New York, NY sells digital voice logging equipment to intelligence agencies. The company says intelligence agencies require sophisticated voice logging systems that enable the recording, retrieval and processing of the information gathered for analysis and evaluation. They require only a small proportion of the large volumes of data gathered, but they are prime users of logging and retrieval systems.

Public Safety Agency Voice Logging

According to Nice Systems, Inc. of New York, NY, most public safety agencies are required to record their telephone calls. Public safety organizations, which include police and fire departments and hospital casualty departments, have traditionally been required to use voice logging systems to record and store voice communications. Historically, the public safety agencies have been users of analog recording systems. Nice's NiceLog system sits between the PBX or ACD at the public safety center. Emergency calls are all recorded – on every line coming in to the center. The voice information is digitized and stores on computer disk. Each conversation or segment is then time-stamped and cataloged for later retrieval.

White House Directory Service

Amcom Software, Inc. of Edina, MN sells the CTI Smart Center for Windows. Amcom's first sale was 15 years ago to a major department store. They wanted an on-line employee directory for their operators.

The biggest sale was to a 300 store retailer. They bought Centralized Attendant Service with Directory and PBX integration for over 100 operators. In 1991 Amcom sold a system for the White House. They provided a directory system for the President and his executive branch. It connected their AT&T switch, Motorola paging system and IBM network.

Hospitality & Travel

Concierge

As many as six or seven specialists can staff the concierge desk. The job of the concierge is to make the guest's stay as comfortable as possible and to "do the impossible" in locating theater tickets, emergency repairs, dinner reservations, or even a baby sitter. Virtually no request is too large or small for a good concierge. The trouble is that the concierge staff is almost always busy. Some guests either have to wait to speak with them over the phone, or they have to get help face-to-face.In effect, the concierge desk is a small call center.

There are several "agents," a small number of incoming lines, and a lot of "will you hold please while I check on that for you..." Voice processing systems can be a great deal of help to these small call centers by providing answers to commonly asked questions. For example, the concierge can indicate that they are really busy and offer the caller a chance to "browse" through the voice processing menu while they wait. The system can dispense information on the best choices for dinner that evening, entertainment, day trips, and other points of interest.

Another benefit is that a voice processing system can help callers judge how long they have to wait before their call is answered. For example, the system could say: "Thank you for calling the Cross Keys Hotline. There are two other callers waiting to speak to the concierge. This will probably take a few minutes. You are free to hold or you can leave a message and we will call you right back..."

Human Resources

Agent Service Observance

Dictaphone Corporation of Stratford, CT has the ultimate solution for Call Center service observance. The Insight quality monitoring system gives agent supervisors 20/20 vision. This client-server system helps call centers automate agent monitoring and evaluation. Insight has monitoring, recording, evaluation and reporting all in one package. And because Insight can capture both the audio conversation and the agent's screen, you hear and see the complete picture.

Here's the scenario. You've often heard recordings while you're waiting for a customer service rep to come to the phone and say: "In our ongoing efforts to serve you better, this phone call may be recorded for training and evaluation of our customer service representatives." And many calls are recorded. But have you ever wondered just what they do with all those recordings? Enter the world of agent evaluation and training. Here, each keystroke an agent makes on his or her terminal whilst speaking to you over the phone is likewise recorded.

Agent supervisors are able to review each transaction (after the fact) and pore over every minute detail. A mirror image of the agent screen is presented to the supervisor. The manager can scroll though the screen data and listen to what was going on during the data input simultaneously. Supervisors can pause these transaction playbacks in order to score an agent's transaction on a separate screen. This gives the manager a more studied and quality approach to making constructive comments and advice.

The insight system is comprised of hardware and software. The Insight server (see figure 5.9) houses voice processing boards. Each port on these boards is associated with a monitoring port on the ACD switch. The Insight Interface Unit fetches data from agent screens and stores this information for each transaction in an archive file. These files are indexed or paired to the voice recordings associated with the transactions.

Figure 5.9 – Insight Agent Monitoring and Evaluation

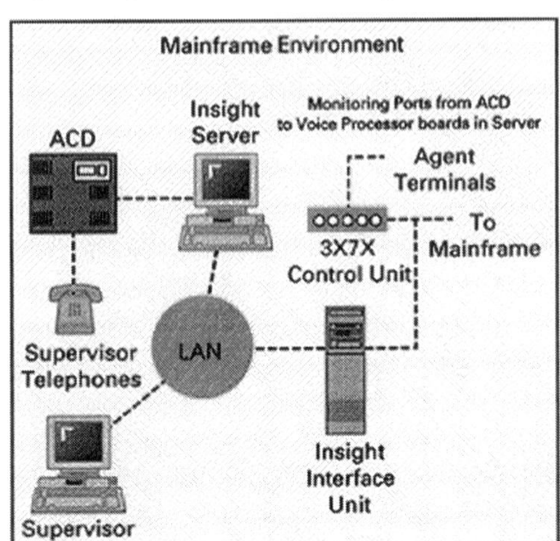

Insurance

Claims Interview

Claims adjusters and investigators must conduct interviews with the insured party in the case of accidents and large claims. Arranging for the interview itself is time consuming, because they take anywhere from five to 30 minutes to complete. Voice processing systems can be used to partially automate the process by asking callers to input certain information via touch-tones so the adjuster can then review the data. For example, yes/no questions can be answered with a "1" or a "2," or several digits can be used in multiple choice questions. The system can ask: "Were you the person driving the car at the time of the accident; or was a police report filed after the accident occurred... Please enter the report number and the telephone number of the police station where the report was filed..."

As you can see, it really doesn't take two people to be involved in these types of questions. The agent can ask certain sensitive questions, or the caller can be prompted to leave an overall statement in the form of a recording for complicated questions.

In this fashion, the voice processing system can take on a conversational tone. For instance, after asking a yes/no question, the system can allow the caller to press a button indicating that he or she wishes to provide additional (verbal) information. Once the claims adjuster reviews the recordings, he or she can call the insured party to follow-up, or decide to accept the data as is.

Claims Processing

Insurance companies process hundreds, and sometimes thousands of claims each day. The actual process of claims reporting requires a lot of data input. For example, there's social security numbers, group policy numbers, coverage classes, and a variety of codes indicating the type of service to be performed (in the case of medical claims).

Since thousands of calls are processed at claims centers, many insurance companies have started to use voice processing equipment to augment the claims agents and other staff. A claims system can ask the caller to input policy numbers, doctor license numbers, and other information required to process the claim. Once this data is entered, the call may be transferred to an agent, or the claim can be "queued" for processing. Depending on the arrangement the insurance company has with the caller (i.e. individual insured, or doctors office), the system may be able to dispense coverage information and authorize the work associated with that claim.

J.N. Phillips Glass is the largest auto glass replacement company in Massachusetts with 27 branch locations and over 1,500 affiliate shops in their national preferred provider network. They receive on average 3,000 calls per week. J.N. Philips contracts with regional and national auto insurers to provide full service auto glass claim processing. With over fifty years experience, they actually have insurance companies' 800 numbers ring directly to them for glass replacement claims. J.N Philips Glass continually strives to maintain the trust the insurance companies have placed in them by providing quality auto glass service and exceptional customer service to their policyholders.

Insurance company customer representatives must handle each client with utmost care, as they are very precious to their business. "Our goal is to enhance the customer service experience. It should be an effortless experience for the car owner," states Bob Rosenfield, President of J.N.Philips Glass. "The customer's first phone call needs to be the only phone call. Therefore, it's crucial that we get all the details the first time around - the information specific to the policyholder's insurance."

In the past, providing such a high level of customer service was a challenge. Originally, J.N. Phillips had 10 lines that rang at several workstations where representatives answered according to call traffic. If traffic was high, a representative would have to put a customer on hold, answer another line, ask that line to hold, then go back to the first caller. While on the phone with customers, a representative would then have to solicit what insurance company they were with so the proper computer screen could be brought up. Customer service wasn't bad, but Bob knew it could be improved considerably.

"We wanted to add an auto attendant with voice mail, automatic call distribution (ACD), and utilize CTI to empower our agents. We wanted our call center to have the capabilities of a mega-call center, at a cost that made sense to us, but still utilize our existing database and Novell Network. I did some research on Comdial and found they offered what I needed at a reasonable price." J.N. Phillips' search for a solution took them to Voice Systems, Inc., a preferred Comdial Dealer in Canton, Massachusetts. Voice Systems suggested a visit to Novell's New England regional office to see a live demonstration of Comdial's DXP switch and QuickQ ACD working with FastCall, a computer telephony application that, among other things, pops a computer screen from a database based on incoming caller ID. All this was shown running on a Novell Network.

"We wanted Bob to see in real-time how effective computer telephony can be," explains Richard Medeiros, President of Voice Systems. "We showed him how with a Comdial DXP, Novell Telephony Services and a Novell Network work together in an ACD.

Figure 5.10 – FastCall-Activated Screen Pop

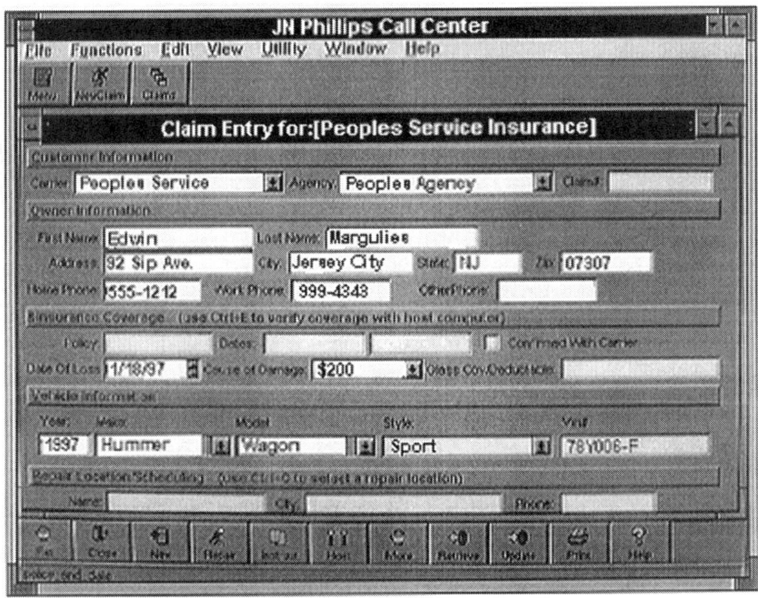

The switch can take a call with a specific ID (an 800 number for insurance in this case), hold it in the queue and transfer it to the next available agent. And in the process, trigger the appropriate insurance computer screen by way of FastCall. With this solution, he was looking at a savings of at least 30 seconds a call, which meant happier customers and lower operating costs." Voice Systems met the budget, provided the features, and installed the system.

Now when customers call in with a claim, the agents already have the pertinent computer screen at their disposal (see figure 5.10). The 800 number dialed links to a database via caller ID, popping the claim form at the agent's PC specific to the caller's insurance company. If all agents are busy, an auto attendant handles the call with a greeting specific to the insurance company's 800 number and transfers the call to the ACD queue. When an agent becomes free, the call is transferred along with the appropriate claim form. Average wait time is about a minute. Operating costs are lower. Customer service is better. Agents are more productive.

Power Utilities

Energy Savings

Utilities are quick to point out how you can save energy, especially if you call about a large utility bill. This is not just a gesture of good will, but also helps to decrease service complaints, late payments, and the hassles associated with service disconnection. One way utility companies overcome these problems is through education. These programs include seminars, school presentations, pamphlets, and special off-peak service discounts. Power company call centers receive many calls having to do with energy saving tips, kilowatt usage of certain appliances, and free publications.

In addition, some companies offer free "home audits." These audits are conducted by utility specialists and often include a visit to the home to point out tips on insulation, water temperature settings, and caulking. Voice processing systems can guide users through a checklist to conduct their own audit. The system can ask callers if they are calling about an energy audit. The application can be developed so that it conducts an interview with the customer before transferring the call to a energy specialist. For example, the system can prompt the caller for their account number so that it can begin the interview with a message like: "Your current bill indicates that your energy consumption is up 20% from a year ago.

Based on last year's average temperature during the same period, only half of this could be attributed to maintaining a constant ambient temperature. Please answer some questions in helping us to complete your audit before we transfer your call to a specialist."The system can then ask the same types of questions that the specialist will ask. These questions will include information on how many people are now living at the residence, working hours, number of large appliances, water beds, pools, additions to the home, and other indicators of energy usage. Based on this profile, the utility can then tell the customer whether or not their energy consumption is within a reasonable range compared to homes fitting the same basic profile.

Figure 5.11 - CTI Integration at Boston Edison

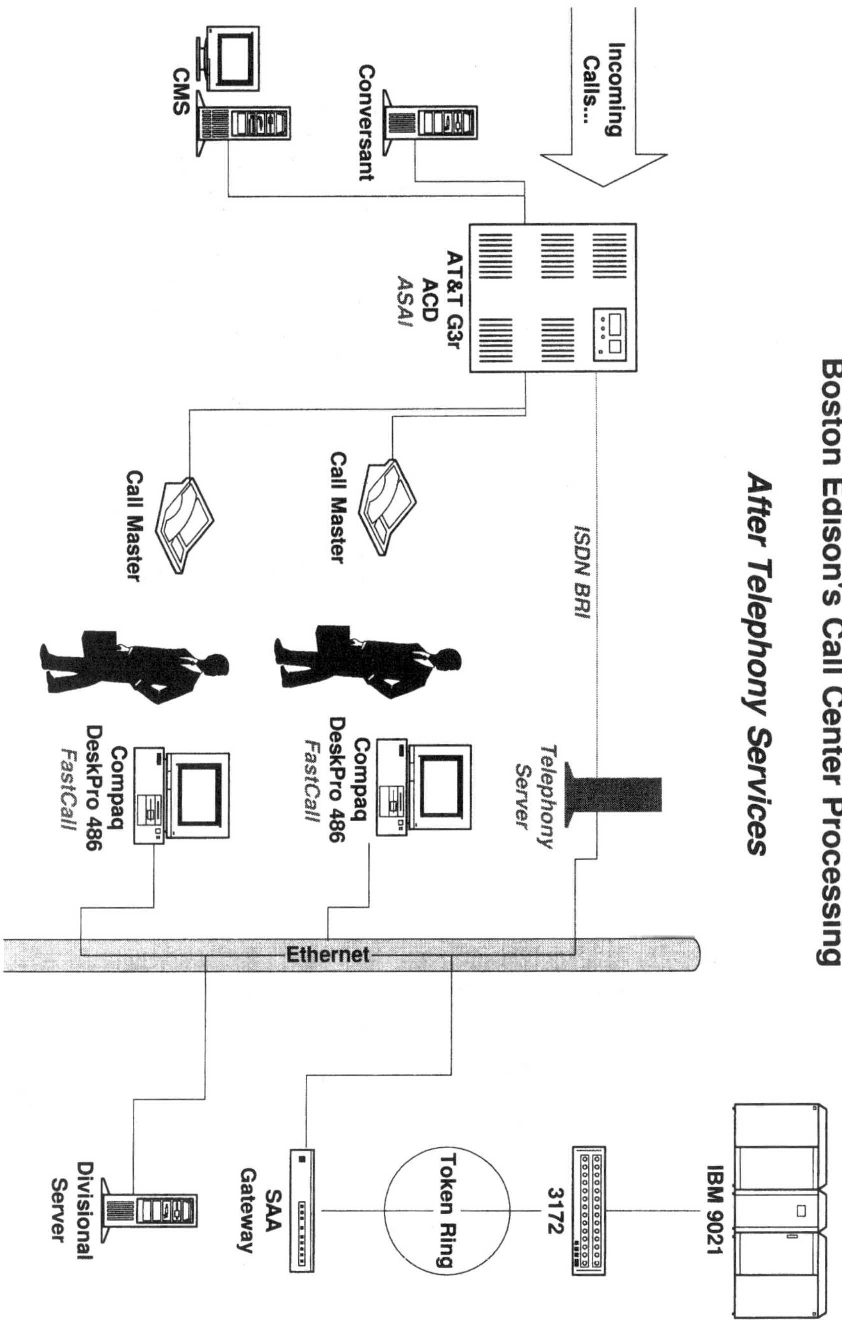

Payment Center

Call centers at most utility companies use a complicated network of computers, terminals, networks, and telephone systems. Take, for example the Boston Edison Company of Massachusetts. As shown in figure 5.11, the utility employs a mainframe computer to house customer account data, a mainframe gateway, and a LAN that connects PCs to the legacy system. In addition, there is an ACD, VRU, and a CTI server. All of this may look like a lot of work, but the company experienced incremental savings and efficiency gains over the past five years.

Call account information can be collected by the voice processing system and then passed to the telephony server. This information allows for the coordination of customer account data and the associated screens to be pulled-up when the telephone call arrives at the desk of the service agent. In this case, Boston Edison is combining the use of voice processing, personal workstation automation and CTI services all in the same system.The end result is that the time it takes to process calls is significantly reduced. According to system planners, the purpose of the automation is not to eliminate the use of agents, but rather to make their tasks more efficient in light of the thousands of calls the center takes each day.

Pre-Service Screening

Utilities are beginning to share customer payment history information both inside the state where they operate, and also with utilities that operate in other states. This information is critical in order to establish service for consumers who need to provide a deposit. In many cases, the utility will ask the new customer for employment information, the telephone number at their last residence, and social security number.

This and other information is gathered in order to establish some kind of payment history.A voice processing system can work in conjunction with the call center telephone system in collecting this data before the call is handed-off to a service agent. If the caller is an existing customer, the call is routed in the normal fashion.

If the caller is a prospective customer, the system can do on-line credit verification. If the caller does not have the information handy, the system can ask them to call back when they have the correct information. Once the phone number, social security number, or other information is input by the caller, the system can put the caller on "hold" while the VRU passes the data to another processor.

This adjunct processor can use data links to credit bureaus or a shared database. If the caller's profile indicates that he or she is a good credit risk, then the call can be transferred to an agent. If a problem is discovered, such as an unpaid balance on a previous utility bill, for example, the caller can be told that this balance has to be satisfied before new service is established. In some cases callers may be asked to provide a deposit. The voice processing application can get this deposit in the form of a credit card payment and then process this payment automatically before transferring the call to an agent.

Print Media

Newspaper Reporter Call Routing

Priority Call Management of Wilmington, MA sells the UNIX-based Oryx system using Dialogic and Gammalink cards to provide single number "follow-me", voice / fax mail, fax broadcasting, voice-rec dialing, conferencing and pager notification. The Boston, a daily newspaper with a 500,000 + circulation, utilizes Oryx to get in touch with field personnel for voice / fax message distributions.

The system uses two 24-port Dialogic cards. One handles ISDN PRI, the other handlcs 24 analog ports (soon being replaced with another PRI card and ISDN PRI line). Tandem (Cupertino, CA); a fault tolerant computer maker, built a switching subsystem. Most roaming cellular calls in the US are routed through Tandem servers. Not surprisingly many really big calling card systems use Tandem platforms.

During message-retrieval calls into the Oryx system, staff can immediately return urgent messages, initiate other calls and gets access to standard call-management stuff, transfer, conference, calling other extensions, etc.

If calls come into the system while an employee is listening to messages, the system allows the real-time call to go through. The Globe also developed an interesting paging application. It relies on two-way text pagers. Not yet rolled out (pending a big investment in lots of pagers) the new app allows dispatchers to send a page via the Oryx platform to a delivery person when a customer reports a paper delivery problem. The paperboy will then confirm delivery with the same pager.

Retail Services

Catalog Sales

Call centers perform a critical function for catalogue companies. Customers often call-in their orders if they want to check on availability, color, delivery options, or to lodge a complaint, for example. This is where a voice processing system can be of great assistance.For example, customers can dial the call center and be greeted first by a system that gives them a choice such as order confirmation, inventory answers, and shipping information.

The system can be programmed to say: "Thank you for calling the catalog center. If you are calling about the availability of items in our catalog, press '1.' To check on the status of your most recent order, press '2.' All other callers can stay on the line to speak with an operator.Since some 20 to 25% of all callers ask about availability or order status, the voice processing system can be used to significantly reduce the traffic presented to the call center.

This way, agents can concentrate on providing good service to new callers who need special assistance. In case the caller has an established account with the catalog company, it is possible for them to order products without operator intervention.

In order to automate or partially automate order processing, the system will ask callers to type-in the catalog number and page number of the item they are interested in. This is necessary, because the caller may be looking at a catalog that is no longer valid. Once this data is collected, the caller can enter the product code number(s) and quantities associated with the purchase.

Installation Help Desk

As consumers become more savvy and technical, unique problems arise. Perhaps the most prevalent is the installation and maintenance of electronic appliances such as computers, remote control units, and the assembly of furniture. While some of the documentation supplied with these items is well written and contains descriptive instructions, there will always be times when consumers need a helping hand.

This helping hand, more often than not, comes in the form of a friendly voice on the phone.Call centers that specialize in helping people with technical assistance or instructions over the phone are often called "help desks." Voice processing systems may ask these callers to input the product number, part number, or model number that they are calling about.

This information can be used to provide an automated tutorial that includes frequently asked about issues, or a list of topics for the caller to choose from. This list can include an option to report lost or broken parts, discoloration, or some other problem that keeps the customer from being able to successfully assemble or install the unit.

Many help desk employees use terminals that allow them to browse through electronic databases of schematic diagrams, parts, and troubleshooting tips. Since this information is pulled-up from a computer, virtually all of it can be translated into the spoken word using speech synthesis. Even complicated diagrams can be faxed to a customer if they have a fax machine.

Software Technical Support

Quarterdeck Corporation of Marina del Rey, CA uses a special IVR system from TTM & Associates, Inc. (Fullerton, CA) to automate technical support calls to their service center. Called AnswerDesk, the system provides automated information delivery 24 hours a day, 365 days a year.

Callers access the system by either being transferred by a technical support agent or by directly accessing the system. Information is accessed in one of 3 ways: 1) "Qwik" look up; 2) Product name; or 3) Menu navigation.

Figure 5.12 – AnswerDesk Call Routing Status Screen

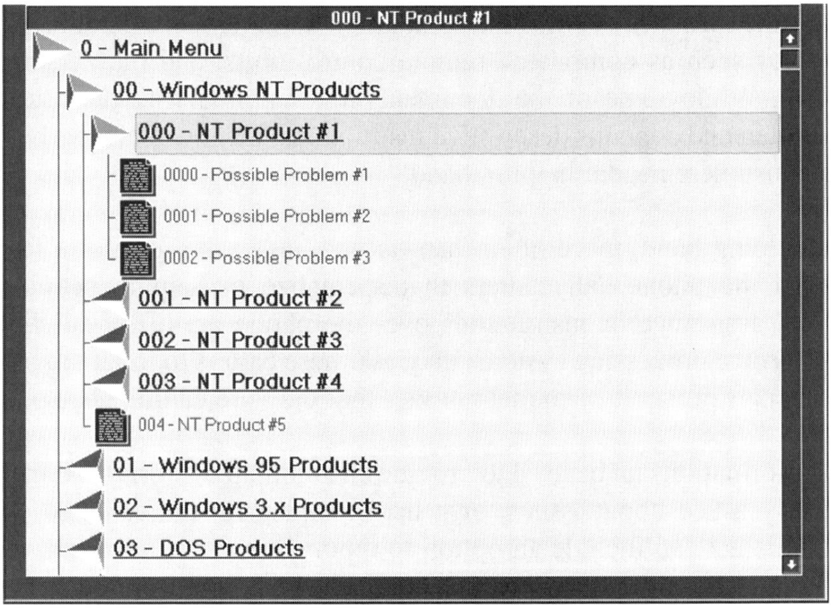

An average technical support representative may be able to service 25 persons a day. With AnswerDesk, this number can more than double, greatly improving the effectiveness of tech support calls. Automating the technical support calls for the help desk frees-up more time to speak with customers who need more personalized assistance. A single screen gives the system operator the ability to do real time performance monitoring and dynamic call routing "on the fly."

Quarterdeck is able to create product names for caller look-ups in real time and add, delete or otherwise modify the call routing in real time. This means technical support calls to both the system and live technicians can continue uninterrupted. Figure 5.12 shows what the operator sees on the routing screen. Here, three callers are making inquiries on the "NT Product #1." The system indicates these callers may be having a problem they need further assistance with. In this fashion, the operator is able to prioritize what calls can be transferred to a technical expert for personalized assistance.

Social Services

Abuse Hotline

Child, spousal, and elder abuse are all hot topics in today's headlines, and social services such as abuse lines are swamped with phone calls from clients. It's hard to know before the phone is answered whether or not the person calling is in danger at that very moment, or whether they are in need of remedial counseling. Once a case profile for the abused person has been established, this information can be used to help automate future phone calls and perhaps even save a life.With the caller's permission, each conversation can be recorded and used as evidence at a later date. In addition, the voice processing system can be used to collect Caller ID or ANI digits in order to help prioritize that call from other (less important) calls. This could be important if the client is calling in a panic and either hangs up the phone or someone else hangs up the phone or disconnects the line. The voice processing system can then be used to call the client back to see if the situation is volatile.

Emergency Services

Municipalities offer a variety of public safety programs and services to help residents report crime, fire, and disasters. The cost to run an emergency call center is usually subsidized by local taxes, surcharges on phone bills, or funded in part by private industry and donations.

Some municipalities share the cost of running a call center by combing the resources of several counties or districts. The primary duty of a PSAP (Public Safety Access Position) operator is to quickly determine the nature of each call in order to dispatch emergency personnel if needed.

The job is difficult if the call center is understaffed, or if there are an unusually high percentage of crank calls or non-emergency calls. The most popular way for voice processing to be used in 911 telephone systems is to detect the ANI (Automatic Number Identification) of a caller. A special name and address database is queried each time a call arrives in order dispatch emergency vehicles.In addition, voice processing can be used to create and maintain a “tape archive” of emergency calls.

This is important so that the conversation can be reviewed later as evidence, used in investigations, and as follow-up to the quality of the PSAP dispatch and call center staff. In some cases, emergency call centers can provide information to residents regarding flood, fire, or other disasters by using a voice response system. The benefit of using an Audiotex capability here is to defray “information only” traffic away from the lines used for true emergencies.

Telemarketing

Telemarketing Voice Logging

Call centers consist mainly of service and sales centers and help desks that support clients by telephone. Such centers typically provide teleservicing, telebanking, telemarketing, teleshopping and teleinsurance. The call center market is growing rapidly and has been increasingly using voice-logging systems. The use of voice logging systems in call centers enables storage of the details of telephone orders and other transactions, supervision of call center operators and campaigns, and evaluation of salespersons’ efficiency, customer service and training.

Nice Systems, Inc. of New York, NY estimates the market for call centers is fully digitized, since it is a brand new segment for logging technology. By 1996, 4.3 million seats are in place, with 2000, 6.76 million seats expected by the turn of the century. Based on the assumption that 10% of seats will digitally record, the market is estimated at $320 million.

Most operational call centers are in the US. In Europe, the number of call centers is still very low and most of them are concentrated in the UK. Total sales of logging equipment to call centers amounted to $20 million in 1995 and $40 million in 1996.

Telephone Companies

Cellular Customer Service Center

Ellen Muraskin, reporter for Computer Telephony Magazine, uncovered this killer application from Scopus Technologies of Emeryville, CA. They helped create a solution for Los Angeles Cellular Telephone (a partnership between BellSouth Cellular and AT&T Wireless Communications). Since passage of the 1996 Telecommunications Act, telephone companies have to actively protect their turf. Customers, particularly cellular customers, demand more sophisticated services and products or they can take their business elsewhere.

Cellular companies, like most other telephone companies, had historically automated different aspects of the business in isolation, on different mainframes and client/server computer systems. This meant that critical customer service applications like sales, billing, customer inquiries, and technical support could not easily share information. In practical terms, this meant that:

1. If a customer called technical support twice in one day for the same problem, he or she would have to completely re-explain the problem each time they called.

2. Depending on the type of information the problem required, the customer would need to contact multiple departments.

3. To communicate between departments, 10 inbound customer calls typically generated 5 photocopies of a document, plus another 5 outbound calls.

L.A. Cellular decided it had to improve customer service by integrating all its islands of automation. The company developed a request for proposal covering more than 150 technical points. Out of several top vendors' responses, Scopus' Enterprise for Call Centers was chosen.

Figure 5.13 - Scopus Call Center at L.A. Cellular

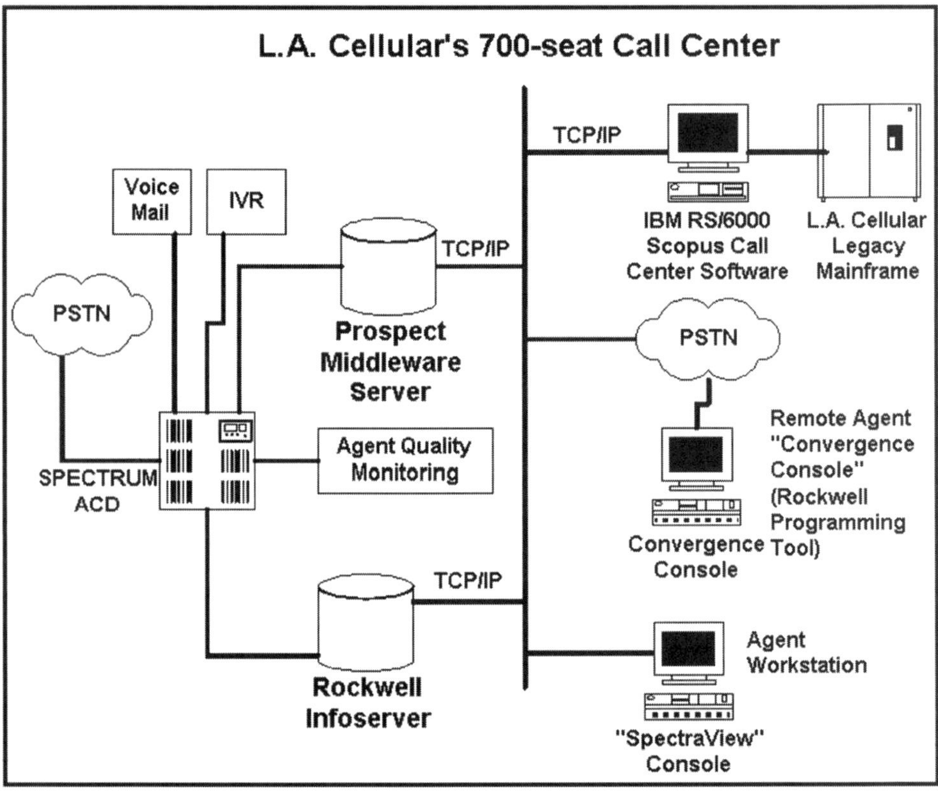

The total solution includes Scopus Corp.'s Enterprise for Call Centers (integrated knowledge base of customer data, case tracking, and workflow routines), coupled to L.A. Cellular's legacy mainframe billing database. Win NT-based clients run Scopus software on Hewlett-Packard Vectra P120 PCs and two IBM RISC 6000 servers run Scopus on the AIX operating system. A separate VRU, behind the Rockwell Spectrum ACD, is used for all IVR inquiries. The computer telephony middleware (used to glue the switch to the Scopus system) is from Prospect Software (San Jose, CA). Figure 5.13 shows how these elements fit together.

Now when customers call in via 611 or 800 number, the Rockwell switch picks up the caller's ANI (Automatic Number Identification) and matches it immediately with a Scopus customer profile. Scopus' system knows which CSR spoke with this customer last and routes the call to that person, or finds another rep whose skills profile dictates a good match. It can route incoming calls to Spanish-speaking service reps, if that's the customer's recorded preference.

It can identify major accounts and make sure that those calls go straight to CSRs dedicated to the high end of the business. It automates incident escalation, sending unresolved calls up to supervisors based on criteria and thresholds established by the call center, taking into account staff skills and availability.

L.A. Cellular can now track and manage every aspect of a customer's interaction. It gives CSRs Windows NT, point-and-click access to all previous dealings; pervious transactions, comments, previous records and screens accessed, and previous company contacts. It provides a friendlier, GUI interface to L.A. Cellular's mainframe billing database, and extracts records from there to its own database daily. Scopus' TeleTeam CT integration piece, communicating with the Rockwell Spectrum switch across Prospect middleware, adds intelligent call routing and screen pops to the mix. On the agent's end, Scopus delivers a screen of customer information that tips him or her off to problem solutions or other selling opportunities, and conducts text searches. Planned enhancements include access to case-based reasoning systems and full text search and retrieval.

CSRs will then be able to tap into an expandable on-line resource of how previous problems were solved, with a cross-reference to products, cases and accounts. Call scripting on the ACD is programmed in-house through Rockwell's own Windows-based graphical development tool.

The new call center, installed in a 157,000-sq.-ft. site in Anaheim Hills, has allowed L.A. Cellular to actively head off potential problems before they can cost accounts. For example, if a customer reaches a set number of call-center calls within a set period, the system be triggered to generate an outbound follow-up call from an agent higher on the command chain. Similarly, requests for specific services can be actively tracked.

L.A. Cellular expects to shave call time 20% and to exceed its goal of answering 80% of calls within 20 seconds or less. Says Jon Klustner, director of call center development for L.A. Cellular: "The benefits of the new system are fewer repeat calls, and the ability to conduct preventative analysis. We have gained consistent customer treatment, more proactive service, and accurate and detailed histories."

Cellular Voice Dialing

Car phone usage has increased over the past few years, and so have the accidents associated with the distractions involved in dialing the phone. While the use of speakerphone technology has gone a long way to solving this problem (in some states, speakerphones are mandatory by law), it's still pretty tricky to dial digits while you're in heavy traffic.

Cellular providers have begun to address this problem by installing voice processing systems that are equipped with speech recognition. These systems are connected as adjuncts to the MTSO (Mobile Telephone Switching Office) in the same fashion an operator call center would be connected. When cellular users are ready to place a call, they simply speak a word or a digit into the phone. This key word triggers an automatic "speed dial" to occur.

Since most cellular phones are not equipped with ASR, the voice processing platform that's installed at the MTSO can be used as an "operator" that performs this function for all of the subscribers

Customer Service

The AT&T MultiQuest Action Center (1-900-555-0900) takes thousands of calls each week. The company upgraded their customer service center capabilities in 1993 by coupling a sophisticated voice processing system to their call center ACD. The purpose of the center is to promote enhanced services including MultiQuest (800/900), DNIS, ANI, and VARI-A-BILL services. Before automating the call center, each caller was queued into the Chicago site's Difinity G3 ACD. Many callers had simple questions or wanted fulfillment material sent to them, but were forced to wait as long as 20 minutes to reach a live operator.

Callers ask for service information, fax documents, or billable technical support. AT&T's challenge was to answer more calls, fulfill more documents and bill automatically for their services. All this with the same number of agents. In addition, the carrier wanted to automate this capability for 24-hours service.AT&T wanted to partially automate their call center and make the system a showcase for their enhanced services. For example, they wanted to show how the cost of a 900 call can be changed "on-the-fly" (VARI-A-BILL) and they wanted to show how real-time ANI can be used in processing and prioritizing calls.Telephone Response Technologies, Inc. (TRT) of Roseville, California worked closely with AT&T to develop a special VARI-A-BILL prototype system. TRT decided that a ProVIDE-based Pro/Switch system would be the basis for solution. As of this second edition, it is still in service.

Pro/Switch systems combine Dialogic, Dianatel, and TRT products for PC-based switching nodes. By taking advantage of other add-on modules, such as Pro/Found, Message Management Service, and I/Fax Plus, the company hoped to automate thousands of calls each day.

Figure 5.14 - AT&T MultiQuest Action Center

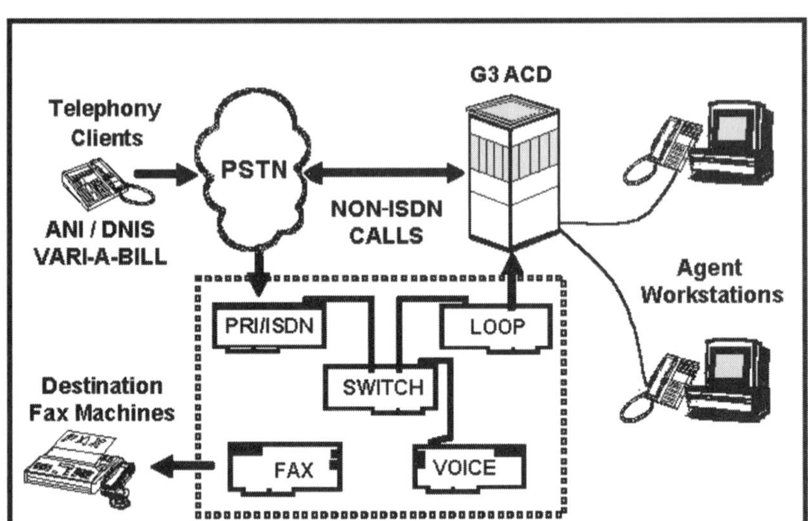

AT&T determined that well over half the calls were for simple information that could be handled by Audiotex. TRT designed an interactive message-taking dialogue for later transcription and fulfillment. For those callers who wanted the same information right away, I/Fax Plus was used to fax data sheets directly to the caller. For callers who needed to speak to a live agent, the Pro/Switch software would insert calls directly into the ACD queue.

The AT&T operators wanted to be able to change the price of the call for callers who were qualified. This was achieved by providing the operators with a touch-tone input (transformed by the Pro/Switch system into a VARI-A-BILL charge). The Pro/Switch system, in turn, would send the charge data over the "D" channel of the ISDN circuit connected to the system.

AT&T also wanted to curb frequent (nuisance) calls into the system by accounting for the frequency of previous calls from the same party. This was achieved by logging the callers ANI (Automatic Number Identification) to a local database. Now, the database is queried at the beginning of each call to determine if tye call should continue.

As the diagram in Figure 5.14 suggests, the platform connects to the PSTN with a Primary Rate ISDN span. In addition to the digital trunk interface, an LSI/120 card connects up to 12 in-house loop start ACD lines. These analog lines are used to transfer calls to live operators manning stations on the ACD.

The system was designed so incoming callers hear a greeting asking if they have a Touch-tone phone. If they answer yes, they will hear the main menu prompt; if not, they will immediately be transferred to a local agent.

The main menu offers 5 announced choices and 1 unannounced choice—the unannounced one is the System Operator menu, allowing an operator to call into the system and retrieve the recordings that were taken from callers.

Main menu choices 1,2: Both of these choices play a single (product and service-oriented) message and asks the caller if they want to hear it again. If not, callers are returned to the main menu.

Main menu choice 3: This choice prompts the caller to record their name/address information, then lets them review and re-record it if desired.

Main menu choice 4: This choice prompts the caller to enter their fax phone number, after which the system validates it and plays it back for confirmation. The caller can reenter the number if desired. A single fax about AT&T' service is then sent.

Main menu choice 5: This choice transfers the caller directly to the live agent queue on the AT&T ACD between 8 a.m. and 5 p.m. (otherwise, a "no agent available" message is played).**Main menu choice 9:** *(unannounced option)* This password-protected option allows system operators to call into the system and listen to each caller message that was recorded in menu selection 3. It also lets the operator hear the total calls that the system has answered during the current day. The MultiQuest Action Center system was put together with off-the shelf hardware components including a 486-33 CPU with 4 MB of RAM.

Debit Card Services

Retail stores, service companies, and financial institutions are beginning to negotiate attractive rates with national long distance carriers by pre-paying for long distance service. This pre-pay arrangement is made possible with the use of debit cards. Debit cards have already appeared in many of the most popular retail chains. PC based systems offer a very cost effective solution to provide debit card services.

Companies and individuals are in a position to achieve significant savings and profit with a minimal investment in the expanding debit card market by using voice processing technology.The operation of a debit card system is straightforward. A customer purchases a calling card from a retail location and dials the 800 number imprinted on the card.

The customer is then prompted to enter the debit card number (usually hidden until the card is sold) and the number that he or she wishes to call. Richard Sachs, President of Interactive Communication Systems (ICS) of Colorado Springs, Colorado, has worked with a number of companies to provide for automated pre-pay of long distance calls. Call-IT Debit Card will search a customer database for the caller's current account balance and calculate the maximum time available for the dialed number. The call is connected and Call-IT will warn the caller at three and one minutes of the time remaining.

Debit cards are distributed to retail locations where they are purchased based upon a preset value. The customer simply dials the access number and enters the debit card number when prompted by the system.

Call-IT validates the debit card account and prompts the caller for the outgoing long distance number. Based upon the rate and the customer's current account balance, the caller is told the maximum number of minutes remaining for the entered number. Call-IT will dial the outbound telephone number and connect the caller. The customer may place another outgoing call at any time by pressing the "#" key.

After the caller is connected to the called party, Call-IT tracks the duration of the call and updates the customer's account balance accordingly.Callers can use options by pressing "*" when making outgoing telephone calls. The customer can set speed dial numbers, select different default language prompts, or increase their account balance with a credit card.

In addition, users can access account balance information, change default language options, and even "recharge" or add additional funds to their account with a credit card. ICS wanted to build-in the ability for flexible control by the system operator (administrator). The options include the ability to restrict cards and users, manipulate rate tables, enable cards with magnetic strip technology, and program various classes of service. The Debit Card Maintenance Screen allows the operator to use "Control Numbers" to access an individual or block of debit cards.

These special numbers are generated when the debit card is created. A block of debit cards is typically distributed to a single retail location. If these cards were lost or stolen, it would be extremely difficult to disable a large number of cards based upon a random customer number. The control numbers allow the system operator to quickly track and modify large numbers of debit cards. The operator also has the ability to execute batch changes on customer records. For example, the system allows the operator to set the dialing restrictions (e.g. no restrictions, speed dial, inactive, or terminated). Speed Dial restricts the card holder to using one of the pre-set speed dial numbers.

The Call-IT system also allows for operator control of country and area code rates, area codes, and prefix codes. When the customer enters the outdial number (or selects a speed dial number), Call-IT searches for and compares the number with the rate ID to calculate a per minute rate. Call-IT will search the rate ID field for a match of the first seven digits. If a match is not found, the system will search the database for a match of the first six digits. This search allows the operator to define both default country rates and rates to a specific area code and prefix.Other maintenance controls include a DNIS / Language table and a password access table.

The DNIS entry screen allows the system operator to assign a default language and greeting prompt to each incoming 800 number. Debit cards can be printed in various languages each with a different 800 access number. When the customer dials the 800 access number, Call-IT will automatically answer with the appropriate language and greeting. Call-IT handles all call processing activities using T1 network interface and matrix switch cards from Dianatel, and speech processing cards from Dialogic. ICS used a variety of Windows development tools for the database portion and the administrative interface for the application.

Administrative workstations handle all account maintenance, billing, and rate table operations based on a standard 486 Windows PC. The units are equipped with a modem for remote maintenance. In addition, a file server can be used based on the type of network and the degree of fault tolerance needed. A credit card server is an optional PC on the network, handling all credit card and account replenishment.

Directory Assistance

As many as 3,000 people are employed as operators in each RBOC (Regional Bell Operating Company), and a number close to that for most larger independent telcos and long distance carriers. These operators handle subscriber calls for coin credit, directory assistance, and toll assistance. With this kind of manpower, it's easy to see why telephone companies look for ways to save money through automation.

In addition to efficiency, the telcos generate revenue by providing premium services. This, in effect, turns a classic cost center into a profit center for RBOCs. Virtually every operator service center now uses voice processing as a means to shorten the "live time" each operator spends on the phone with customers. For example, operators use voice processing to automatically greet callers with their own recorded name and salutation. In addition, voice processing gear is used to play-out the phone number after the directory information search is completed by the operator.

In this case, an operator answers a call for directory assistance and locates a listing by typing commands on a terminal which is connected to centralized SIDB and LIDB (Subscriber Information Database and Line Information Database). These listings are cross-referenced to a telephone number which is then automatically sent to a voice response unit. The VRU then announces the digits to the caller.

Most operator service center VRUs are programmed to repeat the number on request of the caller. Bell Atlantic will complete these calls at a premium if you want to be connected to the number during the same call (called call completion). These automation services have many permutations. For example, US West sponsors a service that allows subscribers to record their own intercept announcement when they have their number changed. This service uses a combination of voice response units, CTI links, and on-line LIDB database access gear to fully automate the recording and playback of these messages.

International Call-Back Service

Interactive Communication Systems (ICS), of Colorado Springs, Colorado has designed a solution for long distance providers called Call-IT. Callers can place long distance telephone calls at significantly lower rates than with their local public carriers. While this differential will probably not last forever, there are significant savings and profits available for companies and individuals that are willing to capitalize on this voice processing solution.

Large multinational corporations and telephone long distance resellers serve thousands of subscribers, many of whom need to make international calls. Calls originating from foreign countries are sometimes costlier than calls placed in the other direction.

For example, there is a substantial rate differential between calls originating from the United States to most foreign countries and the same call placed from the foreign country to the United States.

One example should make it clear just how significant these differences are: Brazil charges $4.88 a minute to call the U.S. The biggest U.S. phone company charges $2.60 for the first minute and $1.08 for each additional minute.Assuming the conversation lasts 10 minutes, the total cost of calling from Brazil is a staggering $48.80 while the cost of calling from the U.S. is only $12.32. This is a 75% savings in the cost of the same call based solely on where that call originated. The trick in capitalizing on this cost differential is to have your international calls originate in the U.S. This might seem kind of tough when you're sitting in Brazil trying to get a call through to your office in Boston, but there is a way to do it using a computer controlled telephone switch called a call-redirector or call-back system.Let's say that a caller in Brazil picks up the phone and dials their special access number. This number is referred to as a "trigger" number because it triggers the start of the call-back process. The caller lets the phone ring three times and then hangs up. Since the phone call is never answered, there is no charge for the call, but the call-back system does know that a call came in and more importantly, knows who the call was from.

Figure 5.15 - Call-IT International Callback System

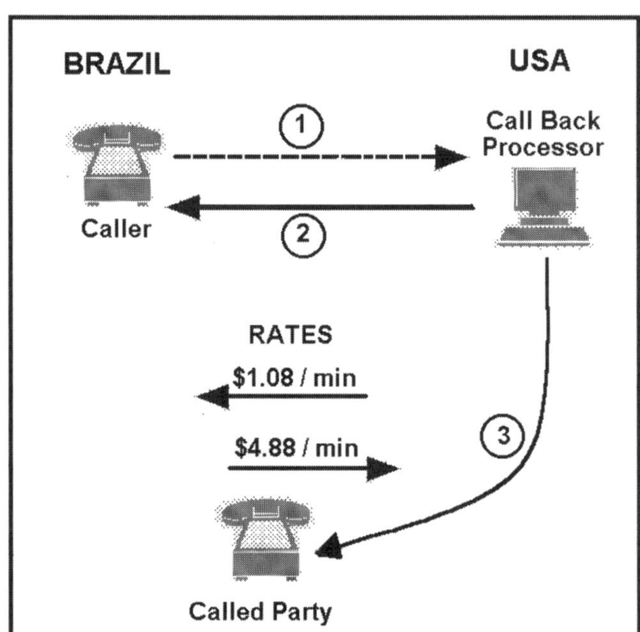

Now the call-back system looks up the customer's phone number and calls back. This time the cost of the call is on U.S. rates instead of Brazil's rates. When the caller answers the phone, the call-back system connects the line to U.S. dial tone. Now the caller can dial the phone number of the party they are calling. Note that the called party could be inside the U.S. or anywhere in the world. The method of originating the "trigger" call can also be customized to support various business requirements. This scenario is represented in Figure 5.15.Call-IT is designed as an entry level to mid-range system, which is expandable from a single T1 node to several nodes depending on the type of local area network (LAN) used. This distributed approach provides optimum response and connect times, operational redundancy, and modular expansion.

Many other solutions are available from systems integrators. Now, call-back systems are beginning to use the Internet as a means to set up and tear down calls. In this scenario, the caller can initiate the call by loggin-on to a Web page, or make a telephone call to a call-back system that signals the U.S.-side VRU to make the call-back.

Toll Saver

Telephone companies are licensed to operate in both small and large geographic regions. In fact, depending on the state or province in question, there could be as many as 30 operating phone companies. Local phone calls are based on a flat service rate each month, or sometimes measured and billed by the number of calls, or by other usage indices. Calls made outside the local area carry an associated toll charge. On the surface, this may not seem to be a big problem, but sometimes the boundaries that are set up between toll and non-toll calls are frustrating for subscribers. A call across the street, for example, could be a toll call for thousands of subscribers. Calling areas that are contiguous do not always require toll charges, so entrepreneurs and local service providers use voice processing systems to "bridge" non-contiguous areas. This helps subscribers avoid toll charges.

A voice processing platform can be placed in the "middle" of two non-contiguous calling areas so that calls made to or from each area from the middle one cost little or nothing to make. Toll Saver systems, then, allow callers to access the voice processing platform, enter their customer I.D. code, and then enter the digits of the phone number in the non-contiguous area they wish to call. The toll saver system then makes the second call for the subscriber and then "patches" the two calls together. Toll Saver service providers charge either a flat rate per month, or a small usage charge per call in order to do this. In some respects, the system behaves in the same way an international call-back system would, except that the calls being made are within the state or province.

Speed Dialing

Voice processing technology can be used to ask callers for information relating to the most frequently dialed numbers. This data can be stored in the same fashion an operator or receptionist would store phone numbers in a Rolodex file. Instead of calling the operator and asking for a call to be placed and connected manually, users simply speak a name or type an abbreviated code on their phone to make the call. Long distance providers like MCI, for example, provide this service and use a voice processing "administrator" to prompt callers for their speed dial information. During speed dial set-up the system asks: "Please enter the speed dial number followed by the area code and phone number you want associated with the speed dial number." Callers are able to enter speed dial database information in this fashion until they have completed a "network-based Rolodex" of their own.

Transportation

Transportation Air Traffic Control Voice Logging

Most countries have legal requirement to record voice conversations of air traffic controllers. The air transportation industry by law requires the recording of voice communications and radio transmissions.

As worldwide air traffic is growing increasingly rapidly, the industry has become one of the heaviest users of voice logging systems. Nice Systems, Inc. of New York, NY has been selected as the voice logging system for 800 air traffic control centers in the US. The NiceLog system is used by air traffic control centers in Israel, Norway, Netherlands, Hungary, Hong Kong, Romania, Germany, Iceland and others.

The market for air traffic control logging equipment is estimated at US $35 million. Nice is planning to gain a dominant share in this market.

In the air traffic control segment, Nice enjoys a strong reputation in many countries, mainly as a result of its US FAA contract win. After the company was awarded its FAA contract, numerous countries followed the trend. In Romania, where seven other logging systems providers were competing, Nice was awarded a contract for 13 sites with total capacity of more than 500 channels. Recently, Nice was awarded a contract by the German Air Navigation Service to supply more than 440 channels.

Transportation Rail Network Voice Logging

Large transportation companies, such as rail networks and road haulage companies, are extensive users of verbal instructions transmitted over the wireless networks. Instantaneous voice retrieval is necessary in order to confirm instructions, review fault reporting and to respond to emergency situations. Nice Systems, Inc. of New York, NY sells these companies their NiceLog system to solve this problem. The system sits between the telephone or radio system and its users. It records all voice information digitally and stores the information on computer disk. Each conversation or segment is then time-stamped and cataloged for later retrieval.

6

Computer-Based Fax

Computer-based fax systems offer the instant delivery of images to any fax machine in response to telephone input. Interactive voice and fax systems increase your ability to service customers.

Customers can select the information they want to receive from menus that can be spoken, faxed, or even advertised in newspapers. In addition, prospects that shy away from traditional sales inquiries order faxes with regularity. On the marketer's end, the automation of routine inquiries has a number of benefits.

First, computer-based faxing allows service representatives to direct their efforts where they are most effective. Secondly, use of the technology increases the quality and efficiency of information delivery to your customer base.

Driving Forces Behind Interactive Fax

The computer-based fax server and fax-on-demand industry has grown to over $50M in sales with a growth rate of about 50 percent annually. There are a number of factors fueling the growth of computer-based fax (CBF). They include:

- Instant Information Demand
- Communication as a Weapon
- Productivity Gains
- Instant Information Demand

Buyers and sellers alike have grown accustomed to getting their information fast. Many business decisions are based on the ability to acquire and manage information quickly. Increasingly, the buying public expects to get product information instantly. The traditional means for conveying product information is susceptible to delay. For example, first class mail or common carrier services are delivery methods with built-in delays.

With fax-on-demand, information can be delivered faster and more accurately because callers can enter simple information on their own rather than verbalizing it to a service agent. The benefits of instant information access include not having to "wait in line" to speak with the next available service agent. These services are also available 24 hours a day, so they can be accessed at the caller's convenience.

Communication as a Weapon

Another driving factor behind CBF is that it offers you a means to competitively distinguish your company. The competitive edge of an enterprise is often measured by its ability to deliver accurate and timely information. This is because customers and prospects alike are looking for clues about how your company is different from your competitor's.

For example, your products may be of equal quality and performance, and your policies and terms could be just as fair. Unfortunately, these are not the only indices of measurement in a buyer's decision-making process.

Prospects use other criterion for deciding to buy from you. For example, they measure your responsiveness, flexibility, and stability. Certainly, by providing access to important product literature around the clock, you can increase your responsiveness to your client base. For example, consider an instance when it is after hours and no technical support staff is available - yet a customer has a problem in the "field." The customer can use a phone to access technical support tips, and will likely perceive this capability as highly responsive.

Customers appreciate having access to your information stores, whether they are made available via computer bulletin board, fax-on-demand, "live" agents, or a combination of these. The image you thus project is that of a strong and stable enterprise. Consider who you would do business with, all other things being equal in your decision.

Productivity Gains

Maintaining this competitive edge is a difficult challenge for most companies, because so many downsizing initiatives are under way. This is perhaps why the most compelling argument for the use of fax servers and fax-on-demand is the productivity of your staff. Most information managers, sales directors, and customer service personnel are delighted with how much easier it is to perform regular duties when this technology is installed.

It's a simple matter of tracing the steps we go through. For most companies, a labor-intensive series of tasks are what's ahead to carry-out a simple communication. For example, consider breaking-down the time it takes to send a fax to a customer without a fax server. First, the document is printed-out. Most of the time, this will be a shared printer that is located outside of the sender's workspace.

By the time you walk to the printer, wait for someone else's job to print, and then gather your document, at least two minutes have elapsed. Of course, this does not take into account any distractions along the way. Unless there is some automated means for creating a fax cover sheet, you then fill-out the form with your name, the recipient's name, and other identifying marks. For most people, this takes about a minute.

Then - on to the fax machine. Here you will either find a machine ready to use, or another job that's in the machine, or worst yet - a line of co-workers in front of you. It may be the case that an individual is assigned to collect faxes and stand in line on behalf of everyone else, although this too, is unproductive. At any rate, by the time you feed, dial, and wait for the machine to transmit your fax, another two minutes have gone by.

Five and a half minutes may not sound like a lot of time, but multiply that by the number of faxes sent each day, and you'll quickly see how much can be saved by automating these tasks. Virtually the same series of events happens with a literature fulfillment request. The only difference may be that instead of faxing the literature, another employee may have to print-out a mailing label or envelope, and then affix postage, etc. The difference in time savings varies from company to company.

Now let's retrace these same steps in a fax server or fax-on-demand scenario. Assume for a moment that you've just completed a letter that you wish to send to a customer. With a fax server, all you need to do is "print" the document to the fax server. A cover page is automatically appended to the communication based on pre-configured data about yourself. Recipient information is input once and then stored in a "phonebook" for future use.

Once the document is printed to the fax server, nothing is the same:

Collecting the document at the printer is eliminated. Manual creation of cover sheets is eliminated. Feeding and dialing the fax machine is eliminated. Waiting at the fax machine is eliminated. Distractions and side-trips are eliminated.

Since the fax server captures the transmission from your workstation, there is no need to go to the printer. Likewise, it is not necessary to create a manual cover sheet. The fax server captures the document electronically and automatically queues the transmission along with other fax jobs. If the number is busy, it will try again and advise you of the status. And of course, since you are in your own workspace while all of this is going on, the distractions associated with walking down the hall and back again do not exist. Figure 6.1 shows the productivity gains you can take advantage of by using CBF at your company.

Figure 6.1 - Productivity gains with CBF

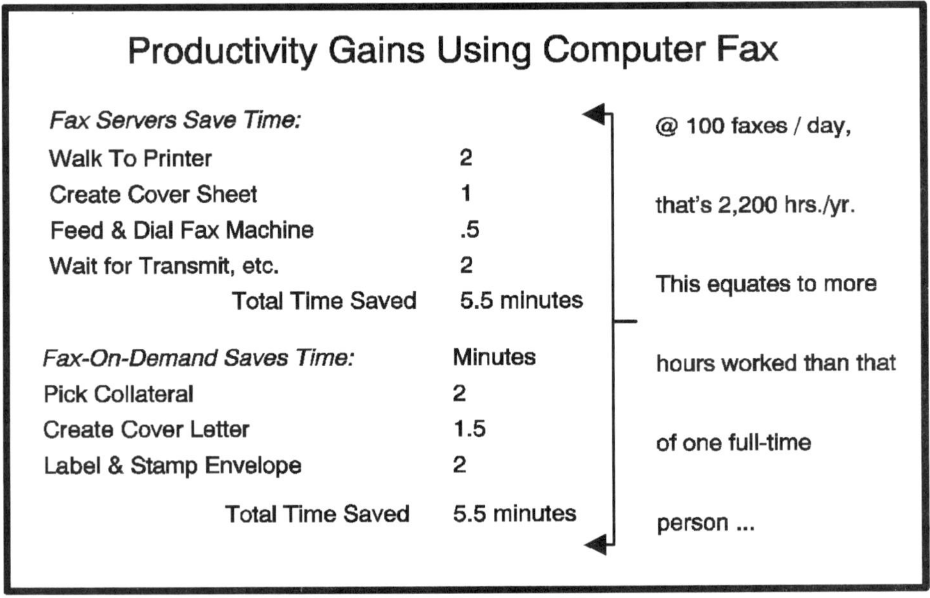

By adding a fax-on-demand system as an adjunct to your fax server, the productivity gains for fulfillment requests are even more impressive. This is because the recipient of the information is the one entering the request for the data. This goes on as a result of an interactive voice response dialogue with callers. A simple greeting message is played followed by instructions for the caller to input their fax telephone number followed by the document(s) required.

If the caller does not know what document number to enter, then the system will fax them a document catalogue. In this scenario, there is no need for intervention by a customer service or sales agent. One shortcoming of most fax-on-demand systems is that they do not provide an alternate means for requesting documents from a workstation versus over the phone. In other words, customers can call in by themselves and input touch-tones to have documents sent to them, but sales agents do not necessarily have the ability to do this on behalf of the customer without having to call into the system themselves. Invariably, the sales agent has to hang up with the customer and call the system after the real-time conversation, or worse yet, ask the customer to hang up and call back.

Figure 6.2 - CBF Link with I/Fax Commander

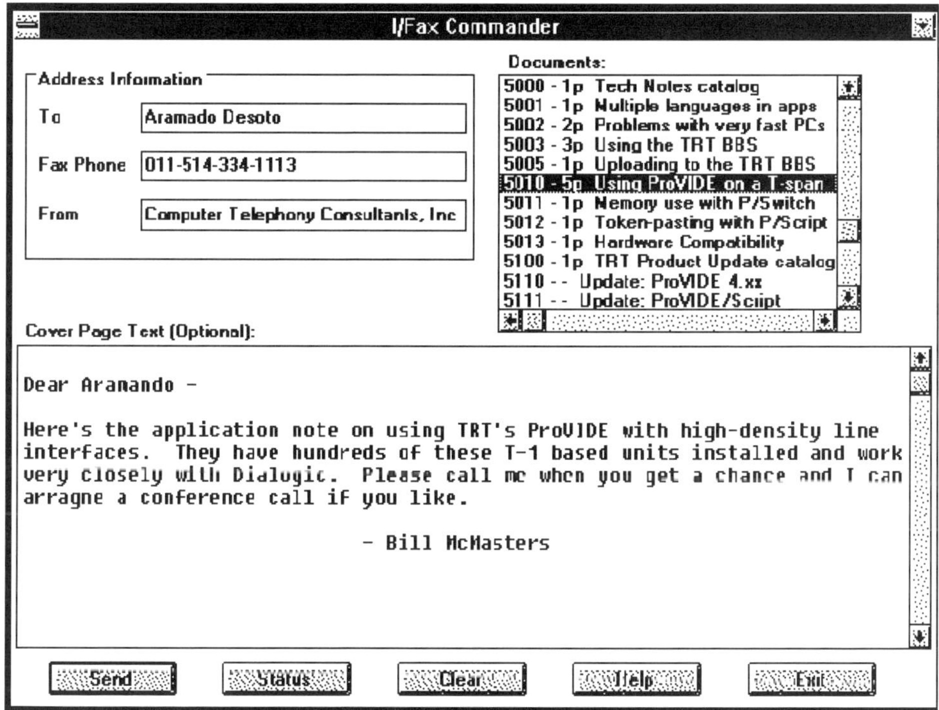

Some fax-on-demand products support workstation access to the same document catalogue with simple mouse commands. In this fashion, agents can fulfill requests for literature to callers while they are still having a live conversation with the customer.

One good example of how this can be achieved is with the use of software such as the I/Fax Commander product from Telephone Response Technologies, Inc. of Roseville, CA. The user interface is presented in a Windows-based screen on agent workstations as pictured in figure 6.2.

Fax-on-demand technology has its basis both in the facsimile and interactive voice response disciplines. The stand-alone fax machine was and still is a great boon to business communications. Among the apparent benefits are the fact that these machines are easy to use and fairly inexpensive. Unfortunately there are some compatibility constraints, and these stand-alone machines do not integrate with computer programs. For those companies still using stand-alone fax machines, the financial impact associated with labor can be overwhelming. In fact, if hundreds of faxes are being sent each day, and there's a wait to get something faxed, then the key instant ingredient is ironically unfulfilled.

Stand-Alone Voice & Fax Systems

Until recently there were only one or two commercially viable ways to get voice response systems to cooperate with existing fax servers. Voice response vendors were convinced they could incorporate fax technology and eventually displace the fax servers that were already installed. The irony in this thinking is that the fax server manufacturers have already done much work here. But most voice response experts are unaware of the years of effort required to get everything to work.

The best case scenario for users of fax servers would be for a voice response system to operate with an existing fax server, thus enhancing the original investment. If, on the other hand, the buyer does not have a fax server, or a LAN for that matter, they could install a stand-alone system. Client-Server architectures have done much to pave the way for a more seamless integration between computer platforms.

For customers who have local area networks, this client-server capability is critical for optimal results. This chapter provides many examples of how CBF is used in real-world applications.

In most of the examples, the CBF application involves a voice processing element, however, there are a few fax-only CBF examples as well.

Advertising

Desktop Fax Routing

"Our worldwide corporate computing plan dictated a migration from a mixed NExT UNIX environment to Microsoft's Windows NT and client/server," observed Tom Morris, Senior Network Engineer at Temerlin McClain Advertising Agency in Irving, Texas, a 526-person unit of Bozell Worldwide. "We wanted NT so the back end could run on NT as a service, while on the front end we looked for a client that supported NT, Win95, and Mac. "

"This responded to our clients' requests: the in-house management people wanted to do their work in Microsoft Office and apps like PowerPoint, the creative folks required the ability to create Web pages, and everybody wanted to be able to send faxes from an application, client software, and MS Mail. Looking further out, we knew we'd need DID capability, too. Omtool's Fax Sr. for Windows NT and Brooktrout's TR114 Universal Port card provided us with a unified solution, the only one that supported all three platforms. They could take us from UNIX to NT and grow with our anticipated needs."

Defining management information needs and testing solutions is an important part of the MIS work at Temerlin McClain, which serves as a "test bed" for Bozell Worldwide. Temerlin McClain designs and implements all corporate technology solutions before they're rolled out to Bozell's offices around the globe.

The network helping the folks at Temerlin McClain do their work is quite complex: 20% of the clients are on Macs, the remaining 80% on NT workstations. The PCs are all Unisys or IBM 90 or 100 MHZ machines and the Macs are all PowerMacs.

There are 11 file servers, two of which are 32-gigabytes each backed up by RAID 5 storage systems. They run two file servers, a fax server, an Exchange server, a legacy MS server for Macs, an SMS server, and a SQL server.

The MIS staff found that they needed to upgrade the backbone sooner than planned as they worked to maintain load balancing among domains as they migrated to NT. They upgraded from a shared flat Ethernet network to a switched Ethernet with an ATM backbone, some 10BaseT and 100BaseT.

Tom and his team assessed telecommunications software in October 1995. MIS knew that the network would come in two phases: (1) temporarily using a manual router, under which a user hits the Forward button to send the fax to the appropriate recipient; (2) eliminating manual routing by phasing in DID, an integral capability of the Brooktrout TR114 board that is part of the Omtool solution. An examination of Microsoft's fax software products list revealed that only Omtool supported all three clients.

"What Temerlin McClain really liked about Fax St. for Windows NT solution," says Craig Randall, Vice President of Marketing for Omtool, "is that you can send and receive faxes directly to the desktop- any desktop. Fax Sr. provides a uniform, easy-to-use faxing interface for MS-DOS, Windows, Windows NT, MacIntosh, and Motif.

And, because Fax Sr. is fully integrated with native applications like E-mail and Microsoft Exchange, users need little or no training, so they can become productive quickly."

"What was important to us," echoed Tom Morris, "was that Omtool was the first to support MS Exchange and deliver MS Exchange compatible server products." these capabilities allowed Temerlin McClain to send mail all from one location rather than separate faxes and MS Exchange messages. (see figure 6.3) "Now, we can create a group composed of mixed types of recipients, but it's transparent to the sender- he just sends the message. Omtool's Fax Sr. was the only product that could do that."

Figure 6.3 - Fax Sr. Exchange Inbox Integration

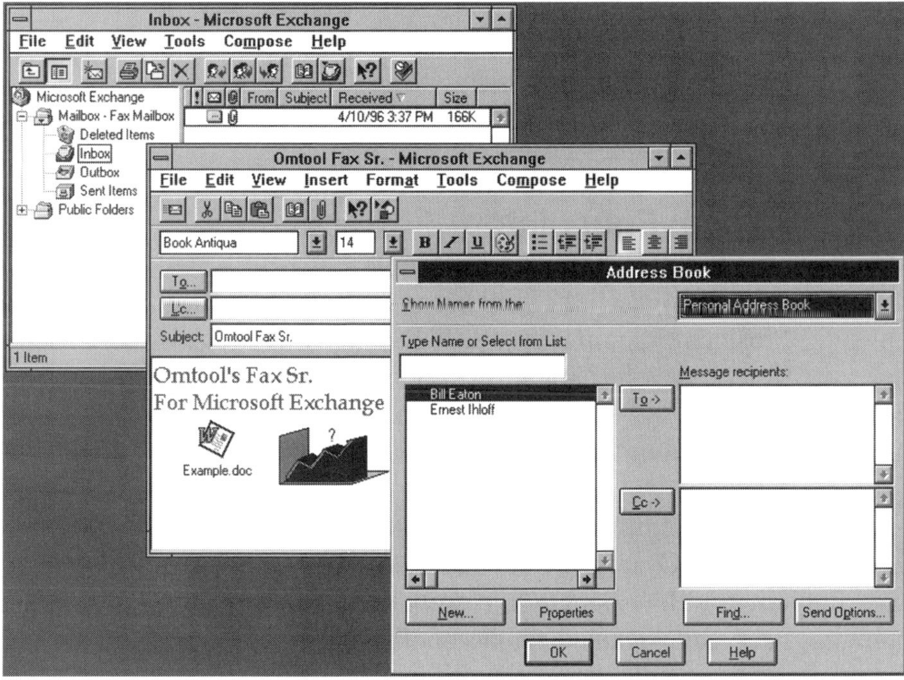

One component that contributes to Fax Sr.'s unique functionality is Brooktrout Technology's TR114 Universal Port boards. The TR114 offers superior performance and reliability and the flexibility that comes from single-source supply, allowing Brooktrout to design hardware/software solutions that meet the system requirements of specific vendors.

" We like to find out from our vendors which hardware they prefer," observer Tom Morris. "Omtool told us they chose Brooktrout's TR114 Universal Port board for its Fax Sr. solution because of its outstanding performance and onboard DID capability. We like the TR114 because its clean and efficient design meant we didn't require a lot of external modems, just a straight cable plugged into the board. "

"In addition, we like to use gear from the same vendor for system consistency. When we looked at inbound messaging we like the Brooktrout approach of putting DID and voice processing on the same board."

DID solved Temerlin McClain's fax routing problem by assigning a fax number to each person so they receive faxes as mail. This also eliminated the problem of routed faxes being seen by unauthorized personnel.

Corporate

Fax Line Voice Enunciator

In reviewing Comdial's PATI3000 for Access/Visual Basic Advisor, Jay Schwartz, President, CompuSolve Data Systems of Monsey, NY, tried to think of different uses for single line telephony products. Says Jay: "After all, if we could put a $99 product to good use in a multi-line office, every one of our clients would want to own one."

Comdial's marketing people positioned this product for the Small Office/ Home Office (SOHO) market. The bundled application, Algo Communications' PhoneKits, allows single line users to make and take calls from their computer, keep call logs and phone books, and pop up caller information when the phone rings.

In analyzing what goes on in our clients' offices, we think the marketing people at Comdial are dead wrong. This product belongs in every office, no matter how many lines they have. Where, you ask? The answer is at the fax machine and modem line.

How many times a day does this happen in your office? The phone rings. "Did you get my fax? Hold on, I'll check." The fax machine has become the water cooler of the 90's with office staff congregating around, waiting for the all-important incoming fax (especially the confidential ones). Hundreds of productive hours are lost each year to this marvelous technology. Isn't this stupid?

We took Comdial's PATI3000 and attached it to a workstation next to the fax line. We then imported the client's customer file and wrote a Visual Basic program to announce incoming faxes, "Incoming Fax, Acme Distributors".

The program reads the Caller ID information, looks up the client, and announces them on the computer's speakers. Anybody (on a LAN) expecting a fax can run a Fax Monitor program, which will announce faxes as they come in. If you like, you can tell the monitor to advise you only when a particular fax comes in. Or, you can attach the speaker output to the office paging system, so everybody hears it. Now, that's a little better.

A variation of this Visual Basic program monitors the modem line with the PATI3000. To prevent unauthorized callers from accessing your system, you can enter a table of authorized phone numbers. Configure your modem to answer on the fourth ring (ATS0=4 for Hayes compatible modems). When the phone rings, the Caller ID information comes down the line between the first and the second ring. The Visual Basic program looks up the incoming phone number. If it is not an authorized number, the program answers the call and hangs up. The modem never gets the call, period.

Document Fulfillment and Messaging

Call congestion, misunderstood or lost messages and "people congestion" at the fax machine are all issues affecting small to medium size companies. These problems are manifest as a result of performing regular communication duties with paper, pen, and fax machine. Rodrigo Almarza, General Manager of CommuniCorp (Santiago, Chile), has worked with a number of his clients to develop a voice processing and CBF system called "ANSWER," to address these concerns.

The ANSWER platform binds workstation functions with telephone traffic. The system provides for the routing of phone calls, voice message administration, and fax-on-demand requests. In addition, ANSWER automates inbound and outbound fax routing for workstations. The system connects to the public switched telephone network either directly, or through integration with a variety of PBXs and Key Systems. In addition, the platform interoperates with a variety of network fax servers including Optus FACSys and Alcom LanFax servers. Figure 6.4 shows how all of these elements play together. CommuniCorp takes a consultative view in designing an ANSWER system for its clients.

Figure 6.4 - CommuniCorp ANSWER

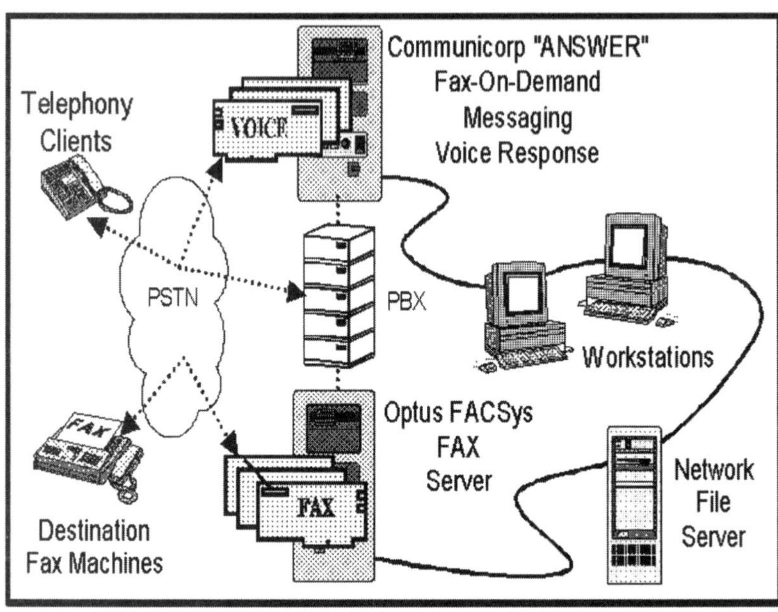

A needs assessment is conducted in the form of a customer survey. The client is interviewed regarding a number of issues:

1) Number and frequency of incoming and outgoing calls (traffic analysis);
2) Number and frequency of fax communications (to size the fax server);
3) The communications needs of each department.

Once this data is collected, a system design is proposed including the sizing of the voice processing platform (voice ports, PBX integration options), the fax server (number of fax ports), and the LAN (suggestions on redundancy, storage of image documents, etc.).

After the system design is ratified, a complete PBX, LAN, and ANSWER upgrade is put into place. This approached paid-off for Juan Pablo Salas, General Manager of the Apple Center in Los Monjes, Chile. According to Salas, the ANSWER system provides his company with numerous ways in which to inform customers of changes and information they require.

Says Salas: "I am delighted by the efficiency with which ANSWER helps our customers to contact not only our company but also the different departments and the personnel in each of them."

Carlos Johnson, the Finance Manager at Representaciones TOPP S.A. also uses the system and is impressed with the efficiency and customer service improvements his company has enjoyed. "We give our customers better service by providing the precise information they require without taking extra time." The ANSWER platform uses Dialogic speech cards and line interfaces, GammaLink facsimile cards, 3COM ethernet network interface controllers, a Novell File Sever, IBM-compatible 386s, an Optus FACSys fax server, and a Nitsuko PBX.

The development platform for the interactive voice response and fax-on-demand portion is Telephone Response Technology's ProVIDE, MMS, and I/Fax Plus tools.

The ability to co-locate fax technology in the same machine as the voice system is a new option being developed by CommuniCorp. Rodrigo Almarza, says that this capability is desirable for those companies who do not have a network fax server.

Employee Benefits

Ask any human resources employee and you're sure to hear about what a problem it is to keep track of the myriad paperwork and forms that are distributed to each employee. You can find these forms in boxes, bins, baskets, file cabinets, and in some cases - on the company's computer network. The management and the distribution of these forms is a big challenge. JetForm Corporation and Coopers & Lybrand jointly developed a project for a "Forms On Demand" application. Clients can now take advantage of a special fax-on-demand capability that will allow employees to access a variety of forms via simple touch-tone commands. Fore example, employees can download faxes such as medical forms, 401K plan worksheets, expense reports, and vacation requests.

With this service, employees can call an 800 number and be given a number of verbal prompts to select the employee benefit or related form they wish to obtain. The system asks for their fax number and then combines the related forms into a single fax transmission. The JetForm company worked with Coopers and Lybrand to develop several methods for managing, changing, and transcribing existing forms into the correct format for fax transmission.

Imaging System Server

A network-based fax server is a communications device (typically a PC), that houses multiple fax cards (or multi-port fax cards), and associated fax communication software. Its purpose is to send, receive, and route facsimile communications on behalf of an enterprise.

With a fax server installed, each workstation on the LAN has the ability to create and send documents over the network for transmission by the fax server. Most network-based fax servers have the ability to process multiple fax transactions simultaneously, because there are telephone lines associated with each installed fax port. In a sense, fax servers behave like a shared printer on the LAN.

In fact, most workstation-based fax server software makes faxing a document as easy as printing it. In addition to acting as a shared resource, a fax server also has the ability to receive and route faxes. Fax servers can be configured to route incoming faxes to the print queue so they are automatically printed.

Alternately, incoming faxes can be electronically routed to administrative workstations, so they can be viewed and then forwarded to the intended recipient. There are also emerging standards for achieving this administrative routing function automatically. Instructing the sender to put the recipient's extension number on the cover page can do this.

This extension number can be understood by OCR (Optical Character Recognition) software that enables the automatic routing.

Another way of routing incoming faxes is the use of DID (Direct Inward Dial) telephone numbers. In this case, the last four digits are used to determine the intended party's extension number.

An example of how this technology is being put to use is the Chevron Corporation imaging system. The company uses a combination of off-the-shelf word-processing software, an Oracle database, an image database, and an Alcom LanFax server. The system automates document requests, corporate communications, and other paperwork.

Employees use the system to quickly assemble customer-related data from the Oracle database and convert documents into faxable images. These images are then "printed" to the Alcom fax server and transmitted to the intended party. Figure 6.5 shows how all of these elements work cooperatively over the company's local area network.

Figure 6.5 - CBF at Chevron Corporation

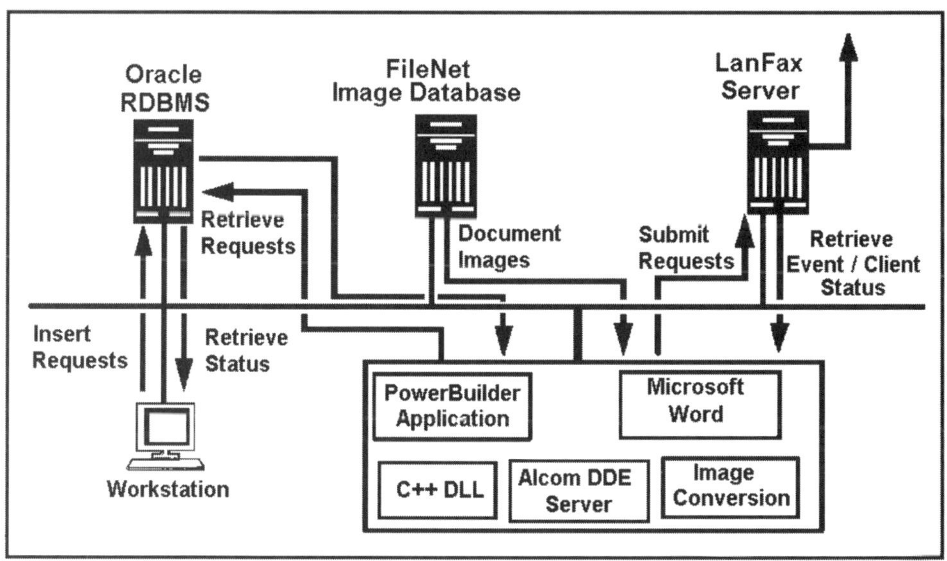

Intra-Company E-Mail to Fax Gateway

From the beginning, Lotus Notes developers viewed the inclusion of a fax gateway as an integral feature of their LAN-based group communication product, destined to become one of this Fortune 500 company's most important offerings. GammaLink fax boards and technology are a primary technology resource for Lotus in its ongoing quest to remain aggressively competitive in its industry.

Lotus Notes gives people working together a comprehensive environment in which to create, access, and share information using networked PCs. It develops and deploys applications such as customer tracking, status reporting, project management, information distribution, E-mail, and collaborative freeform discussions of all kinds.

It can also integrate text, numerical data, and graphics, including photographs. this makes it ideal for geographically dispersed organizations, where the composition of work groups must be fluid to respond to changing business opportunities.

When Lotus added a fax gateway to the Notes Mail Router, they selected GammaLink's GammaFax CPi board because they knew it was designed for high-volume, multi-network environments. "We consider it to be the best in the area of reliability," said Cindy Lavoie, Notes Product Manager.

"Its flexibility and programmability for our customization purposes was also a principal consideration during the evaluation process." According to Lavoie GammaFax's international certification in over 20 countries was another plus in influencing Lotus' decision to standardize on GammaLink products.

The Notes Fax Gateway allows users to send faxes from individual workgroup surroundings and receive faxes from a Calera Optical Character Recognition Server, a Kodak Imaging subsystem, or a Notes user's workstation. The Notes Fax Gateway contains the software supporting its operations.

The software falls into three categories: fax gateway programs, GammaFax programs, and Notes databases forms. The GammaFax software works as a subset of the control software that comes with the GammaFax board. These programs have been preconfigured for a standard fax gateway setup. Once the Notes Fax Gateway routes Notes mail to any fax machine worldwide. Because it is an extension of the Notes Mail Router, sending a fax is as easy as sending mail.

The user simply composes a document and addresses it by name and fax number, using the system's standard mail memo form or a fax form. The fax gateway programs convert the message into a bitmapped image and the GammaFax software residing in the server that controls the fax board and manages transmission sends it along.

The Notes Fax Gateway enables users to send messages or documents from any application to fax machines worldwide. These compound documents can then be sent through the Notes mail system to another user or users, to other mail systems, or through the fax gateway to an individual fax system. The system not only provides an outbound fax gateway, but an inbound one as well. People can send a fax from their location and have it routed directly to the Notes server or to an individual workstation, not a server.

According to Lavoie, when members of workgroups in different locations have needed to work together on a project, they have had to exchange information by telephone, supplementing this with E-mail, fax, and perhaps electronic file transfers.

"There are significant limitations inherent in all these methods," she explained. "First, whatever information is exchanged during a project cannot easily be organized, except perhaps chronologically. Second, it is difficult to establish and maintain a storehouse of information for use by the entire group when that information is stored in individual mail files."

In this case, members of a group cannot easily track the history of a project, or view shared data from different perspectives. Third, maintaining security and confidentiality in such a haphazard information-sharing environment is nearly impossible.

Notes is viewed as a major advance in intra-group communication because it stores all information users generate in databases designed to be shared. Each database has its own manager, who controls access to it. All documents within databases (both historical and current), are indexed, so they can be displayed in a variety of "views." Information is stored on servers independent of individual users' electronic mail files so that it can be structured, accessed, and maintained over time. Finally, Notes provides privacy, access control, and encryption features that no other PC-based group communications software package offers.

Voice Augmented Fax Transmission

For Hong Kong Telephone, fax services represent an area of business with unlimited growth potential. Personal computer fax boards and fax technology from GammaLink are helping Hong Kong Telephone's DataCom Services Unit take advantage of that potential.

With over 200,000 businesses operating in the small territory-a highly dense, international business center where much of the world's import, export, manufacturing and financial trading takes place-Hong Kong has been touted as being the world's most "fax-conscious" community, second only to Japan.

"Both Hong Kong Telephone and GammaLink have been pioneers in developing and using fax technology," explains Roy Ellyatt, manager of DataCom Services. "GammaLink was the first to develop fax boards that you could install in a PC, and Hong Kong Telephone was the first to offer networked fax services to its customers."

Those services, under an umbrella name called Faxline 100, have over 85,000 subscribers today and are growing at a rate of over 2,000 installations per month. Faxline 100 customers can choose from six separate Faxline 100 options, each offering unique facsimile communications.

In Hong Kong, the fax has already alleviated much of the need for personal messenger services for many Faxline customers. Executives who have personal fax machines in their offices have found that real-time fax transmission through the DataCom Faxline services has radically changed the way the fax is used.

David Connelly, General Manager of the telephone company's services branch says confidentiality is no longer an issue with this type of fax. "People can send handwritten notes and bypass the telephone, eliminating 'telephone tag.' Real-time fax also allows a document to be transmitted during a telephone call to augment business communications." In fact, in many businesses, physical mailing and other information systems are being replaced by fax.

In addition, the phone company offers Infofax, an information retrieval system for customers to retrieve financial, airline, and other information in both English and Chinese; Multifax, a broadcast service which allows customers to send out a fax to a large distribution list; FOLDS and Chekfax, a Fax on-line diagnostic service and a checking system for testing fax terminal operation and obtaining diagnostic information such as signal power; Datafax, a PC-to-fax service which guarantees the fax will be delivered to the recipient if the first attempt to send the document does not go through; and Telefax, an operator service which provides assistance to Faxline customers on dialing procedures and service operation; and a directory service which lists over 83,000 fax machines.

"A lot of telecommunications companies treat facsimile as a terminal on the telephone network," says Paul Honey, engineering manager for DataCom and one of the key members of the division's initial project team. "But we treated it differently. We made sure that the facsimile communication was treated as data, sending it through digital, not analog systems."

To accomplish this, they purchased a number of PC chassis and installed five GammaFax CP boards in each machine to accommodate the kind of traffic the Faxline services require.

Today, approximately 200 GammaFax CP boards provide more than the throughput needed for the over 20,000 calls each day.

"By using PCs you can develop applications, get them to market, test them, and determine whether you have a viable product. You can't do that with more expensive systems." Honey adds. Besides which, the level of fax technology provided with the GammaLink products is currently unavailable on mini and mainframe machines.

Construction

Employees who work for companies that rent and sell condominiums and apartment units wear many hats. For example, workers describe floor plans over the phone, mail brochures, and answer payment status and financial questions for hundreds of callers each day. Such was the case for Dasfin Industries of Seoul, Korea.

Due to rapid growth of their construction business, the company was faced with hiring five new employees to fulfill information requests and to maintain good customer relations. Locus Corporation, a Dialogic Corporation Open Advantage developer, was asked to create a unified voice messaging and fax-on-demand system that would automate these tasks.

According to Heywon Lee, Associate Marketing Manager at Locus, prospects would make calls either after visiting a model home or browsing through a brochure. The status of construction, floor plans, and pending contracts is all stored in a database, so for a company employee to access the data was straightforward.

Doing this quickly in light of the many telephone calls and other duties became a bigger problem over time. Based on their V33/F voice and fax processing system, Locus developed an integrated solution that combined fax-on-demand (for floor plans, Audiotex (for general information), and voice messaging (for non-real time answers on payment information) to solve the communication challenge.

According to Lee, the number of applicants for newly constructed units increased by 30% after the Locus system was installed. Says Lee: "The new system plays a distinctive role in reducing personnel expenses. In fact, the five new employees we were going to hire do not need to be brought on board for that purpose." Locus used Dialogic speech and fax boards, several LAN-connected PCs, and a small PBX to put the system together.

Entertainment & Sports

Sports Arena Luxury Box Reservations

Jim Kelly, President of Syscom Services, Inc. of Silver Spring, MD designed a killer productivity application for the sports industry. According to Jim, productivity means mixing solutions. He looks for transaction-intense initiatives or seasonal and "spike" transactions to identify problem areas for his clients. In many cases, Syscom has found that IVR and Fax work well together. IVR does a good job of handling data input and self-navigation. You can use IVR to unify disparate information stores. It also helps cut down unnecessary phone traffic.

With IVR, you can do instant record retrieval without the assistance of an operator or agent. Syscom mixes the fax element with IVR to handle the fulfillment part. Says Jim: "With fax, the visuals are striking, details can be added on-the-fly and the turn-around is instant." One great example of this is an "event booking" solution he created for the Core States Center in Philadelphia.

Arenas face a huge problem in communicating availability and reservations for their luxury boxes. Corporate sponsors get first dibs on the luxury boxes, but this right of first refusal means it's the arena's burden to contact the sponsors first. This means hundreds of phone calls each time an event is scheduled.

Syscom figured they could solve this problem by using a fax broadcast to communicate the seating availability and reservation information to the sponsors. Now the sponsors are contacted virtually simultaneously. There is a minimal staff requirement because the fax system delivers all the information automatically. The fax shows forms with all the available seating and box reservations. Clients can call the IVR system and use it to accept or reject their box assignments. Figure 6.6 is a sample form used for the Core States Center.

Figure 6.6 - CoreSates Center Form

24410

CoreStates Center

JANUARY
FLYERS CLUB BOX
TICKETS
Please return form 48hrs
Before day of game
Fax form to (215) 389-9791. Thank you.

Remember. . .
Please be sure to fill in all selections completely and accurately to ensure that your order be processed in the quickest possible manner.
If you do not receive a confirmation within 48 hours, please call(215) 389-9585.
Seating is done on a first come first serve basis and there are no refunds or exchanges on tickets.

CONTACT

ACCOUNT

EVENT
0 1 0 7 2 8

TICKETS @ $XX.00 PER TICKET
$ X.00 SERVICE CHARGE
$XX.00 PER TICKET

Remember:
If you wish to order tickets for multiple games, copy the blank form and use a separate form for each game.

Game Dates
January 7, 1997
January 9, 1997
January 11, 1997
January 14, 1997
January 21, 1997
January 25, 1997
January 28, 1997

Specify Delivery:
Will Call
U.S. Mail
Courier (additional cost)
FedEx (additional cost)
Fed Ex Number

Leave @ Will Call in the Name of:
FIRST NAME
LAST NAME

Payment (Credit Card)
Payment: Primary Credit Card on file
Secondary Credit Card on file
Other (Fill in Credit Card Number ONLY if you are NOT using your Primary or Secondary Credit Card on file.)
Visa
MasterCard
Discover
American Express
Credit Card Number

Does a member of your party require ADA seating?
Yes
No

NOTES

Expiration
Month Year

SIGNATURE:

24410

Ticket Refunds

No one likes it when a concert or special event is canceled due to rain, sickness, or some other unforeseen event. What makes canceled events even more tedious is the refund procedure for all of the disappointed patrons. Voice processing and fax-on-demand can help to alleviate this hassle by providing easy access to information. Let's say for example, that a sell-out concert gets canceled and 50,000 tickets have to be refunded. When those customers try to get a refund over the phone, the call center is swamped with calls from thousands of other patrons. A voice processing system can answer most calls and greet callers with an up-front menu to deal with the cancellations. The greeting can say "If you're calling about the canceled event, press '1' now.

All other callers can stay on the line for the next available agent. The system can then provide information on re-scheduling or refund instructions. By incorporating a CBF capability, instructions for refund procedures, a schedule of alternate events, and other information can be easily faxed to customers. The overall effect of providing this capability is that it can easily cut in half the number of calls that have to be dealt with manually. In addition, the voice processing system can take calls 24 hours a day, therefore reducing the need to put extra call center agents on-line for second and third shifts.

Financial

Mortgage Rate Broadcasting

G&H Consulting of Tolland, CT sells Technically Speaking boards and Microsoft NT and SQL Servers. Their VoiceTrac product is a full-fledged fax mail, voice mail and fax-on-demand package. It handles 48 lines. Their biggest sale was to Telefax, Inc. in Westlake Village, CA. They're a fax broadcasting service bureau. They've got 500 ports with plans to go to 2,000. G&H helped them design a Windows-based client package that's installed on remote PCs.

Mortgage industry executives use it to fax rate sheets in bulk. They fill out a form on their PC and transmit it Telefax via modem. The bureau assembles the fax electronically from templates and sends them with a personalized cover sheet.

Financial Retail Credit Approval

The Monogram Retail Credit Services, Inc. (MRCSI) had a problem: the number of credit applications it had to process daily from the more than 300 Montgomery Ward stores across the country, as well as other client companies, was increasing. Although extra personnel had been added, processing time per credit application at the MRCSI end had expanded beyond the ten-minute-mark. Since the number of daily applications being processed averaged between six to almost eight thousand per day, this was an unacceptably long time. When added to an outside credit bureau's processing requirements, this meant customers were kept waiting for their in-store credit approval far beyond what Montgomery Ward considered as a reasonable time.

"It became very obvious to us that something needed to be done-and soon!" recalled Werner Marschall, vice president of information systems for MRCSI. "As the number of in-store credit applications that we had to process kept increasing, so did our turnaround time. The situation was rapidly reaching a point which was becoming critical both for us and the client."

It was then that MRCSI called in T4 Systems. The Little Rock, AR company studied the problem, and provided the solution. "We installed our T4 MULTILINK fax server to do all the credit application-processing for Montgomery Ward nationwide, through GE Capital," said Frank Carollo, T4 System's vice president of sales and marketing.

T4 installed a 40-line system in a configuration that enables them to do 12 to 15 thousand faxes a day. At present, credit approval procedures for the others are still being done manually, but they expcct this will change soon, based on the success of the installed systems.

Figure 6.7 - Montgomery Ward Credit Approval

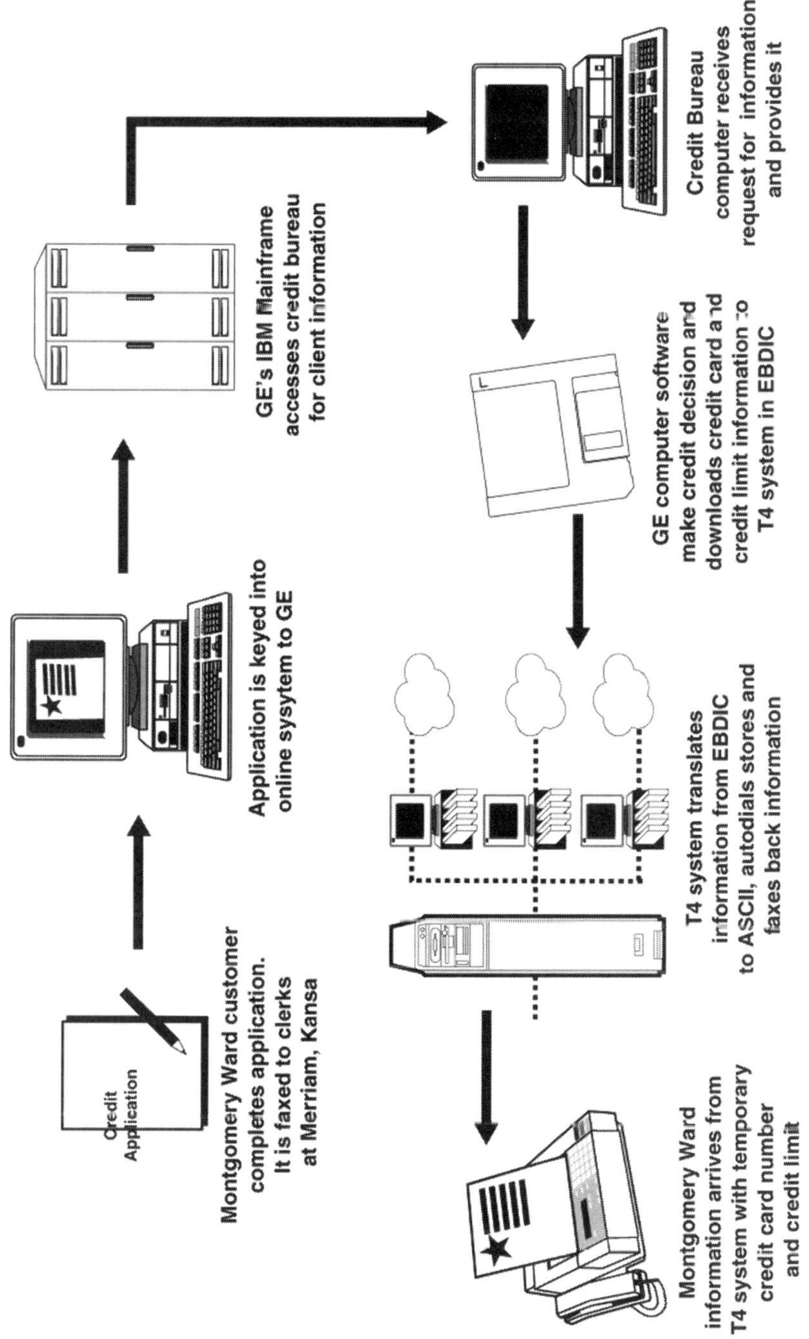

Carollo explained how the T4 MULTILINK system works, using the Montgomery Ward configuration as an example. "When someone goes into a Montgomery Ward store and makes an application for credit, he or she fills out an application form. That form is hand-faxed to Merriam, Kansas. There they have data entry clerks who get the fax and input the information into the on-line system, which travels over the public telephone network to Stamford, CT, to GE's 3090 IBM computer." This is illustrated in figure 6.7.

From this point, there is no longer a physical document. The computer's program accesses for information a credit bureau like TRW, and based on its software parameters makes a decision on the applicant's credit worthiness using the credit bureau's data. The decision may be to reject, accept, or queue (the latter goes to a human being for evaluation).

If it is either of the first two, the IBM mainframe downloads the response back to the T4 MULTI-LINK in Kansas City, as if it were a remote printer. The mainframe information is contained in an Extended Binary Coded Decimal Interchange Code (EBDIC) file, which is a standard computer character set coding scheme used to represent 256 standard characters.

Since IBM mainframes use EBDIC and PCs ASCII coding, the T4 system first translates the file information into ASCII. It then uses the information contained in the electronic document to autodial the store the credit application originated from, and faxes the decision and the information back to it.

Assuming everything went well at the credit bureau, the electronic document notifies the particular store that the credit has been approved. The software also provides a temporary credit card number and credit limit, and the new Montgomery Ward customer is then ready to begin charging merchandise immediately.

The complete process has to take no more than ten minutes. With the T4 system, from the time the customer hands in the application to the time a response gets back, less than ten minutes have passed.

According to Marschall, it would be impossible to process anywhere near the number of applications the T4 does, using current personnel levels. "We reduced our staffing considerably," he said, "once we felt comfortable with the system's high level of reliability and capacity. It has enabled us to do automatically, almost 24 hours a day, seven days a week, what would have taken a considerable staff of full-time employees."

Stock Market Information

Investors and brokerage houses face an interesting challenge in accessing time-sensitive and accurate information. Institutional and private investors pour over hundreds of pages and listen for all types of clues to know when to "hold" and when to "buy" stocks or mutual fund securities. What's most difficult is the consolidation of the data into a form that can be easily consumed and interpreted by the investor.

For example, there are historical price charts, previous and current day sales data, First Call earnings estimate reports, and a variety of insider trading reports to consider. Investors using news wires, electronic mail, paging and newsletters collect these disparate bits of information. And of course, there are the frequent calls to brokers and financial planners who have access to this data as well. As the saying goes, investors are "drowning in data, but starved for information."

Even if the data could be easily interpreted by the reader, the sheer number of minutes or hours spent in acquiring the data is a commitment in its own right. The time lag involved in accessing accurate information is more acute when a lot of market activity is happening. In this case, each small delay could mean thousands of dollars earned or lost for the investor.

James Tanner, President of Wall Street By Fax, Inc. (WSBF) of New York, NY has solved these problems by developing a special financial service bureau application that provides instant access to stock updates. Subscribers can access the service by dialing a toll-free number and entering a PIN code.

After a ticker symbol representing the stock or mutual fund is entered, the caller is prompted with a selection of reports (dial 1-800-938-5555 to get one free report via fax). These reports are then dynamically assembled by special imaging and database access servers at the service bureau location in New York. Once the reports have been prepared, they are automatically faxed to the subscriber's fax machine. Subscriber profiles are also stored in a special database that triggers an outbound fax with alerting information.

This occurs when one or more of the securities in that person's profile experience market changes beyond a defined threshold. WSBF recognized that there were several key elements to answering the challenge of the investor and wholesaling investment houses. First, it would be necessary to "multiplex" and manage various news sources and electronic feeds and manage the resulting data in a centralized database.

This would solve the problem of collecting the decision support data for any stock transaction. Secondly, it was determined that most investors still do not have computers, e-mail links and special pagers - but that most of them had access to a fax machine and everyone had access to a telephone.

Therefore, James Tanner and his development staff developed a fax-on-demand "front-end" to the securities database. By developing a voice processing platform with Telephone Response Technologies, Inc. Pro/Switch software, WSBF was able to build a system that serves 24 callers simultaneously without operator intervention.

The platform would also be designed to "insert" calls into the call center telephone system for subscribers who need operator assistance. Lastly, it was determined that the database of financial information had to be massaged in such a way that reports on any security could be dynamically created.

This would avoid the cost of either reading-out bits of information over the phone to clients, or cutting and pasting the information manually onto a fax. Tanner developed an imaging server that merged critical information from the financial database onto easy-to-read charts, reports, and graphs.

Figure 6.8 is a sample of a Standard & Poor's report that can be accessed via simple touch-tone commands from the WSBF system. This is one of dozens of reports that are available to subscribers.

Figure 6.8 - Sample Report from Wall Street By Fax

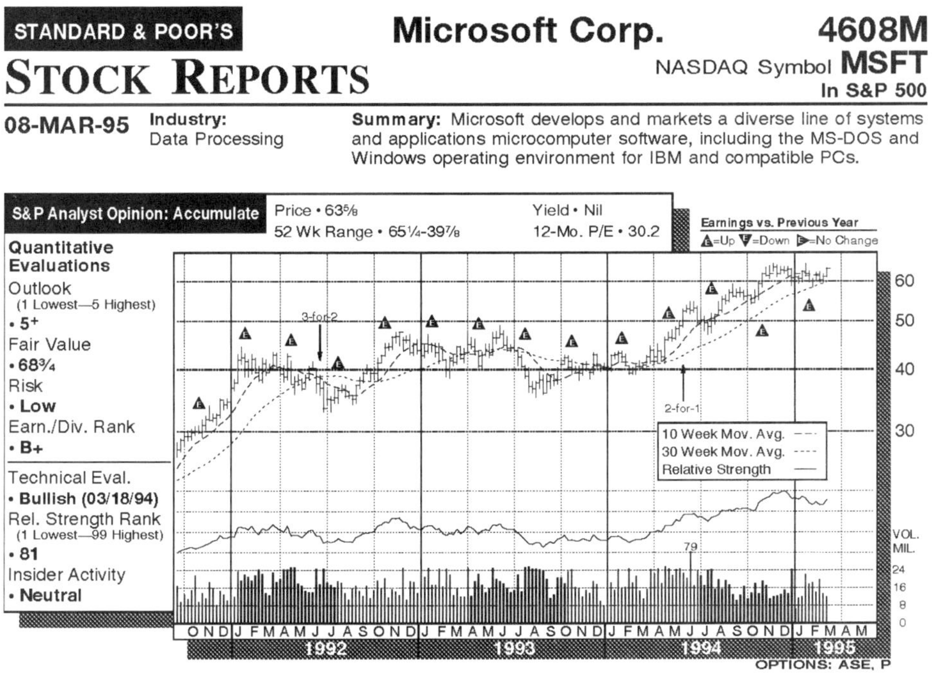

STANDARD & POOR'S

STOCK REPORTS

Microsoft Corp. **4608M**

NASDAQ Symbol **MSFT**

In S&P 500

08-MAR-95 **Industry:** Data Processing

Summary: Microsoft develops and markets a diverse line of systems and applications microcomputer software, including the MS-DOS and Windows operating environment for IBM and compatible PCs.

S&P Analyst Opinion: Accumulate

Price • $63\frac{5}{8}$ Yield • Nil

52 Wk Range • $65\frac{1}{4}$-$39\frac{7}{8}$ 12-Mo. P/E • 30.2

Quantitative Evaluations

Outlook (1 Lowest—5 Highest)
• 5+

Fair Value
• $68\frac{3}{4}$

Risk
• Low

Earn./Div. Rank
• B+

Technical Eval.
• Bullish (03/18/94)

Rel. Strength Rank (1 Lowest—99 Highest)
• 81

Insider Activity
• Neutral

Earnings vs. Previous Year
Up / Down / No Change

3-for-2

2-for-1

10 Week Mov. Avg.
30 Week Mov. Avg.
Relative Strength

60
50
40
30

VOL. MIL.
24
16
8
0

79

O N D J F M A M J J A S O N D J F M A M J J A S O N D J F M A M J J A S O N D J F M A M

1992 1993 1994 1995

OPTIONS: ASE, P

The Wall Street By Fax System is accessed by thousands of subscribers on a daily basis. It would literally take hundreds of service agents to deliver the same kind of information that is being dispatched automatically by the system 24 hours a day. Clients of Charles Schwab, for example, have worldwide access to the system, which is marketed under the Schwab name for their investors.

Clients of CNBC, Standard and Poor's and Morningstar are especially fond of the ability to access the system with a regular telephone. The system provides a primary vehicle for accessing securities information that is more timely and accurate than financial newspapers.

The WSBF system uses a variety of voice, fax, and network interface cards from Dialogic and Dianatel Corporation. These include D/121B speech cards, SS-96 switching technology, and a CO24 PEB-to-Loop converter for PBX system interfacing. The computer telephony platform is a rack-mounted industrial grade PC with direct digital telephone trunk connections to the PSTN (Public Switched Telephone Network).

Statements-On-Demand

Sometimes it's not convenient to wait for your monthly checking, savings, or IRA statement to arrive by mail. While most major banks and other financial institutions offer customer support over the phone, getting a full statement on demand is rarely an option. Considering the capabilities of voice processing, this can easily be achieved by adding a fax-on-demand capability to any IVR (bank-by-phone) system. For example, callers can be prompted to enter their account number followed by a security code. As an extra measure of security, the system can be programmed to only fax statements to a pre-determined number.

Since the information for statements is typically housed on a mainframe computer, terminal emulation technology can be used to download the customer's current financial records. An image server can manipulate this information. The image server can be programmed to "pick and place" portions of the customer record into a pre-set form that resembles a regular bank statement. This data can then be converted into a faxable file and assigned a unique file name. The CBF system can then "pick-up" the file in question and fax it to the requesting party.

Trader's Commodities Futures Charting

RamPage Publishing of Sunnyvale, CA created a network-based fax-on-demand system. Built with Parity Software's (Sausalito, CA) VOS CT app gen tool, it's not your average FOD fare.

"Our system does market charts for commodities futures traders," said RamPage's Dick Milewski. "With data on more than 600 futures contracts changing daily and support for spread charts that compare the performance of two different contracts, there are more than 60,000 possible charts available, each updated daily. Such a larger and changing library wouldn't let us create the fax images in advance and load them into a conventional fax-on-demand system.

"VOS let us do dynamic page make-up. This means the fax image isn't created until the incoming call arrives. After we know what the caller wants. We pull it out of our library, customize it on the fly and then fire it off." RamPage also implemented a Telepublishing system on the network that makes available more than six million different photocopier and fax machine spec sheets.

DID information is used to decide which VOS application to spawn. Most applications write the caller request to the Telepublishing queue database. The Telepublishing engine runs on another machine in the network. It takes requests from the queue, creates fax page images and hands them off to the network fax server.

Version 2.2 of the RamPage Telepublishing engine is implemented in MS Access so any OLE-2 capable Windows application can be used to create page images. The Telepublishing engine is modular so user written modules can be added. It has access to data in dBase, Paradox, MS Access, FoxPro, SQL-Server, Oracle, Sybase and many others.

Government

Flood District

The National Rivers Authority of the United Kingdom is a government agency that tracks, among other data, the state of flood emergency nationwide. The agency recently installed a special CBF system to assist in the role of emergency information delivery to associated agencies.

The system was prototyped and delivered using Pronexus' VBFax and PureData SatisFAXion modem technology. Weather stations all around the UK maintained by the National Rivers Authority are used to monitor rainfall. These measurements are collected on a minicomputer at a central site. Depending upon the amount of rainfall in an area, the system becomes active. High rainfall triggers an automatic transmission of a fax to the relevant emergency service agency (police, ambulance, and fire crews).

The fax contains a map that illustrates the projected size of the area that could be affected by the rainfall. The faxes are transmitted 30 minutes before projected damage would occur for the area in question. In this fashion, the system is proactive in warning emergency service personnel so they can prepare for the situation. Once triggered, the system continually faxes the flood progress information until the flood danger subsides.

Permits

We need permits to parade, build, burn trash, register pets, sell food, and conduct a variety of personal and business-related functions. Both small and large county government offices dispense thousands of pages of forms associated with these permits each week. This usually takes a personal trip to the agency, or a phone call to request forms to be mailed. Considering the number of government offices that issue permits, the manpower associated with sending out forms is staggering.

There's at least one person in each office that is assigned to serve this function. Voice processing can go a long way to consolidating this effort by acting as a "general fulfillment" front-end to an entire county's forms database. For example, forms for marriage licenses, building permits, hunting licenses, parade permits, construction, and pet licenses could all be managed as part of a general county database of forms.

Each county office can be assigned a prefix number associated with the forms available from their office. A fax-on-demand system can provide all of these forms from one platform, providing fulfillment services around the clock.

A system like this can accept new forms as part of a document replacement function. This would allow county office workers to simply fax the replacement forms into the system. This would be enabled by dialing a series of security codes from the fax machine being used to send the new form into the fax-on-demand system. In this fashion, both the fulfillment of the forms themselves and the updating of the document database could reduce each office's work load.

Tax Collection

DATATAL AB of Klintehamn, Sweden is an Expert Systems VAR. They sell a voice mail system called DataTal Professional (DataTal means computer-voice in Swedish). It integrates with Ericsson and Northern Telecom switches. It handles 32 lines under DOS. It also works with the Interception Computer System (Computerized Screen Pops, Directory Services and Employee Status Package).

Datatal sold a fax-on-demand system to the KFM Agency (part of the Swedish government). They are the Swedish IRS deadbeat office. They collect unpaid taxes and put liens on your house. The system tells you what houses are up for sale (the ones that have been seized by the government). The system started out with one fax line and one voice line. It grew to eight lines.

Health Care

Immunization Record Keeping

Ellen Muraskin, reporter for Computer Telephony magazine, recently interviewed Mark Nickson, president of DAC Systems in Shelton, CT. He developed an immunization record-keeping solution for a State Department of Public Health. Health departments must maintain immunization records of all children in the state from birth to age two.

It's a pretty big record-keeping headache, as it comes from a wide range of health-care providers and covers a long list of shots, vaccine manufacturers, and lot numbers. It must make this information accessible to health care providers statewide.

The Department of Public Health has to exchange this information with state health-care providers and keep it current. The department decided that the easiest way to do this was via fax and IVR. It wanted the system to be easily maintained using industry-standard tools, and wanted it to run on Win NT. DAC Systems won the job, beating, says Nickson, vendors proposing all-inclusive, proprietary solutions.

Twin mirrored systems, written in Microsoft Visual Basic with Visual Voice controls from Artisoft (Stylus Product Group, Cambridge, MA) allow healthcare providers across the state to consult and update health department records of children's immunization histories. Health care providers can update the SQL Server database through voice recognition or touch-tone IVR, or by faxing in forms. ICR (intelligent character recognition) identifies the form and strips out the handwritten fields.

Each system uses DAC System's rack-mounted SMC 3000 VRU for IVR and broadcast voice mail. The units house Dialogic VR/160s for 12 ports of voice, including Voice Control System's Voice Recognition and 6 ports of fax based on a GammaLink CP/6. In addition, the OPTUS Software (Somerset, NJ) Facsys fax server is used to send and receive fax forms and information. NCS OmniReader Intelligent Character Recognition is used to interpret hand-filled forms. The Dept. of Public Health's MS SQL Server 6.5 database sits on a DEC Alpha Server running Windows NT.

Healthcare providers can now update patients' immunization histories via IVR or via fax, using one of five different request forms supplied by the health department. They enter the information by hand, check off boxes, and fax it. In DAC's system, NCS OmniReader intelligent character recognition software strips the handwritten fields and uses them to update patient records in the SQL Server database.

In an option pediatricians should especially appreciate, users can even screen the database to retrieve a faxed-back list of all those Johnnies and Jessicas overdue for their last vaccination. The system also faxes back blank input forms (or takes requests to have them mailed), broadcasts urgent messages ("An outbreak of Hepatitis B has been reported in ____") from the health department, and offers a fax-on-demand list of generic immunization information.

Nickson says the job took six months, using Visual Basic and Visual Voice. The hardest part, as usual, was integration with the existing customer database. Viewed as a pilot in this state, the CT solution is being eyed for adoption by others.

Lab Results

Cardiff Software, Inc. of Carlsbad, CA sells Teleform for Windows. It's a data capture solution for fax communication. With Teleforms, users can fill out a form, fax it to a Teleform server and have their data automatically read by the system. It's an alternative to using touch-tones for remote input applications.

The companies' fist sale was a hospital healthcare sale in Denver, CO. They were originally running off of a single PC, but needed a better solution -- one that could handle jobs in multiple environments. The customers needed results from labs and patient evaluations and found that Teleform helped organize and ultimately solve the problem.

Medical Records

Medical personnel store patient records both in paper and electronic form. The management of these records is often critical, because they include information that could mean life or death in critical situations. For example, patients have information about what they are allergic to, what complications they have had, and what problem they are in the hospital for.

Voice processing and CBF technology can be used together to provide instant access and updates to all kinds of medical information. This is especially useful if records need to be transferred from one medical facility to another. Take, for example, a product called WillFax2.

Developed by Digital Technologies, Inc. of Broad Axe, PA, WillFax2 is a Windows application that provides a fax-on-demand capability complete with Audiotex and IVR features. Digital Technologies built the application using Pronexus' VB Voice Toolkit for Visual Basic. WillFax2 was developed for Advance Choice, a company specializing in the preparation of Living Wills. Advance Choice provides a consultancy wherein Living Will documents are prepared, archived, and distributed to authorized parties. The distribution of the documents is achieved using the fax-on-demand capabilities of the WillFax2 system. Living Wills have recently come into the public spotlight. For example, both Richard Nixon and Jacqueline Kennedy Onasis wrote Living Wills to cover directives to family and physicians regarding terminal medical treatment.

WillFax2 automates the process of filing and storing documents, tracking customers, and responding to customer inquiries. In addition, the system faxes documents to customers, generates reports, and tracks customer follow-up. If an Advance Choice customer becomes hospitalized, physicians can call the service at any time to receive a copy of the living will. The WillFax2 system is a good example of how some off-the-shelf packages can be used alongside voice processing development tools. In addition to VB Voice and the VOX-FAX GOLD application, other elements include Goldmine for Windows, Crystal Reports, and File Magic Plus.

Help Desk

Fax-On-Demand Help Desk

The Dragon PhoneQuery Crystal Info application allows callers to access information by voice over the telephone using natural dialog.

This approach employs "word spotting" techniques for grammar-free interactions with the system. Dragon PhoneQuery is ideal for telephone based applications or systems such as database queries, Fax-back services, information retrieval, scheduling/shipment queries, etc.

Dragon PhoneQuery is a key emerging technology that will revolutionize data access. It will allow users to instantly access information by using a phone from anywhere in the world. It uses natural dialog to interact with the user to select the specific information that they need with no need to remember exact commands. Continuous speech and word spotting technology contribute to a state-of-the-art product that takes speech recognition to the next level.

Dragon PhoneQuery technology is first being featured in future versions of Crystal Info, a leading enterprise reporting and analysis system from Seagate Software. It will provide multi-line remote telephone access to corporate information and reports contained within the Crystal Info software. The Dragon PhoneQuery software obtains information about the reports available to each user from any number of Crystal Info Automated Process Scheduler (APS) systems located on the LAN. Once the report is identified, it may be sent to a fax or other gateway for delivery to the supplied user destination. Seagate Software announced its partnership with Dragon Systems and showcased a future version of Crystal Info incorporating this unique technology at Computer Telephony Expo 97 in Los Angeles.

One killer app enabled by this technology is the ability to fulfill technical support documents and diagrams over the phone via fax-on-demand in a help desk environment. In this scenario, a caller in need of help on a computer repair, for example, could call the PhoneQuery system and speak naturally about the nature of the problem. The system can use "word spotting" to hone in on unique words. For example, the system could prompt the caller by saying: "Please tell us what the product is and what the nature of the problem is." The caller could say: "I have a XLT model 5 unit here, and the power supply is making a humming sound."

At this point, the PhoneQuery app searches the key words and makes queries to the Crystal Info database. The application finds several matching documents. These documents include diagrams on dismantling the power supply to make modifications to it. These documents are then faxed to the caller.

Dragon Systems technologies are available in hardware and software configurations ranging from single-line PCs to applications supporting hundreds of lines running on multiple platforms including Windows NT servers as well as solutions using Dialogic Antares and National Microsystems DIVA DSP cards.

Hospitality / Travel

Cruise Line Agent Broadcast

What started with "The Love Boat" has now become a boom industry-vacation cruises. Competing cruise lines are locked in hand-to-hand combat for available vacation dollars. This particular battle is waged in great part within the country's 36,000 travel agencies, where group and individual tours are booked.

So how did one cruise line win the hearts and minds of the nation's travel agents (and the travelers' dollars)? For Royal Caribbean Cruise Line-the industry's second-largest cruise line-the answer was GammaFax PC-to-fax technology provided by GammaLink. The Miami, Florida-based firm had been using traditional technology to communicate with its network of travel agents across the country--direct mail, advertising, and brochures.

The problem was that its "Cruise-A-Grams," once printed, still had to be sent via the regular mail service. This meant delays in getting crucial information out to travel agents regarding fares, promotions, and itinerary changes. "We would even have to pull our district sales managers out of the field and put them on the phones in some cases," says David Hancock, Vice President of Field Sales for Royal Caribbean.

Even more pressing a problem for Royal Caribbean was the need for swift confirmation of reservations. In the parlance of the travel industry, the initial traveler's reservation is "taking out an option." Many such reservations are made months in advance, and may eventually be cancelled, changed, or simply ignored by the traveler.

Royal Caribbean began exploring ways to speed up the communications process with the assistance of Cambridge Technology Group, a systems integrator based in Cambridge, MA. Cambridge is a 10-year old firm specializing in systems integration, rapid prototyping, and executive education, with particular expertise in rapid systems development.

Howard Kolodny, vice president of software products at Cambridge, says that while the overall system configuration for Royal Caribbean was complex, the actual solution in terms of hard copy output was simple: a total of 16 GammaFax CP boards installed in two NCR 80286 PCs.

"We decided that PC-to-fax represented the quickest, most cost-effective way to link Royal Caribbean with its agents," explained Kolodny. "And GammaFax CP was the only PC-fax board on the market that met our needs for multi-board support on a single PC chassis."

"CruiseFax" incorporates reservation information from an IBM AS400 to enhance Royal Caribbean's communications with the travel agent. An NCR Tower is connected to the AS400 via LU6.2; the Tower in turn is linked via Ethernet to the two NCR PCs which serve as the communication gateways to Royal Caribbean's agents.

Cambridge Technology used its proprietary Surround architecture and tools to develop the software that links the IBM and NCR systems with the PCs. The software then takes the data received and formats it into the appropriate documents. Using GammaLink's communications software, the 16 GammaFax CP cards then broadcast a wide variety of documents to approximately 1,200 agents throughout the United States.

Royal Caribbean was so committed to PC-to-fax that they purchased nearly 1,000 facsimile machines which they in turn provided free to travel agents, all to ensure receipt of faxed communications.

The benefits of the new system were immediately apparent to Royal Caribbean. "We now have instant verification of reservations," says Royal Caribbean's Hancock. "We have created a two-part confirmation form; one part is kept by the travel agent, and the other is given to the client. It increases the client's commitment and improves the level of service they receive from us and the travel agent."

Just as important as reservation confirmations are such things as marketing information-new cruise packages, pricing changes, and the like. "It's made a big impact," says Hancock. "We broadcast fax-based communications every week to our agents. They love it, and it means a constant flow of communications.

Kolodny of Cambridge Technology reports that the entire system cycle, from planning to final installation, was very rapid. "We had our initial consultations with Royal Caribbean in April, with a pilot system installed in May. By the end of July, the installation was up and running."

Royal Caribbean's Hancock is enthused about system performance. "We have increased our visibility with a key audience, and the agents are pleased with the benefits they receive with instantaneous communications. On a scale of 1 to 10, this is a 9.5 for us."

Fare & Schedule Hotline

Keeping track of bus schedules, tour lines, and ferries is tough enough for local users, but even tougher for people traveling on business or on vacation. Since virtually all hotels and resorts now have fax machines, it's easy for information to be transmitted to guests with a simple phone call. A centralized Hotel Association or Chamber of Commerce can sponsor a city-wide fax-on-demand system that can be used by visitors all over town.

One approach for such a system would be to post a marquis in the lobby that lists the codes for show times, fee schedules, and day trips. Callers can enter digits to indicate where they are staying, and then choose from a verbal menu of schedules and fares. For example, the system can be programmed to say: "Press '1' if you want a current Whidby Island ferry schedule. For the Patriot's Tour Bus schedule, press '2,' or press '3' for the Airporter schedule for this month."

Travel Package Confirmations

When the autumn leaves fall and the first signs of winter begin to appear, thousands of Canadians flock to their travel agents to book vacations in sunspots and ski centers around the world.

Air Canada Vacations sells 150,000 vacation packages a year. During the winter get-away season, traffic reaches 4,000-plus packages per week, even more during the peak Christmas season. Travel agencies book the majority of these packages either by phone or through direct access to the Air Canada Vacations reservations computer.

Air Canada Vacations used to print all invoices on laser printers and delivered them by courier from the Montreal accounting center to travel agents across the country. This was costly and caused a two to three day delay between the time the vacation was booked and the invoice delivered to the travel agent.

Air Canada Vacations wanted to improve the turnaround time for several reasons. First, travel agents require quick confirmation of rates and schedules so that they can invoice customers. Second, customers want their confirmation as soon as possible so they can plan their vacations. And finally, the airline requires payment in full before it can issue tickets.

How to close the gap between the receipt of the reservation and the delivery of the invoice? Answer: computer-based fax (CBF).

Since July, 1995, CBF has allowed Air Canada Vacations to save those two to three days in the booking, billing and payment cycle. According to Roger Rouse, Director of MIS at Air Canada Vacations, "We didn't do it to save money, but to save time. But obviously the saving in manpower and other costs of printing 4,000 or more invoices a week, stuffing them into envelopes and delivering them to individually by courier was a worthwhile bonus."

Vacation packages include not only air travel, but hotel accommodations and such options as car rentals, side tips, meals, lift tickets and, often, connecting travel through other carriers. Because of all these add-ones, the information systems people at Air Canada Vacations must produce more complex invoices than their colleagues on the airlines side of things, who typically sell only air travel. Vacation package invoices include not only prices, but flight schedules, agent's commissions and all the other details of the package. The switch to CBF didn't require Air Canada Vacations to change any invoice-generating software in its UNIX-based accounting system. Said Mr. Rouse: "What makes the system feasible for us is the Faccent software created and supported by Montreal's Elan Technologies."

This software includes a HP LaserJet III printer emulation, which means that any document that can be printed on a printer can seamlessly output to a fax. "We didn't have to rewrite any of the programs that generate our schedules and invoices. Our computer can output to fax or printer; it doesn't know the difference, and it doesn't care," said Rouse.

Faccent also generates an electronic image of the invoice form. It then automatically overlays all the invoice information received from the accounting database directly into the form's predefined information fields. The result? The only change we made from the printed form," said Mr. Rouse, "was to eliminate the decorative halftone areas to speed up the transmission time The resolution is comparable in clarity to our printed invoices. Even the smallest type comes out sharp and clean on the receiving fax machine."

The CBF system, with its four-port PureData 14.4 Kbps SatisFAXtion 4000 fax card, was designed and implemented by ChronoFAX, a VAR that specializes in CBF technology. If necessary, several cards can be placed in a single PC for large scale applications.

The system uses Traffic Software's Object-Fax server solution as its principal fax engine. The server is essentially responsible for managing the outbound fax queues, load balancing between the fax card's four ports and generating activity reports. The Object-Fax server also doubles as a departmental solution for sending and receiving faxes directly from users' desktop PC applications such as word processors and E-mail.

We calculated that four dedicated fax lines, each transmitting approximately one invoice per minute, would be more than sufficient to meet our needs and those of our other departments that also distribute documents by CBF," said Rouse. "We queue our invoices to transmit during the regular business day. Of course, we could take advantage of cheaper night-time phone rates, but that would delay receipt of the invoice until the following morning. For our travel agencies, this same-day service by fax is a major improvement and a very important advantage."

At the moment, some 60% of travel agencies have requested their invoices and confirmations by fax. The others continue to receive their material by courier. But Rouse is convinced that as more and more travel agencies become aware of the speed and added convince of invoices-by-fax, they, too, will want to do it.

Import / Export

Fax Routing Platform

A California developer is using VBFax by Pronexus (Carp, Ontario) to create a custom fax and E-mail integration application to speed up order processing and customer service for a large Import/Export business.

The business has large clients that require special attention from individuals in the company assigned to specific customer accounts. When a customer faxes in a transaction request, it must be determined who has responsibility for that customer. Copies of the fax are then made and distributed to the individuals assigned to that customer.

Using VBFax, an automated system is being developed that will automatically distribute incoming faxes to the appropriate recipients via the company's e-mail system.

VBFax allows for easy e-mail integration through the open architecture of Visual Basic and the large library of tools that are available. Once the customer is identified through Caller ID, DID or fax header information, Visual Basic can be used to query a customer database to determine the list of e-mail recipients. E-mail integration tools for Visual Basic can then be used to broadcast the fax to that list of recipients through the company E-mail system.

The system saves the company money by eliminating the manual copying and distribution of the faxes and improves customer service by shortening the transaction processing cycle.

Internet Service Provider

Internet Service Provider Internet-Enabled Fax

Universal Interactive Systems (UIS), is a Los Angeles, CA based manufacturer and provider of WebMessenger, a unique e-mail to fax relay service. The service lets Internet users send Internet e-mail to any fax machine or any alphanumeric pager worldwide. Internet Service Providers can offer this service on a monthly or yearly basis. UIS handles all the administration, actual fax and pager relaying and invoicing back to ISPs for final billing to their subscribers.

UIS bills ISPs on a wholesale basis - the service providers then markup the service when billing individual subscribers. A newer addition to the WebMessenger is called FAX2FAX over the Internet. FAX2FAX provides fax transmissions over the low-cost, high bandwidth Internet communications highway. It gives the international clients the ability to send faxes throughout the US and all over the world at extremely low rates simply by sending their faxes to a local FAX2FAX system and specifying the destination number using touch-tones.

Here is how the system works: The user calls FAX2FAX automated IVR (Interactive Voice Response) system from a conventional fax machine having his fax ready for sending. Using the touch-tone keys on the phone pad, the customer identifies himself by user ID and password and enters the destination fax number. Then the IVR system prompts the user to press the "start transmit button" on the fax machine, and it will proceed with sending the fax using the same line.

Using the destination fax number, the local FAX2FAX system checks its routing tables to determine which is the remote FAX2FAX system responsible for that long distance area or International access code. Then the fax is sent over the Internet from the local to the remote FAX2FAX system. On receipt by the remote system, the fax is transmitted to the destination by a local phone call, and a confirmation E-mail message is sent to the originating FAX2FAX system. The user can check the result of the fax transmission by calling the IVR that plays back the date, time and duration of successful send or failure after a certain number of attempts. The IVR can also be configured to notify the user whenever a fax confirmation is received, or to send by fax on a daily basis status reports with the fax transactions for the period.

Subscriber Technical Support

V*Channel, Inc. of Santa Clara, CA specializes in interactive voice, fax and call processing technologies. FAXLink is their big seller. One of V*Channel's biggest sales was to America Online with a FaxLink system.

AOL had only 14 documents on their system, but they had a tremendous amount of calls coming in. They send out over one million pages of information (of those 14 documents) on an annual basis. And when FaxLink was installed, it offloaded a large amount of their call load. The system is open 24 hours a day, technical questions can be answered and there's always a hard copy to work with.

Legal

Temporary & Expert Services Finder

Law offices are often in need of other professional services including process servers, investigators, career counseling experts, and doctors. These services are sought in order to do research or provide expert testimony in certain court cases. Most of these experts are hired on a temporary basis and are promoted through word-of-mouth.

One way to help lawyers in their search for a professional service or expert is to provide their offices with access to a daily schedule and resume service via fax-on-demand. Let's say that a lawyer needs a career counselor to interview a client involved in a wrongful termination suit or divorce. If the regular or alternate choice for a counselor is not available, this could introduce delays or at the very least uncertainty in getting a good reference.

A voice processing system can provide simple touch-tone access to a database of local or regional professionals so the lawyer can call one centralized number to search for help. Professionals who are listed in the database would be responsible for forwarding updated resumes and for updating the system with information having to do with their fees and availability. Such a system would allow lawyers to access a menu that prompted them for the type of help they were looking for followed by a choice list based on availability and fees. Once a good match has been found, the law office can type in commands to receive the expert's resume. This type of system can significantly reduce the amount of time associated with conducting a processional search.

Military

Military Maintenance

Millions of parts, hundreds of suppliers, and thousands of procedures must be managed by the military. The task of maintaining current records on parts inventory and manufacturer procurement is staggering - it takes thousands of military employees to do this. When it comes time to fix a piece of equipment, maintenance personnel often refer to documentation from a maintenance database. The trouble is that specifications change, procedures change, and products are updated all of the time.

A voice processing system with CBF capability can go a long way to simplifying the entire procedure. Take, for example, the repair of a computer. The maintenance person can call a general document fulfillment center, input a series of digits for the computer in question, and then answer questions having to do with the problem on that machine.

The answers to these questions can be linked to a centralized document database that "tags" image files. The files tagged in each phone call can be assembled by the fax-on-demand system and then prepared for transmission as a single fax. Contractors and manufacturers can help to keep the centralized system updated by faxing-in new parts sheets and instructions on a regular basis. These same pictures and diagrams could also be scanned-in manually by a system administrator.

Power Utilities

Power Utility Service Center

Utility companies experience huge call volumes, causing customers to wait for long periods of time to get service. This is especially true of special call centers, such as the PG&E's Smarter Energy Line in San Jose, California.

PG&E serves millions of users and sponsors a special help desk to assist customers in the conservation of energy. The solution is based on a product called FAXLink, developed by V*Channel, Inc. of Santa Clara, CA. FAX-Link aids the Smarter Energy Line service agents by greeting callers, as well as providing interactive voice and fax options. The system thereby reduces the number of callers that the service agents have to assist over the phone directly. It also helps because callers do not have to wait in order to get information.

Approximately 25 - 30% of all callers elect to help themselves to the information by selecting various options provided by the automated information system. The automated attendant menu allows callers to choose to transfer to a service agent, listen to information on rebate offers or energy saving information, or to order fax documents with the fax-on-demand portion of the application.

Figure 6.9 - PG&E Smarter Energy Line

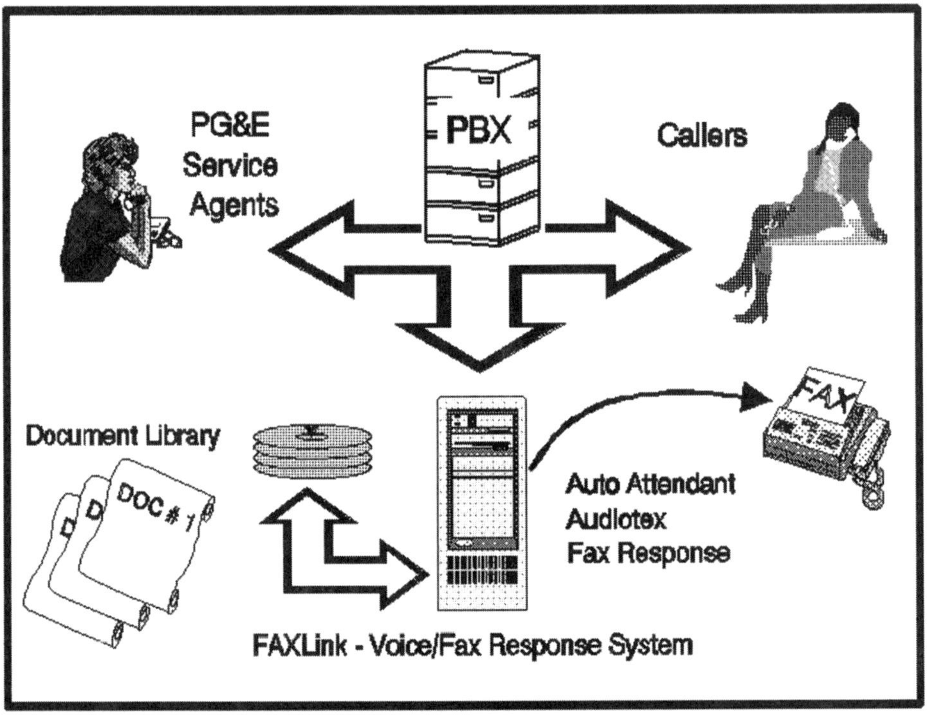

V*Channel first surveyed the types of information that were being requested by the callers. The company determined how much information to put on the system in order to achieve the budget "off load" percentage. Then, V*Channel designed the interactive messaging for use by the callers. According to Herbert B. Gong, Jr., PG&E Smarter Energy Line Technical Assistant: "The FAXLink feature has given PG&E's Smarter Energy Line customers a valuable option to receive literature immediately via facsimile. Customers are pleased and surprised with this new and upcoming technology. It's fast and easy to do. It's very efficient and convenient."

V*Channel used Dialogic and GammaLink speech and fax cards to interface with a Rolm 9000 ACD as pictured in figure 6.9. In addition, MCI's TA and 800 services are used to connect the callers via the public switched telephone network. Both customized software from V*Channel and software modules from Telephone Response Technologies, Inc. were used to develop the application.

Print Media

Article Reprints and Extensions

Publishing companies are always on the look-out for new and stimulating ways to maintain readership loyalty and increase circulation. Some information industry planners have used Audiotex, CD-ROM, and the Internet for this purpose. Perhaps one of the best examples of how CBF can help to stimulate readership is an application developed by a St. Louis, MO company called Nuntius.

Pulitzer Publishing's St. Louis Post-Dispatch has been promoting the use of its "PostFax" service since the summer of 1994, and now the system receives over 1,000 requests a day. Every day, the newspaper publishes a menu of approximately 35 items in the business section that readers can inquire about via PostFax.

For example, selections can include IRS forms during the tax season, EPA car mileage listings, local bank performance data, mutual fund information, or even Dave Letterman's popular "Top 10 List" for the day.

In order to access the system, readers dial a local number and are then prompted for the document choice they have made. In addition, the system prompts callers to enter their fax number. PostFax then delivers the document to their fax machine within seconds of the request. Figure 6.10 shows how the system uses a client-server topology in providing this service.

PostFax offers an "extension" that provides customized newspaper delivery. During the pilot test of the system, over 2,000 calls in one weekend were made for late-breaking baseball scores and a collection of the best recipes published during the previous year. The Post-Dispatch is also selling advertising space around the perimeter of each fax that is delivered, so the system is a revenue generator for the paper.

Figure 6.10 - PostFax at St. Louis Post Dispatch

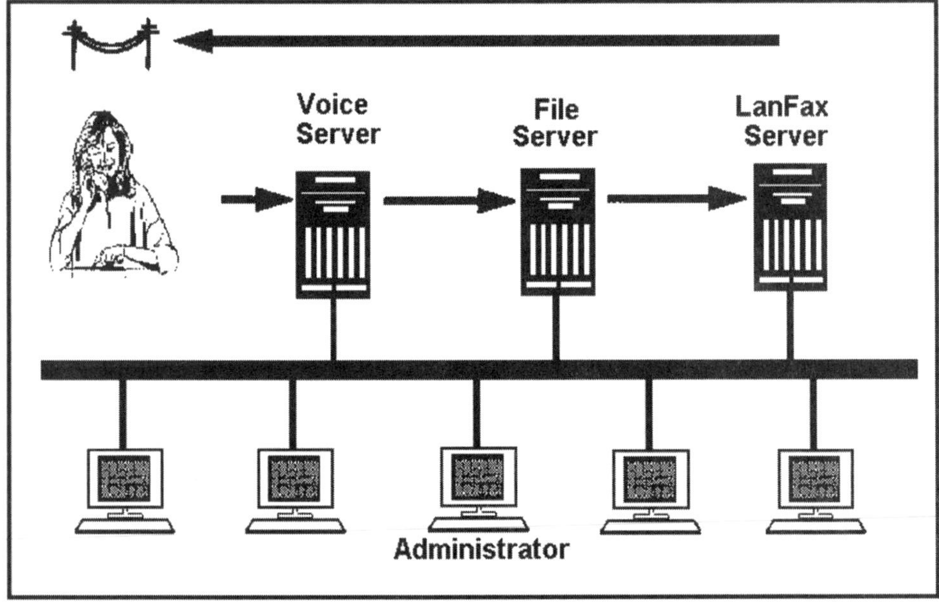

Public Relations

Fax Broadcasting

Microware, Inc. of Anchorage, AK is a Telephone Response Technologies VP Express VAR, Microsoft Certified Professional and an Artisoft Premier VAR. They sell mostly custom apps, but use TRT's VP Express as the base platform. This includes fax-on-demand and fax broadcasting.

Microware started with Delrina WinFax Pro for a SOHO Public Relations company. They bought a 486 DX2 along with a Practical 14.4 fax card and the single-line WinFax broadcaster. The company made money in service, because the solution was very inexpensive. At the time, a VP Express Broadcaster was not available -- it handles 10 fax lines. The customer acted as a miniature fax service bureau and sent out bulk faxes on behalf of his PR clients.

A bigger sale was Northwest Strategies. They're the second largest PR firm in the state. They represent the largest native corporation and a major oil company. They had to get their "consistency notices" out to as many as 600,000 people within a few days. They were doing mass mailers. It was slow and expensive. They bought VP Broadcaster and two four-port GammaLink boards. They are now able to fax out over 20,000 faxes in two days with one PC. $22,000 includes installation and training.

Retail Services

Catalog Sales

Merisel, the second-largest computer products distributor in the United States had an information bottleneck: Its customers, mostly computer stores, needed fast access to thousands of pages of current product information from the hundreds of manufactures that market through the El Segundo, California distributor.

The only way to meet this requirement seemed to be using already limited sales resources. This proved unsatisfactory, however, since it soon became obvious that the time and effort being expended solely to meet customer's information needs was hampering the actual sales effort. According to Lorraine Watkins, Merisel's VAR marketing manager, keeping several hundred customers a day abreast of the newest developments in products and services had become a very severe problem. "We were unable to give our customers the type of service that we believed we should," said Watkins. "Before we got our fax-on-demand system in, we were not doing much to meet their information needs, and what we were able to do we certainly were not doing in a timely manner."

Watkins indicated that whenever a customer needed information, sales was the only contact and resource available to provide it. "This meant that a salesperson had to take time away from the phones to go to the library and locate the information, usually a spec sheet, then walk to the fax machine and send it. Sometimes there might be a line at the machine, which meant the information could not go out right away and he would have to attempt to send it later, if he had the time. Meanwhile, a customer is waiting and a potential sale delayed."

Merisel sales people did not like having to provide product information to their customers in this time-intensive way. They were, in a classic Catch-22 situation: Although getting information of this type to the customer was certainly a part of the sales strategy, it took an inordinate amount of time away from the actual selling process. Besides, the effort often yielded small results. "The volume of information needed was overwhelming. What we could provide was trivial and often outdated, and the client base we could provide it to was not nearly large enough to justify the time it took our sales personnel away from the actual process of selling," admitted Watkins. "It was a very inefficient way of doing things."

Faced with a situation that deteriorated further every time manufacturers sent in additional and new product information for Merisel to pass on to its customers, Watkins came up with the idea to automate this part of the operation using a fax-on-demand system. It was then that Ibex Technologies, of Placerville, California was called in.

Merisel's problem had a straightforward solution. Where it differed from the commonplace was in the magnitude of the database that the system would have to handle. Also, because of the intensive, continuous use they planned for it, reliability was a paramount consideration."

Ibex Technologies was certain they could provide the required reliability. "This is one reason we use GammaLink products exclusively," Mattos explained. "They are robust, and offer the best tools to easily and quickly develop systems that customers can really use. We have had clients come over to us from their original developers because we use GammaLink products and nobody else's. Another important aspect for use, particularly when dealing with customers who operate worldwide, is GammaLink boards are approved for use in well over twenty countries. This enables us to put together systems that will operate across international borders without any difficulties or unpleasant surprises."

The Ibex FactsLine integrated voice/fax response system configured by Ibex Technologies for Merisel uses twelve GammaFax boards, and 12 Dialogic voice cards. The Dialogic cards are the 21D, 2-port card; the 41D, 4-port board; and the 120, 12-port board. All GammaFax boards are network-ready, providing multiple phone line support, broad software support, and are compatible with any PC-based LAN.

When Ibex Technologies was through, Merisel had a system consisting of 12 voice lines and 16 fax lines. Merisel then scanned in the literature they received from the manufactures, compiling a database of over 3,700 different products fact sheets, about 19,000 pages or six gigabytes of information. According to Watkins, this was the most difficult part.

"Initially, scanning was a problem because we had no previous experience and the volume of material was so staggering. We had people working in shifts from 7:00 a.m. to 11:00 p.m. It took about 200 hours to scan it all in." Once the original inputting was completed, updates were no problem and are presently handled by a single full-time worker.

When the database was ready, Merisel ran ads in the trade press and carried out a direct-mail campaign to let its 30,000 customers know about the service and information that was now a phone call away, 24 hours a day, seven days a week. "It was instant gratification," recalled Watkins. "Where before a customer might get the information he requested in a few hours if he was lucky, or a day or more if he wasn't, now he gets it in a matter of minutes." At present, just a few months later, the system is fielding an average of almost one thousand calls a day. The setup is so formidable and obtains such consistently excellent results, that it merited coverage by The Wall Street Journal.

Although primarily intended for use by distributors, VARs, and resellers, the way the system is set up also allows end users to call into the database. "If the caller has a Merisel account number; that is, he is one of our customers, he inputs it when he calls in to request information," explains Watkins. "If at the time we happen to be running a promotion on that item, he automatically receives both the information he requested and the promotion with the pricing information. If there is no account number the system omits the promotional and pricing information."

Merisel also uses the system to broadcast regularly schedule electronic newsletters they call Fax Bulletins. These go out every Friday to some 6,500 top customers. Practically every day of the month the distributor also does what if refers to as unscheduled fax broadcasts. These take place whenever one of their clients introduces a new product and wants this information broadcasted right away to Merisel's customers. This has opened a whole new profit center for the company since, essentially, what it does is sell ad space in these broadcasts.

Because of the added capabilities the new system has brought with it, Watkins finds it difficult to quantify its costs. "If we go by bare numbers," she indicated, " the system more than paid for itself in under one quarter. However, if you factor in what the cost would be of doing everything we can do now, using outside contractors, the payback period then becomes ridiculously short."

Custom Quotation System

Ellen Muraskin, reporter for Computer Telephony magazine, interviewed BSC of Waltham, MA. The customer sells computer, fax and copier supplies to businesses over the phone. Dan McLean, marketing VP at BSC, needed a much cheaper fulfillment strategy. At $10 a crack, mailing catalogs to prospects is pricey. And slow. By the time customers got them, they'd already bought elsewhere. Customers need information fast. Who needs to thumb through a whole catalog to find one case of print cartridges?

And broadcast faxing through a service bureau was expensive. A bureau's fax minutes couldn't be discounted along with voice traffic from BSC's own carrier. Merchandise returns were complicated. RMAs differed from supplier to supplier.

Enter the FaxFacts system from Copia International, Ltd. of Wheaton, IL. The FaxFacts system is an outbound, telemarketer-driven fax-on-demand system. It uses FaxFact's FFMerge modules for customized quotes using fax-on-the-fly. BSC hooked two Novell servers together, marrying fax-on-demand and fax-merge to a DOS sales-and-marketing database. Now sales reps send prospects tailor-made product and price information on the spot, in two pages or less. BSC was also able to take broadcast-fax in-house, stretching the company's marketing budget and increasing control. Ready-made fax drivers for MS Word (for templates and documents) and MS Access (to transfer the merged fields) simplified the job.

"With printing and postage, it's $10 to get a catalog out the door. At 5,000 mailings, you've spent $50,000. But a two-page fax is $1,000 to the same number of people. And I can send information tailored to that individual customer - immediately."

BSC uses Copia's FaxFacts faxserver software in three basic ways: to send customized quotes to individual prospects, to send canned product and material safety sheets, and to distribute fax broadcasts targeted to specific customer niches.

McLean is self-taught in Netware, MS Access, Word and FaxFacts. He built Access and Word hooks into the DOS-based sales and marketing application already running on sales reps' workstations. Sales reps, while on the phone with customers, pick out the part numbers and quantities they need from the sales and marketing database.

Six or eight times a day, the quote records are downloaded into BSC's Access database. Since Copia has a plug-in print driver for Win95, the system can take a Word document template, plug in the Access database record of name, fax number and quote fields using Copia's FFMerge module, and print it right to the fax queue.

At day's end, Access and Copia's logging file tells McLean which faxes did not reach their targets. These are manually printed and faxed. The company's library of 40 to 50 fax documents is geared to specific vertical markets. "True Value hardware stores all have a specific hardware setup," says McLean. "After you've sold to 100 True Value dealers, you know the needs of your 101st." McLean broadcasts hardware store faxletters with information on just the products they tend to order. Telesales reps can select these faxes from a pop-up window to send to individual callers.

Another Copia module makes it simpler to tell customers how to return merchandise to any of BSC's 50 different supplier warehouses. Fax merge both notifies the warehouse of the return and tells the customer that supplier's RMA requirements. The system can even fax the customer a bar code or a shipping label. McLean creates all faxes using Word and other off-the-shelf graphics and desktop publishing software. He says it took him two days to upgrade the system from a version he'd installed six months earlier, using analog Hayes fax boards.

"We decided to upgrade our outbound capacity and go digital, so we went from eight lines to 24, and got a T1." He bought a new computer from CompUSA and stuffed it with Win NT, (that was the hard part, he says) three Brooktrout TR114 voice-and-fax cards and a Dianatel T1 interface card.

Of FaxFacts, he says, "It's simple. You just identify the configuration parameters the setup program asks for, and just like any other off-the-shelf software, it'll configure the program for you for fax broadcast, fax-on-demand and fax merge."

Bringing fax in-house also gained McLean more bargaining muscle. "We said, 'Why should we pay exorbitant fax service bureau fees and not get credit for the usage time with our primary voice carrier?' The company was also getting bumped by higher-volume customers, and not learning which faxes had failed to arrive. Now McLean gets that information from the FaxFacts log, and he can negotiate long-distance rates with carriers based on voice and fax minutes combined.

McLean says that the fax merge, fax-on-demand system has measurably improved his company's sales productivity. It's allowing him to cast a wider net and diversify his customer base. He can send a whole campaign of messages over a period of weeks, and his sales cycle shrinks from up to six weeks to a matter of days. "I can reach more people in less time," he says simply.

The payback can be quantified. The cost of reaching customers goes from $10 for a catalog to 20 cents for a fax. The sales cycle shrinks from up to six weeks to a few days. Voice and fax long-distance charges can be combined, allowing BSC to negotiate better deals with carriers.

Floral Orders

Nothing beats making a trip to the flower shop so you can see exactly what you're buying. It's a lot easier to make comments on a certain arrangement, choose the best-looking flowers from the refrigerated section, and watch as your arrangement is prepared. Of course, this luxury cannot be duplicated if you are calling an 800 number to order flowers for someone far away. A voice processing system can go a long way to alleviating this problem and also help to promote the purchase of more products for the vendor.

For example, a voice processing system can ask flower buyers a series of questions having to do with the price range, color, type of flowers, and the occasion the purchase is being made for. Based on this criterion, the system can then narrow-down the choice for the caller.

The system can be programmed to say "Based on what you have entered the Floral Rainbow and the Mumtastic arrangements would be a great choice. If you want to order them now, press 1 or you can press 2 to get pictures of these arrangements faxed to you within seconds."

Telephone Companies

Fax Gateway

Companies such as MCI have popularized sending out many faxes to a group of customers or fellow employees. The idea behind a fax gateway is to distribute information by fax to hundreds of recipients simultaneously. Of course, to do this manually would require dozens of fax machines, as many people, and several solid days of work.

Computer based fax technology has made this a reality, and now it is possible to upload both the fax you want sent along with a list of recipients and fax numbers in one simple data call. MCI provides PC-based software that automates this procedure. Voice processing can also be used to schedule the transmission of faxes to pre-defined list of recipients. In this case, the system administrator assembles distribution lists and associates a code number with each list. These code numbers are provided to the subscribers that have authority to distribute the faxes. This capability is sometimes referred to as a "fax mailbox" system.

Fax Store And Forward

Managing confidential messages is a special challenge for those who travel extensively. For example, if you are traveling and a fax "comes in" for you - you need someone else to intervene and re-transmit it.

The fax can then be viewed by someone else, lost, and degrade in quality during re-transmission. If the communication is important, a serious problem presents itself if you can't get someone to re-transmit the communication. The receipt of a fax can also be difficult for small business owners or consultants who share telephone lines for multiple purposes such as e-mail, voice calls, and fax.

The same problem arises if your line is busy and the sender quits trying after a few attempts. Voice processing and CBF come to the rescue for those meeting this profile. The answer lies in providing worldwide access with fax store-and forward capability.

For example, SkyTel (a nationwide paging company), offers their own "SkyFax" service to subscribers. With SkyFax, subscribers are issued unique 800 numbers; each of which can be published as a dedicated fax phone number for that business. When subscribers receive a fax, their beepers indicate that a fax is waiting.

The beepers are able to display the number of fax pages waiting the users' fax mailboxes (beepers are also used to alert users of e-mail and voice messages as well).Users of the SkyFax service then dial a special access number to hear about the messages they have received. The system asks them for their PIN number and security code and then announces how many faxes are in queue, the length of them, and where they came from.

Users then have the option of storing the fax, having it sent to their fax machine, or having it sent to any fax machine in the world. The system operates much the same way a voice messaging system would, with the exception that the message is a fax instead of a recorded voice message. In fact, the fax can be "pointed" to a distribution list, so that it can be re-faxed to a list of other individuals. In this fashion, the SkyFax system can be used to totally control the time and place when faxes are sent. This is useful to the extent that lines can be made ready to receive faxes on demand, and the communications can be kept more confidential.

7

Interactive Voice Response

IVR systems are popping up everywhere. This can be attributed to the popularity of the PC platform and the ease with which databases can now be accessed with programs like FoxPro, SQL servers and host gateway programs. In years past, accessing computer data with voice processing was a great feat. Older IVR systems used speech processing cards custom designed for the proprietary backplanes of mainframes and minicomputers. This is true of the DEC PDP series, the IBM Series 1, the Unisys A Series, and a number of other legacy-type systems.

Today, it is a straightforward business to use terminal emulation cards, xBASE access modules and PCs to achieve the same result for a few thousand dollars. For example, there are Visual Basic extensions that provide complete voice processing access to ODBC data. These tools can be purchased for less than $1,000.

Add a PC and a speech card, and you're ready to do some IVR. Regardless of development tool and hardware cost, the sophistication of some IVR solutions is impressive. There are systems that talk to several computers, interface to a variety of switches and automate transactions for thousands of calls each day.

This chapter provides examples of how IVR has been applied to a number of industries. In some examples, the components and software used are described. The makers of these systems can be contacted in order to purchase turnkey versions of the software, or perhaps to negotiate a cross-licensing arrangement. With the IVR industry growing at over 25% each year, there's plenty of room for private label agreements, technology transfers and strategic alliances.

Automotive

Parts Order Status

Most auto parts managers use computers to search a parts database or log into a network to retrieve information on behalf of customers. On many occasions, parts are ordered and customers receive a call when the part comes in so they can schedule an appointment for the installation. If the item is out of stock, there may be a long wait. Since most of this data resides in an inventory database, there's no reason why voice processing can't be used to retrieve the data, and make it available to customers over the phone.

The voice processing system can use terminal emulation in the case of a mainframe database, or it can be connected to a file server. Customers can access the system and type-in an invoice number or telephone number in order to "locate" their record.

The system can then provide information such as backorder status, estimated delivery date, and an option to cancel the order. As an option, the IVR system could be programmed with an outdialing capability, so that it calls the customer when the part arrives.

Repair Status

Over the past decade, automobile manufacturers have worked closely with dealerships to develop ways to streamline repair procedures. This has been enabled with computerized diagnostic equipment, on-board microprocessors, specialized training, and even voice processing. Intellivoice Communications, Inc. of Atlanta, GA was one of the first IVR vendors to design voice processing systems like this.

How voice processing is applied is straightforward: When customers deliver their cars to the dealership for repair, they are provided with a repair order number. Customers are then instructed to call a local telephone number in order to check on the status of the repair, authorize work, and to find out about the final cost and completion time. In order for the system to operate efficiently, it is connected directly to the dealership's invoicing and repair order system.

In addition to the repair status being available to callers, special voice messages can be left on the system by the technician. This is especially helpful if the car is picked-up after hours when the person who fixed the car is not available. For example, the repair person might leave special instructions, or answer a question that the car owner had when the car was dropped-off earlier in the day.

Valuation Hotline

Finding out the market value of a car is as easy as calling your bank or looking into the official "Blue Book." All you need is the year, make, and model of car along with mileage and information on the options in your car. All of this data is taken into consideration when applying a value to the wholesale and retail value of the car.

These books are used by auctioneers, dealers, and financial institutions to establish interest rates, insurance premiums, and trade-in values. If the bank is closed, or you don't have a book - it's hard to assess the value of a car.

It's always a good idea to have this information before engaging the sales cycle, no matter what side of it you are on. Voice processing can provide universal access to this information. An IVR system that's hooked-up to a database of automobile valuation data can make the search for information a pleasure. These systems have been in use on premium service lines (976), and lending institutions has sponsored some. An IVR system of this type prompts callers to select car types, foreign and domestic makers, and then the year and model of the car. Once the data has been assembled, the system can announce the value of the car.

Cellular Telephone

Cellular Telephone Company Alarm Dispatch

Andre Leveille is a system analyst at Rogers Cantel in Quebec, Canada. He is thrilled with his CallStream Communications (Oakville, Ontario) IVR system. Rogers Cantel is the largest national cellular company in Canada. They provide coverage for 85% of the population with over 1 million subscribers. The service area spans Victoria, British Columbia to St. John's Newfoundland (five time zones). The carrier operates 1200 cellular sites and repeaters to provide service. Rogers Cantel uses NetExpert from OSI Systems to monitor their site activity. Net Expert is based on the Sybase database. The system sends alarms to the Rogers Cantel Control Center (RCCC) for dispatch of field personnel.

When field staff enter a cell site, an alarm is sent to a RCCC's NetExpert display. A staff member then calls to confirm the identity of the person in the site and the reason for the visit. Rogers uses VoiceStream IVR running on a Sparc 5 with Linkon FC-4000 cards to automate the process. Now field staff call in, identify themselves, and enter a maintenance code to update the NetExpert system. VoiceStream IVR updates the Net Expert Sybase database with this information using UNIX shellscripts to downgrade the alarm. According to Andre Leveille, Rogers' Operations Support Systems Analyst, this represents about 300 calls per day. This has freed up one full time position for someone at RCCC to work on other projects.

Charitable Organizations

Auction Events

Christie's Auction hosted a special charity event to raise over a billion dollars for medical research and these series of auctions were broadcasted live on television to enable viewers to participate in the bidding process via the bidding IVR.

The configuration consisted of Windows 95 and Workstation clients, including a voice box with Dialogic T-1 boards for network interface and voice processing. ActiveX objects were distributed among the clients and Windows NT Server.

The Dialogic hardware took the touch-tones of bidders and fed back real-time responses indicating whether or not the bid was valid.

Construction

Job Status

For small construction businesses, job status is an obvious fact, since foremen and employees are on the same job. The task is not so easy, however, for management employees who are tracking as many as three dozen construction sites all over the country. One way to help manage the process is to have foremen and job supervisors regularly update a centralized IVR system.

For example, at the end of each shift, employees from each site can call an 800 number. The IVR system can prompt the foremen for their site I.D. number, and status codes for different aspects of the job. These codes include progress reports on permits, material, and percentage of completion.

All of this data can be written into a database, so that management reports can be generated each day in an automated fashion. In addition, the system could double as a voice mail platform in order to trade important messages between field employees and the home office.

Materials Sizing

Wholesale suppliers have all kinds of instructions and aides to assist architects, builders, and general contractors to size jobs. There are calculation wheels, books, pamphlets and even help desks. The sizing of materials is perhaps one of the most important aspects of estimating the scope of work and cost for a job. It's no wonder that only the most experienced workers are counted on to do this.

Concrete, plywood, bricks, and siding can all be sized by a computer if the correct temperature, drying time, depth, height, weight, and other data are supplied. All of this information can be input over the phone. For example, an IVR system can say: "Thank you for calling the Wholesale Materials Hotline. Please choose the material code you are interested in. Please enter the length of the space in question. Use feet and inches with the "*" as the decimal. Now please enter the width in the same fashion followed by the depth... You need 2,000 cubic feet. If you want to order this, please enter your account number." In this way, an IVR system could simultaneously serve dozens of construction sites for an entire metropolitan area.

Time Reporting

It's difficult for subcontractors and construction foremen to track and report paid time for employees. This is especially difficult if there are many people on the job who are dispatched to different areas. It is typical for an individual's working hours to be logged manually and then later transcribed for the preparation of payroll checks. The process is time-consuming, and often susceptible to fraud.

Voice processing can solve much of the time reporting problem by turning the IVR system into a "punch clock." A database of employee numbers, job sites, associated telephone numbers, and payroll information can be accessed and updated by the IVR system. Each employee or foreman can be instructed to call into the IVR system each time an employee punches-in or punches-out.

By using a combination of speaker verification, ANI, and Caller ID, the task of collecting this information is made easy. In addition, this technology makes the process less susceptible to fraud and mistakes.

Corporate

Benefits Enrollment and Fund Liquidation

Access to benefits information, including the liquidation of funds, is a time-consuming affair for benefits workers and the employees they serve. The liquidation or conversion of funds comes about as a result of changes in corporate ownership, regulatory changes, or by employee choice. The benefits department usually distributes letters instructing employees to fill-out a form. These forms are transcribed and the data is used to disburse checks or transfer funds. In many cases, temporary employees are hired to handle these transactions.

Much of this process can be automated with voice processing. For example, Automatic Data Processing (ADP) has developed a system based on Talx Corporation voice processing components and the ADP Client Server Series (CSS) Human Resources Management System (HRMS). This system allows for interactive voice response for benefits enrollment, attendance reporting, and other human resources functions.

By integrating the HRMS with interactive voice response, the users of the system have telephone access to virtually all databases and services supported by ADP.

This includes benefits coordination, payroll, and human resource databases. The system is capable of alternately routing the same information to authorized users by fax or e-mail.

Direct Inward System Access

Most larger PBXs have the ability to connect outside callers to system resources for automated business communications. This is especially helpful if a company wishes to avoid the use of calling cards for long distance calls made by employees. Instead, employees can dial a local number, enter a special code, and then have access to virtually any feature that a single-line PBX phone has. The difficulty in administrating this feature is that it is simple for fraud and theft of service to occur. This is because Direct Inward System Access (DISA) requires only a sequence of touch-tones to activate.

An IVR system can curb the abuse of DISA lines by using speaker verification, or by using a call-back mechanism. For example, authorized users can record their voice prompts and have the system administrator store the voice prompt in the IVR system. When users call the system, they need to speak the same word in order to gain access to the system. Once the speaker has been verified, the system can hang-up and call back at a pre-determined phone number, or another level of security can be used. This type of authorization could be used for computer network access.

Order Entry & Order Status

Voice processing systems routinely automate the process of checking the status of an order. Each order is assigned a number that is entered when a status report is needed. Users are prompted to enter part numbers, or they can enter a phone number that is associated with the order.

These IVR systems interface with database files that contain inventory, backorder status, and delivery information. All of this data can be read aloud to callers with text-to-speech technology.

For example, the Radix II company of Oxon Hill, Maryland developed enhancements to the Defense Logistics Agency (DLA) DESEX system with IVR. The DESEX upgrade is based on the Radix II MULTEX platform. The system uses a dual 486 platform, offering component-level redundancy.

MULTEX networks with the DLA's Ethernet LAN, providing connectivity to workstations and other nodes. In addition, the company developed a mainframe access capability by using cluster controller emulation cards that perform SNA data access to the order entry host. According to DLA manager Shelly Broussard, the DESEX IVR enhancement has reduced the operating expense associated with each call from $1.48 to about 60 cents. After the installation of the IVR system, the agency experienced a 200% increase in overall call handling capability.

Employment Verification

It is a burden on most large employers to provide employment verification to state agencies, prospective employers, or financial institutions. For some companies, there are literally hundreds of these inquiries made each day. In most cases, employers try to keep the transactions short by limiting the amount of information provided. For example, some employers will only verify employment dates, and refuse to provide information on salary or the reason for termination.

Voice processing can be applied in order to announce and verify employment based on a cross-reference to an employee's social security number and dates of employment. This information could either be maintained separately in the IVR system database, or the data can be retrieved from a "live" human resources computer.

For example, the system could say to callers: "Thank you for calling the employment verification system. Please enter the social security number of the person you are inquiring about... This person worked here starting June of 1993 and is still employed on a full-time basis."

Training Centers

Large corporate training centers operate in a similar fashion to hotels and universities. For example, there's room reservations, course selection, and scheduling that has to be organized by the training center staff. Training center employees prepare instructor contracts, process registration records, and also maintain course materials and inventory. Managers can add a fax-on-demand capability so students can get prospectus data and other information instantly.

Correctional Institutions

Visitation Scheduling

Prison administrators have both an internal and external constituency to serve. First and foremost, there's the prisoners themselves. To a certain extent, the same officials act as social directors in arranging for visitation from family, friends, and court-appointed specialists. The task of managing these schedules is difficult and time consuming. The schedules have to be arranged around activities, inmate privileges, and visiting hours.

An IVR system can sufficiently automate these tasks by providing touch-tone access to the same schedules input by prison officials. For example, a database of inmates, their privileges, and "open" hours can be shared by both the IVR system and an administrative workstation. When visitors call to make visitation arrangements, they can be asked to input their desired time based on prompts played-out to them after entering the inmate's I.D. number.

The caller can be asked to speak their name and select the time they would like to schedule. Each day, the system administrator can generate a printed report and then post or distribute the information in a way they're accustomed to doing so. Optionally, inmates may access the system from "house phones" or a payphone to hear about their visitation schedule.

Education

Course Registration

College course registration is a time-consuming and frustrating task. Just ask any student or college administrator. It is typical for the entire student population to fill the school gym so that course selection and availability can be negotiated on a case-by-case basis. Registrants are armed with a booklet including course numbers and are also given an entry form marked with the students' selections. If courses are no longer available, or if credits are miscalculated, entries have to be re-submitted. Even if all goes according to plan, it is typical for students to have to make multiple passes before completing the process.

Voice processing systems have become a popular way to alleviate some of the concerns associated with college course registration. One good example is the University of Quebec in Montreal (UQAM). By working closely with MediaSoft Telecom, of Mount Royal, Quebec, the school was able to automate their entire registration procedure.

Yves Bouchard, UQAM's Information Systems Manager, was in charge of integrating the MediaSoft system with the school's VAX minicomputer, HP Netserver, and Oracle database.

With the MediaSoft IVR system installed, students are asked to call the system at alternating time intervals in order to spread-out the traffic presented to the system. Students enter their I.D. number first, so that their authorization to use the system can be verified with the Oracle database. Once this is done, students enter their semester and program codes. The system then allows them to continue as long as they are logging-on during the set-aside time intervals for that group of students.

Most students pre-register, so the system is programmed to confirm registration and accept cancellations as well.

For example, students can hear a list of the courses they registered for, cancel a course, and subsequently register for new ones. The VAX system contains all of the data for the students' records, course timetables, academic programs, and course availability. After the host gets information uploaded to it from the IVR system, it returns codes so the voice processing gear can announce messages confirming or denying access to each course.

The system is also programmed to deliver specialized messages to groups of students. The teaching staff and the registrar can record these messages. According to Bouchard, UQAM is considering an upgrade to the system that will also allow students to access registration via their own PC. Overall, the IVR capability has significantly decreased the labor and time involved in administering the registration process. The school says that the system handles as many as 5,000 calls a day during registration.

Dormitory Registration

Students and college administrators who handle dormitory assignments can use interactive voice response. For example, all students who are registered and have a valid I.D. number can access the system before arriving on campus.

The system can let students know what their room number, hall name, and scheduled arrival time is. In addition, the IVR system can allow students to leave messages for their new roommates and resident managers.

More sophisticated systems may be programmed to allow students to choose from a list of rooming options before the assignments are finalized. In this way, seniority, cumulative grade averages, or some other merit system can be used to give preferential treatment to some students.

These assignments could also be made according to the location of certain classes. In this case, the system could assign rooms based on the location of the closest classrooms.

Financial Aid Hotline

The financial aid office at most colleges and universities is perhaps one of the busiest places on campus. Students, parents, and school officials frequent these offices to get financial aid packages, forms, and to attend interviews. One of the constant themes in answering questions over the phone is the status of financial aid eligibility and approval. Some universities have greatly automated this work by installing IVR systems.

For example, New York University has a financial aid hotline that handles calls for the bursar's office, collections, dormitory and housing assignments - as well as financial aid. All financial aid status inquiries are now referred to the IVR hotline system. Students access the system by dialing a local number. They're then asked to enter their social security number. The IVR system then checks the database for a correct match and then speaks out a status message to the student. If a financial aid request hasn't been processed, the students are told that they are not yet on file and to call back later.

In addition to the IVR database access capability, the system provides basic Audiotex functions. Information dispensed includes tips on how and when to apply for financial aid. The IVR system also allows students to record requests for paperwork, or transfer to a financial aid officer. NYU's system also provides student employment office information.

Report Card Hotline

There are inherent delays in the distribution of report cards due to the mail, and the processing time required to generate them. In addition to the delays, students unfortunately try to change the grades on report cards in order to avoid punishment from their parents. This disrupts the process for corrective action or tutoring if problems are not caught in time. Teachers and administrators alike can use voice processing technology to speed-up the process of reporting a student's grades and also to avoid fraudulent changes to the reports. Parents can access this information over the phone at any time from home or at work.

IVR system access can be limited to authorized users with a special access code. Parents can "log-on" to the system with their access code, the student's I.D. number, or by calling only at a predetermined time from a certain phone number (ANI or Caller ID can be used in this case). Callers can choose from a menu of classes, or opt to get an entire report subject by subject. An IVR system of this type can also be programmed to provide specialized voice messages to the parent in the form of comments, congratulations, or advice on a subject-by-subject basis.

Student Housing

SOCR of Ontario, Canada worked with Voysys Corporation of Fremont, CA to develop a combination Web page and database input form with IVR. The system is used to help university students secure campus housing.

Universities have a shared problem during the period just prior to the beginning of each semester or quarter when students are enrolling. The problem is that these students all need housing. There is generally a wealth of property owners who would like to rent housing to them, but the problem is connecting the two together in an efficient way that gets roofs over the students heads in an affordable manner.

Housing rental information is often unavailable to out of town students. Most landlords do not have access to the Internet to retrieve or submit rental information. They need a communication device accessible by anyone, anywhere. The solution to this problem according to Voysys, was the use of an IVR application to enable renters to list property on the Internet. This lets landlords use IVR to activate and cancel property listings by using their touch-tone phone.

The Internet/SOCR application is a combination of an Internet (Web page) database input form and an IVR. The way it works is this: a student visits the Web page and fills out a form that includes such information as amount willing to pay for rent, how many bedrooms are required, are pets required, geographical area, duration of rental need, etc.

That information is placed in a database and a search is conducted that provides the student with a list of candidate landlords and rental properties. The candidate landlords "list" their available properties using the IVR section of the application. They call Internet/SOCR in advance of listing their properties to make billing arrangements (there is a per-listing fee) and to obtain an identifying PIN. The following is a description of the call-flow for the telephone portion of the application.

1. Greeting. The caller is greeted and advised that they have called the Internet/SOCR automated student-rental listing system.

2. ID Request. The caller is asked to put in a valid PIN if they wish to add new listings. The caller enters his or her PIN.

3. Geographical. The system asks for a geographical area where the listing is located input: (areas surrounding the university or within reasonable commute distance. This is a multiple-level menu system that breaks down the geographical area to more granular options as the menu levels are negotiated). The caller identifies the geographical area.

4. Listing Type. The system asks the caller to press 1 if this is a single-family dwelling, 2 if it is an apartment, 3 if it is a boarding house.

If the caller responds with either 2 or 3 the system then asks how many units are available (from 1 to 9). Note that in this case each unit is considered a billable unit in terms of the listing fees the candidate landlord pays to Internet/SOCR.

5. Number of Bedrooms. The system asks the caller how many bedrooms the listing has and instructs the caller to press from 1 to 9 to identify how many bedrooms the unit has.

6. Number of Bathrooms. The system then asks the caller to input the number of bathrooms (again, from 1 to 9)

7. Number of Renters. The system then asks the caller how many students can occupy this Renters unit. The caller responds with a number between 1 and 9.

8. Pets allowed. The system then asks the caller to press 1 if pets are allowed, 2 if pets are not allowed.

9. Rent amount. The system then asks the caller to use the telephone keypad to input the rental amount (per student) in rounded dollars.

10. Contact info. The caller is then asked to input a telephone number, including area code, where the student can contact them.

11. Confirmation. The system then repeats back to the caller information they have entered, and asks the caller to press 1 if the information is correct, 2 if the information is not correct. If the caller presses 2, the entire sequence is repeated. If the caller presses 1, they are given a total amount to be billed, thanked, and the transaction ends.

This information is now posted to the database and made available to students searching through the Internet/SOCR Web page.

Internet/SOCR application is complete and functioning for the use of candidate venture capitalists on a demo basis.

Student Registration

Students are busy people. Whether they are full-time carrying 15 credits or part-time carrying 6 credits, during the hours that they are not at work or dealing with family responsibilities, time is precious. At Onondaga Community College, some of the time that students in the past used to traverse the cold Central New York weather only to wait in long queues during the registration process has now become quality time for them to spend as they wish.

Onondaga Community College, located in Syracuse, New York, is a two-year college with a student enrollment of about 6,600, evenly divided between full-time and part-time attendees," says James J. Sotherden, associate registrar of the college. "We are a part of New York's SUNY system and cater primarily to the needs of a diverse student population, the vast majority of which have commitments beyond their educational activities at the college.

Most have full-time jobs, and many are working and going to school while tending to the needs and demands of a family life. Needless to say, time is a precious commodity to these busy people. Having them stand in long lines to register for classes was a poor use of valuable time, and when you consider the cold, rather harsh climates of Central New York, the ability for students to register from the comfort of home was an attractive alternative.

"After hearing of the positive results similar institutions have had using voice response technology, the college decided to implement a voice response system," Sotherden says. "A Periphonics (Bohemia, NT) voice response system was selected as a result of our evaluation process, and in April of 1994 a 20 port system was installed and commissioned."

The system was given the name STAR, for Student Touch-tone Assisted Registration. The students are the primary benefactors of the system, enjoying the ease of telephone registration coupled with the significant increase in the time available to schedule courses.

The system also has had a significant impact on office productivity at the college, because it permits valuable staff hours to be redirected to other priorities and demands. It has eliminated the need to hire temporary workers once required for the traditional 'arena style' registration. As a result, the expense for registration related to temporary staff and staff overtime has declined by a significant 79 percent.

"The STAR system consists of a Periphonics model VPS 7500 equipped with a total of 20 ports that are connected to our Fujitsu 9600 PBX system," says Sotherden.

"Three of the system ports are reserved for the staff who support our 'help line.' If a student attempts to register on the system and experiences some difficulty STAR automatically transfers that student to our office. The student can also request such a transfer by pressing the Operator key at any time during the registration transaction. After providing the student with the necessary assistance, the staff member frequently transfers the student back to the system. The three reserved lines are for that purpose, so students do not have to call back or wait for a line to become free. This is especially beneficial during peak calling periods. Student records are maintained on the college's Unisys A6 mainframe. Burroughs Poll Select protocol is used for communications between STAR and the host. Periphonics developed the custom application programming to meet the unique needs of our host application, after which they provided application development training for a campus computer programmer. We did our own in-house recording of the system script using the facilities of the college's Radio/Television academic department and a student voice. Student involvement has added an additional dimension of interest to the system."

"The primary applications presently running on STAR are the student registration application and the grade reporting application," Sotherden says. "The system easily supports the college's registration load by offering a generous registration period. Students currently attending can register during the interval between November 12 through January 26. To further control the system load, we have structured the application so that during specified periods; registration is restricted to particular groups. That also allows us to offer potential graduates first crack at the courses. Our current plans for additional system features to be implemented include the capability to process credit card payments via touch-tone entry. We also plan to offer students access to their account information."

Sotherden concludes, "Voice response technology is very well suited to the process of student registration and course scheduling, especially in the community college campus environment where all students commute to classes."

Electronic Media

Cable Pay-Per-View

Cable TV companies face a major challenge in providing pay-per-view (PPV) services to hotels. This challenge arises from the fact that regular subscribers are charged monthly for PPV activity on their cable bill. Cable companies usually contract for their billing services from companies whose mainframe computers handle many cable companies' billing cycles.

Standard PPV systems use terminal emulation cards in the voice response unit in order to transmit the billing information directly to the billing service. The billing service then automatically inserts the PPV charge into the subscriber's monthly bill on behalf of the local cable company. Since smaller hotels can't generate enough use for dedicated systems to justify the cost, PPV is not provided.

Figure 7.1 - Great Lakes Data Systems PPV

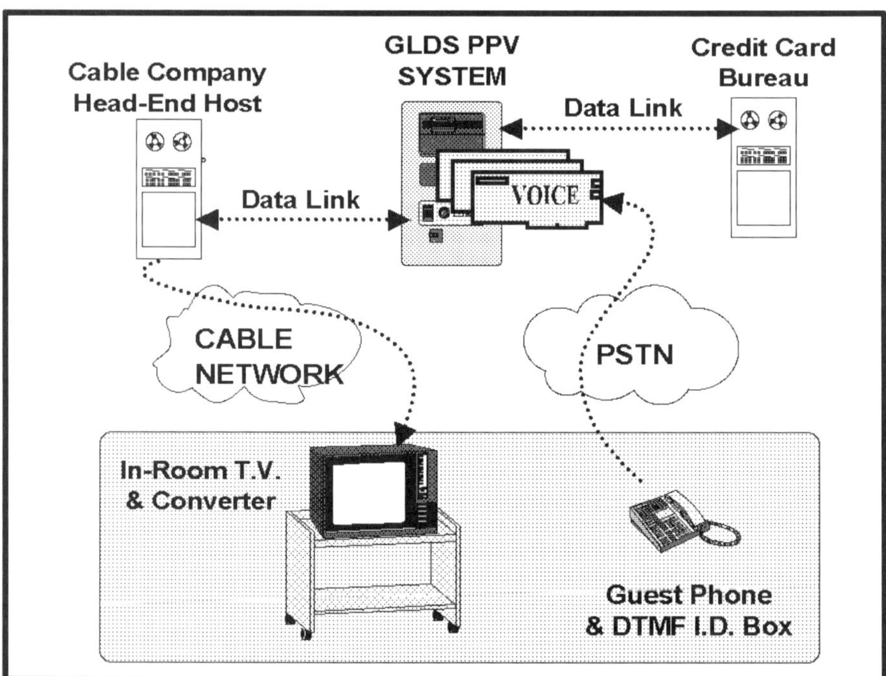

J. Alonzo Rosado, Vice President of Great Lakes Data Systems (GLDS) in Encinitas, CA, developed a system that is the first PPV system that enables profitable operation for properties with less than 300 rooms. The solution is called The Great Lakes Data Systems Pay-Per-View (GLDS PPV) system. Users can access direct ordering capability from their rooms using a combination of touch-tone input based on menu prompts on their television screen. Figure 7.1 shows how the system is connected to the billing system and to the televisions in guests' rooms.

Before the GLDS PPV system, expensive on-site equipment would have to be installed at each property. The GLDS PPV system uses addressable converters on each TV, so the computerized playback gear used for dynamic access can be located centrally. This helps smaller businesses to achieve an economy of scale not realized with on-site systems. The system features around-the-clock access for PPV customers and assures proper billing. The GLDS billing system is off-site from the property, so the cost of dedicated equipment is eliminated.

GLDS developed a means to provide credit card transaction processing services over the phone. Authorized transactions are linked to the billing system automatically. This enabled the PPV system to provide service to hotel guests without having to collect for the service at the front desk. The company designed the application so that I.D. Boxes installed on hotel phones would precede each caller transaction with the associated room number. This data was used along with the caller's credit card number to construct the PPV transaction. The touch-tone input is converted into an ASCII message that is transmitted to the credit card authorization center.

A Telephone Response Technologies-based system using ProVIDE was chosen for the IVR node because of its ability to handle serial communications gear. This eliminated the need for host emulation software, terminal emulation cards in the PC, and real-time links to the cable company's mainframe. The use of individual credit card processing helped to greatly reduce system cost for the hotels and their guests.

The GLDS system communicates with the cable company computer that controls converters on each T.V. The system captures the guest's movie selection along with their room number from the DTMF box. At that point, it's simply a matter of transmitting this data to the controller so the T.V. will display the movie.

According to George Blake, Vice President of Intelenet in New Braunfels, TX: "The GLDS Pay-Per-View system is the only way in-room movies can be provided profitably for smaller hotels." Blake works closely with smaller properties to help them provide the premium services usually associated with larger hotels. GLDS uses 486 PCs, Dialogic speech cards, and network interface controllers for the IVR system nodes. In addition, serial modems, PC-based video character generators, addressable converters, DTMF I.D. boxes, and DigiBoard intelligent serial controllers are used.

Home Shopping

Shopping by phone for items advertised on television has been with us for a long time. It was not until about ten years ago, however, that entire channels were dedicated to this function. Perhaps the most popular are the Home Shopping Network and QVC.

The home shopping network uses voice processing and has personalized the machine by naming it "Tutti." Tutti was designed as a large network of PC-based VRUs based on Dialogic speech processing technology programmed by Precision Systems, Inc. (PSI). The system now has over 3,000 ports of IVR and can handle 50,000 transactions in several hours.

First-time callers are routed to service representatives who get credit and billing information from the caller. Once and account has been established, callers need only to call the system and input their account number. Items that are for sale can be automatically delivered in this fashion. The IVR system can transfer callers to service agents if they arc in need of assistance.

Interactive TV

Dozens of new companies, cable concerns, telephone companies, and universities have been designing interactive systems of this type for years. For example, Bell South Advanced Networks, and New York University to name a few. Rating services like Neilson, for example use "black boxes" in selected viewer's homes. These devices log the results of viewer reactions and then transmit the data to a computer for compilation.

A voice processing system can be used in conjunction with addressable converters, which are installed in great numbers all over the world. An IVR system can be connected to the "head end" controller for the addressable converters in order to enable on-the-fly changes to interactive programming.

For example, by viewing certain prompts on the screen, viewers can call a toll-free number and input touch-tones in order to rate a show, ask for a different show, or even view bank balances and order products. In this fashion, the simple and prolific telephone instrument is transformed into a "multi media" order entry terminal.

Interactive Videos

Companies like Cox Cable, and some telephone companies have been planning on video-on-demand services in order to increase network usage and service levels. Video-on-demand allows viewers to select from a library of titles and then view any movie with the touch of a button. These systems are popularized by in-room movie services offered by major hotels. Voice processing systems can be used to enable the technology on a large scale, however, by retrofitting addressable converters with telephone access.

For example, the billing, selection, and viewing times for movies can all be made over the phone with touch-tone input for homes that are not equipped with special response units. This has the effect of accelerating the use of such a service by several years. This enables service providers to roll-out modern offerings without having to wait for the distribution of special equipment.

Program Polling

Ratings companies have devised a number of elaborate schemes to monitor home viewing of television programs. For example, a selected set of viewers can push buttons, change channels, and answer interactive questions on the screen of their television with the aid of special converters that download information to a centralized site.

This data is compiled in order to determine the popularity of certain programs and to analyze the demographics of viewers. The data is used by advertisers for targeting their promotional efforts and by networks to establish fees.

Unfortunately, the ratings industry comes under heavy fire from producers, viewers, and advertisers for their selection of "private viewers." Some say that these viewers do not represent the likes and dislikes of the majority. One way around this problem is the use of IVR-based polling for broadcasts.

An interactive voice response system can easily use ANI (Automatic Number Identification) to limit the counts derived from callers based on the number of times they have already called. In addition, the results from a telephone-based system can be compiled instantaneously and provide an input mechanism open to anyone with a telephone. This eliminates the need for special equipment to be installed at a viewer's home.

A system of this nature can be advertised during the airing of certain shows and instructions for its use can be given verbally when people call the system. In this fashion, literally thousands of responses can be collected each hour.

Radio Contests

Contests can generate an incredible number of telephone calls. In fact, there have been occasions when certain contests have literally "choked" the local switching network and significantly degraded service and thus the safety of other subscribers.

Voice processing systems can be used to distribute the traffic more evenly and cut down on the service staff required to answer the phone. For example, let's say that each day, a radio station wants to post clues on their IVR system for a "scavenger hunt." Contest players can call the system for clues, and then call the system back in order to answer a series of questions.

If these questions are answered correctly (via touch-tone input), then the caller could be eligible to win. Only those callers who are eligible would then be transferred to a special telephone number that is answered by the radio station personnel. These IVR systems can be accessed via 800, 900, or some other type of number, so that access to the contest is not concentrated at the local telephone exchange.

Radio Station News Clipping

The Radio Station News Clipping Service (RSNCS) is an automated IVR/Fax Retrieval application built by ACCESS Conference Call Service, a telephony service bureau.

The system was built using MasterVox Telephony Application Development software from MasterMind Technologies, and operates on Natural MicroSystems telephony hardware. Fax features are enabled via MultiFax from Commetrex (built-in to MasterVox).

The RSNCS system was developed by ACCESS to provide daily news clippings in audio and hardcopy form to radio station program managers (up to 1500 subscribing radio stations nationwide).

New professionally recorded news clips (including special radio station recorder signal tones) are remotely downloaded to the system very early in the morning each day from one or more studios. A MasterVox IVR menu of these clips is then dynamically updated automatically.

Radio station program managers call in to the system when they arrive at work later in the morning (e.g., 4:30 AM) and listen to the new menus.

The program manager can then choose to listen to, retrieve (i.e., record for re-broadcast), or obtain a faxed copy of one or more selected news clips, for use in a radio station broadcast. The system can also perform a broadcast fax in the middle of the night to all radio station subscribers, featuring highlights of the next day's news clips.

Emergency Services

Air Ambulance Caller Information System

In Europe, the Austrian Air Ambulance of Vienna uses a voice-controlled application to direct the flow of telephone calls for administrative purposes, and to handle emergency calls. Since 1976, Austrian Air Ambulance has conducted thousands of life-saving missions using specially equipped Lear-jets and helicopters. Using the same medical equipment that might be found in the intensive care unit of a large clinic, the Ambulance crew provides care to patients as they transport them to hospitals in Austria and abroad.

As an emergency air service that utilizes the latest technologies, the company wanted a highly- sophisticated, automated information system that could respond to calls of a critical nature, and help separate the emergency calls from the administrative ones that pertained to the day-to-day operation of the business.

The solution, offered through Siemens Austria, is an auto-attendant system that uses MOSCOM (Pittsford, NY) products and technologies. Initially, the system offers three main choices to the caller, and then either provides information or takes details of the problem, and directs ambulance personnel to take some action. Callers use verbal commands to choose from several menus of information or to leave a message. Perhaps most importantly, callers can transfer to an agent for more information or to report an emergency, at any time during their call.

Figure 7.2 - Austrian Air Ambulance of Vienna

Entertainment & Sports

Golf Tee Scheduling

Globestar Systems, Inc. of Pickering, Ontario is a Value Added Distributor for Norstar. They do call accounting, ACD, and IVR solutions. Their most memorable sale was an IVR product. The application automated tee-off bookings at a golf course. It lets callers book tee-off times.

Golfers prefer to play in teams and dislike having to wait for partners at tee time. Some country clubs encourage singles to meet and get to know one another at the clubhouse in order to put parties together on the spur of the moment, or when someone cancels. Club workers do their best to match people to the appropriate skill level, suggest handicaps, and make phone calls to complete a party. For service personnel, this takes a lot of work and diplomacy. An interactive voice response system could help golfers find the right partner at the right time by keeping a database of preferred golfing times, skill level, and other membership data. Golfers could call into the system, indicate their membership number, or act on behalf of a guest. The system can be programmed to automatically match people together and then to suggest tee times and arrival times.

As an option, the IVR system could have access to a common database for scheduling and therefore update reports or calendars that are regularly printed. In addition, the system could be programmed with an outdialing function to remind golfers of their appointed tee times.

Off-Track Betting

Many states now offer off-track betting facilities in addition to lottery and other games of chance. Depending on the governing body, bets can be placed with cash, credit card, or personal checks. In the case of credit card orders, and IVR system can be used to automatically process bets and the charges for same.

The gambler can select all information about post times, horses and odds. The system can then ask the caller to enter his or her credit card number and expiration date. This data is then checked for accuracy using a checksum algorithm. The next step is to query a local database for fraudulent and stolen cards.

Lastly, if the first two steps are O.K. - the system transmits the credit card data. The clearinghouse will issue an authorization code that can then be written into the caller's transaction record on the IVR system. The charge is then associated with the time and date of call. Systems of this nature will sometimes require a signature to be on file before the caller can place a bet.

Ticket Sales

Call centers for concert and event tickets employ hundreds of workers, and the closer to a certain event, the busier the telephone lines are. Some of the telephone traffic associated with popular events can be pushed away from the call center by giving customers the option to purchase tickets in an automated fashion. There are several ways to achieve this with a voice processing system

As with home shopping, the callers can have an established account and order via touch tone for a block of tickets, or a grouping of several kinds of tickets. This would be typical of a travel agency, ticket broker, hotel, or corporate marketing manager, for example. Users with established accounts can be given verbal or faxed confirmation that their tickets are reserved.

Callers with no account can establish one by either using a credit card, or a bank ATM card. Depending on the type of on-line credit card or ATM verification service used by the ticket agency, it is possible for several hundred credit card orders to be automated in a single hour. This would have the effect of cutting in half the number of agents required for that shift.

Financial

Bank Teller Voice Verification

Sue Giuseppetti, Marketing Communications Manager, MOSCOM Corporation (Pittsford, NY) and distribution planning consultant Elizabeth Boyle submitted this killer application about the Customer Functionality Group at Chase Manhattan Bank. The Group is responsible for identifying cultural, organizational and industry barriers performed by 600 branches. Part of this focus is re-engineering system processes within the bank. For two years, the group has developed approaches to make branch offices more efficient. After reviewing several areas where improvement was needed, Chase focused on the process of customer identification. They wanted to ensure quick identification of customers and thus improve their satisfaction with the bank by eliminating approval processes. This was driven by the idea fact that customer identification is time-consuming and it reduces teller productivity. The bank was also interested in reducing fraud.

The first step was to establish and assign weights to the selection criteria for an identification system. The criteria consisted of four areas: customer impact, effectiveness, strategic consistency and implementation.

For example, the strategic consistency criteria specified the ability to identify customers not only at the branches but at all banking channels (i.e. over the telephone).

In addition to voice verification, photo, fingerprint, and dynamic signature identification systems were researched. Each type of system had its advocates within the research team. After conducting research and tests with consumers, voice verification clearly came out ahead. In fact, voice verification achieved the highest score: 9 out of 10.

One of the deciding factors that put voice verification over the top included that it has a very low false rate of rejection. Also, a person's voice produces a very high quality biometric "print" which meant that not only could customers be enrolled over the phone, but their voice print could be used to identify them for various other channels such as ATMS, and PC and home banking.

Fingerprint and photo identification usually have to be done in person, making those technologies unwieldy for fitting a large established customer base. In cases where people were asked to provide their own photo, research showed that the photos were unusable approximately forty percent of the time. Signature identification looked promising, but had a slightly higher incidence of false positives. Considering Chase's plans to enroll a customer base of two million customers into the identification program, these factors were crucial in the decision to go with voice verification.

The current MOSCOM voice verification system consists of magnetic stripe readers, telephone handsets and printers connected to voice cards in a Pentium PC. Eventually the architecture of the verification system will be expanded so it is hooked into the main teller system on a server. All identification would then go through the verification server, simplifying the process bank wide for both tellers and bank customers. The verification is done with MOSCOM's model 2400 voice board, which includes a proprietary algorithm and chip.

After the success of the first pilot, the bank trialed voice verification with more than twenty thousand customers at two more branches, one in Manhattan, and the other in the Bronx. The customer enrolls in the system under the guidance of a coach by repeating the phrase *"Verification by Chemical Bank"* into a telephone handset. Customers have responded very positively to the verification system, and because they are being cooperative with the enrollment process, the pilot will be completed even faster than expected. In fact, within the first three weeks of these pilots, over eight thousand customers enrolled.

Potential applications for verification within the bank include foreign currency exchange, vault access, home banking, and improvement of internal banking processes. For example, voice verification can replace the process of faxing transaction approvals from the branches to the main office -- a process which today increases the amount of time customers have to wait in line.

Financial Bank Transaction Security

MOSCOM Corporation of Pittsford, NY helped Chemical Bank deploy voice verification systems for use in customer identification trials. Chemical Bank has ordered two verification systems that will use spectral voiceprint analysis to verify the identity of customers at two Chemical Bank branches.

The two trial sites used voice verification technology to identify customers prior to performing their banking transactions. The verification stations consist of a magnetic stripe card reader, key pad, printer and standard telephone. Prior to using the system the customer is "enrolled" either at the branch or over the telephone from home.

The enrollment process takes a little over a minute. The customer enters a bank debit or ATM card number using the magnetic card reader at the branch or from home using the touch tone key pad on a telephone. The customer is prompted to say a pass phrase three or four times. The resulting "voiceprints" are stored for use in later verification of the same customer.

When the customer visits the branch, verification is fast and simple, taking about 12 seconds. The identification number is entered into the system by sliding the debit or ATM card through the card reader. The customer then picks up the telephone handset and says his pass phrase. Upon successful identification, the verification station notifies the teller to process the transactions. If unsuccessful, the customer is directed to proceed to the customer service counter. According to Chemical Bank, the program is designed to enhance customer satisfaction by providing an extra measure of security while simultaneously decreasing the time it takes to process a transaction.

Cash Management

The transfer of cash receipts from one account to another seems like a mundane and boring task, but it can add up to thousands of dollars in savings. Retail outlets make bank runs on a daily basis, and then communicate cash receipt and deposit information to a centralized accounting office. This information is then analyzed in order to determine whether or not cash can be invested, or transferred to another account. By using an IVR system, the status of deposits and cash receipts can be ahead of the bank's data. For example, as each store manger can make his or her deposit after closing-out a shift, and then input the data into the IVR system.

Users enter their store number and the cash amount in question. The system would then send this information either to a host computer for processing, or the data can be used to update a standard report that is run each day by the finance personnel.

Check Clearances and Authorization

Jackson Hewitt Tax Service (Virginia Beach, VA) provides income tax preparation and refund check services nationwide. Each year, the company processes thousands of tax returns resulting in $14 Billion worth of refund checks this past tax season alone.

The major portion of all verification activity is squeezed into a two-week period of time. In the past, this required the hiring of part-time employees to manually verify social security numbers, check numbers, and amounts when tellers or bank managers made inquiries.

An interactive voice response system was designed to provide instant access to the Jackson Hewitt refund check verification database for authorized users. By requesting the same input from callers that the live operators would have asked for, the IVR system automatically queries the financial database and matches social security numbers with check amounts and check numbers. This is pictured in Figure 7.3.

It is necessary for the company to handle a large number of calls in a very concentrated time period, so it was decided that the IVR system should handle as many as twelve calls simultaneously. Originally, Jackson Hewitt had single lines installed for a manual operation. These existing lines were re-installed onto a single 486 DX2/66. In addition to the telephone lines, the application eliminated the need for 12 workstations as well as the manpower to operate them.

Jackson Hewitt used Telephone Response Technologies' Pro/Found database access server. This IVR software enabled the PC to mimic the keystrokes of all twelve workstations in retrieving the check verification data from the host database. With this automation, the company was able to operate 24 hours with no scheduling difficulties. Since banks from all over the U.S. were users of the system, this was especially helpful owing to the time zone constraints of a nationwide service.

According to Sharon Harwell, Director of Banking, IVR has significantly enhanced customer relations for the company. Says Harwell: "The TRT system allowed Jackson Hewitt to provide quick and accurate responses to banks needing authorization for checks. The system was one more application that allowed us to provide the best customer service in the tax preparation industry." With the new system, Jackson Hewitt verified $14 billion worth of checks in a four month period.

Figure 7.3 - Jackson Hewitt Check Authorization

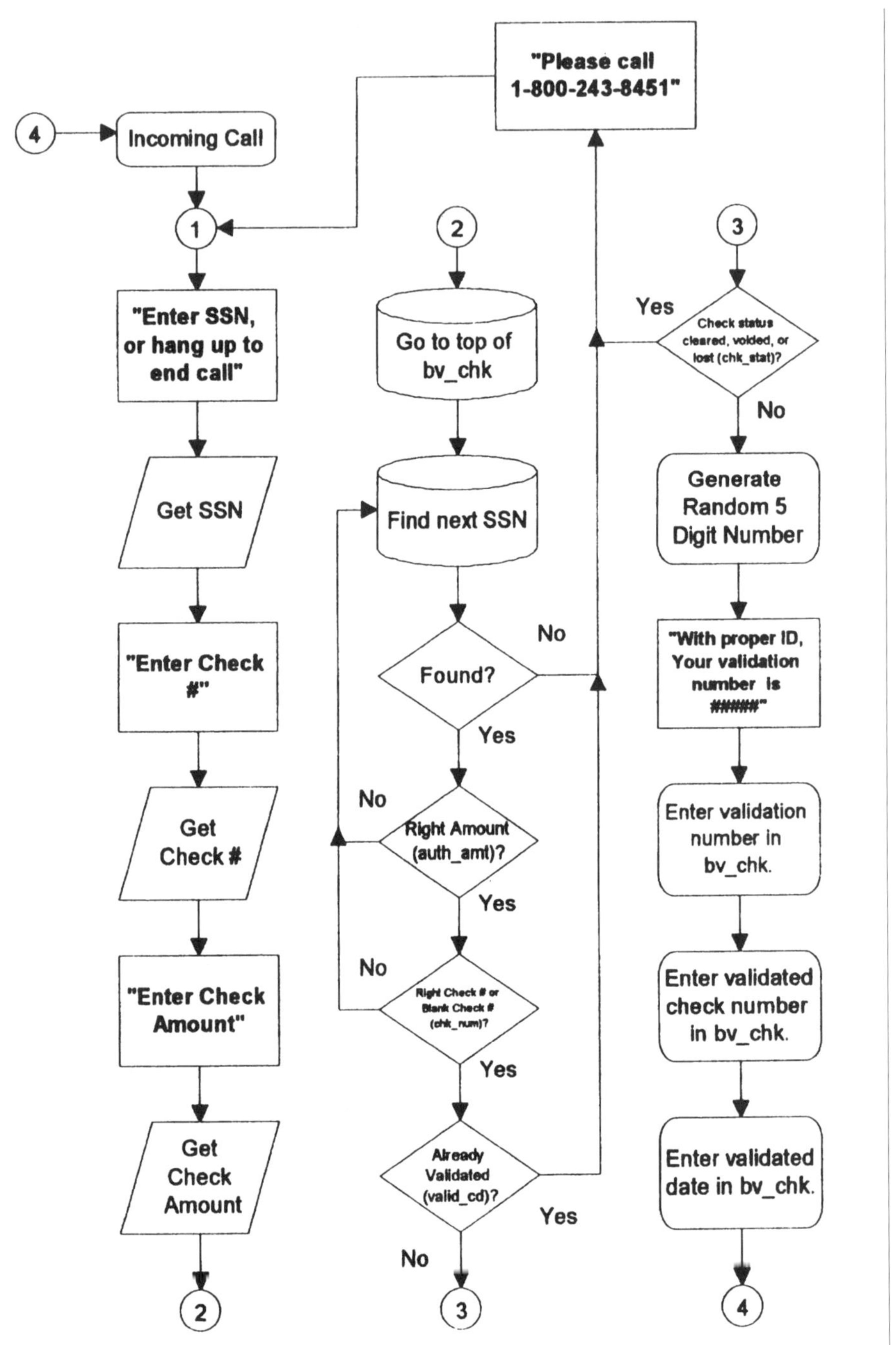

The company is considering making the verification service available directly to their customers so they can learn the status of their refund check. The company is also thinking about using the system to route calls to "the closest office" so they can use a universal 800 # in their advertising. Other ideas are to implement a telemarketing department, and use the system to collect customer survey data.

Checking and Loan Balances

Syntellect of Phoenix, AZ developed a financial IVR application called MortgageWorks. It interfaces homeowners with information on their mortgage balance and other information. The application is part of the company's ApplicationWorks line of IVR products. Information such as loan balance figures and prior year's tax interest are among the data dispensed by MortgageWorks. After callers interact with the system with touch-tones, they can get a verbal response, or the system will prepare a customized fax document and transmit it after the caller hangs up. Developers can also customize administrative screens based on the utilities packaged with the MortgageWorks application.

Credit Bureau Subscription

Credit bureaus such as TRW offer a variety of subscription-based services. These services appeal both to commercial accounts who are extending credit, and also to consumers. The TRW Credentials service offers a consumer-oriented appeal. Members can order credit reports, get advice, and use special personal profiles that are prepared by the company for use in loan applications.

Armed with a membership number and access code, TRW Credentials members can access an IVR system 24 hours a day to order reports, forms, and make special inquiries. In most cases, reports and forms are sent within days of the request, and require only a few seconds to order. Considering the alternative of having to request reports by mail or waiting to speak to an agent, the IVR system offers the quickest and most efficient way to process simple requests.

Commercial clients can benefit from a variety of IVR-based solutions. These systems conduct credit searches and inquiries based on the entry of social security numbers. This eliminates the need for a computer, so credit managers don't need special software to access the information. As an option, the system could provide reports in the form of an e-mail message or fax.

Credit Card or Account Information

Farmers Bank & Trust of Blytheville, AR had an overloaded bookkeeping staff. They were swamped with telephone calls. Bank customers were constantly calling the office regarding bank balances and other information that interrupted the work flow of their regular duties. While the staff wanted to maintain a friendly and helpful attitude towards customers, it was nonetheless difficult to maintain a high level of productivity and service customer inquires at the same time. Being a small bank, the cost associated with additional staff or a more expensive "on-line" system was out of the bank's budget.

Craig D. Hutson, President of Tara Systems, Inc. (Irving, TX) met with the executives and accounting staff at Farmers Bank & Trust to develop a cost effective way of downloading information from the mainframe computer. The idea was to get the resulting information into a multi-line voice processing platform. The cost associated with establishing a real-time link to the mainframe were out of reach for the bank, so Tara Systems developed a means to implement a batch download routine, which assembled data from the mainframe for use on a daily basis. This batch file includes daily account and transaction data.

The INQUIRE24 system was then developed to provide access to customer records over the telephone. Callers can enter their account number and personal identification number in order to hear their balance, and to learn of account activity such as transfer of funds from checking to savings, for example. In addition, a special "rates and services hotline," a "community bulletin board" feature adds a layer of Audiotex to the platform.

Tara Systems planned the application so that the batch file could be converted into a FoxPro database record. A Telephone Response Technologies package called Pro/Found was chosen because of its ability to work with xBASE files. Direct access to the data from the downloaded file was thereby enabled. This eliminated the need for host emulation software, terminal emulation cards in the PC, and real-time links to the mainframe, thus greatly reducing system cost for the bank. This is pictured in Figure 7.4.

Figure 7.4 - Tara Systems INQUIRE24

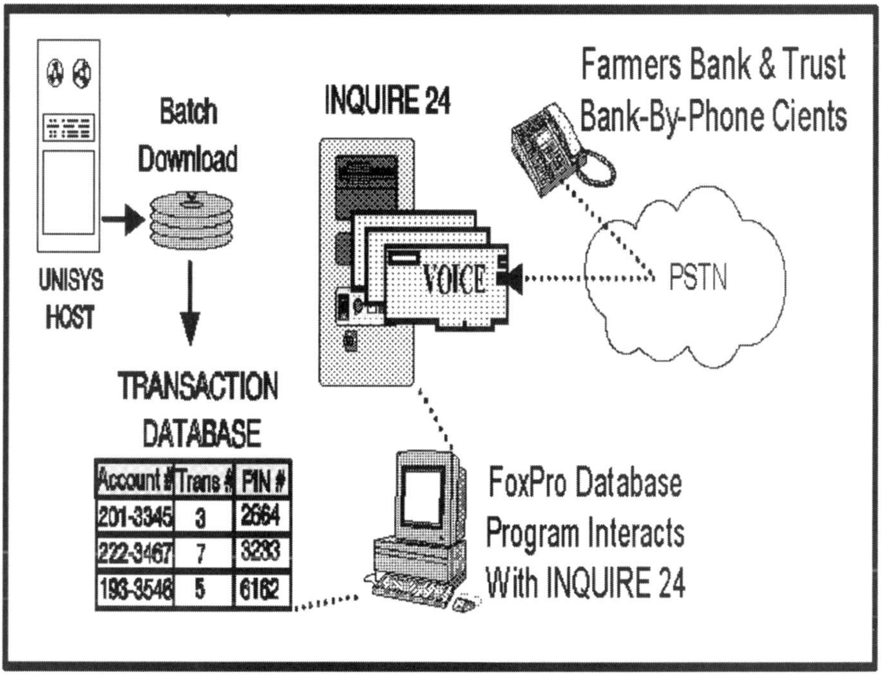

According to Glenda Eubanks, Vice President of Farmers Bank & Trust, the bank researched a number of automated IVR bank-by-phone solutions. Customer response to the IVR service has been "very rewarding." Tara Systems used a Dell 486 DX 33 MHz PC, with a Dialogic D/21D speech card as the basis for the IVR platform.

A variety of enhancements have been discussed including "loan by phone" capability, in which callers can get automated details on the cost of loans, inter-account transfers, and merchant inquiries.

Employment Verification

TALX Corporation of St. Louis, MO is bullish on their "Work Number for Everyone" line. It's a national employment verification service that uses interactive voice response, fax and EDI (Electronic Data Interchange) to let lenders verify employment information in an automated fashion. This saves participating companies money and time and helps lenders close loans faster.

Companies using the service include Hewlett-Packard, McDonnell Douglas, Motorola, Ralston Purina, USAir, Wal-Mart, and Walgreens. Lenders traditionally contact employers by letter or telephone to verify information on an employee's application. They ask for length of employment and salary. Each manual request can cost as much as $6 to process -- taking days or weeks due to the sheer number of calls companies receive.

At a minimum. most companies receive one verification request per employee each year. The Work Number simplifies the process by providing the lender with a single telephone number to call. Lenders use an authorization code provided by the applicant to access a database. Employment information is verified within minutes and can be faxed or transmitted electronically to the lender.

Information in the database is updated each pay period so that information remains current and secure.

The service is virtually no-cost to employers, and TALX Corporation maintains the service and provides the necessary equipment. And lender grant approvals significantly faster while keeping processing costs low. Verifiers using the service include local and regional banks, credit unions, leasing companies, rental companies and credit card issuers. The service can eliminate days or weeks from the application process.

Health Care Credit Information

Creative Advanced Technologies, Inc. of Forth Worth, TX is a VAR for Voysys Corporation (Fremont, CA) developed a killer application for the medical profession. Doctors and clinics are using this IVR service to check information on the payment record of patients and to enter information on "deadbeat" patients. Doctors and medical clinic billing personnel call this service to check on the payment history of perspective patients and to enter information on known "deadbeat" patients. The system uses the patient's SSN as an identifier. The service is on a subscription basis and with 12 lines running the call volume is averaging 400 per day.

These callers are primarily clinics and doctors calling in to check on the payment history of patients who have come to them for treatment. The remaining callers are doctors or clinics calling to enter into the system information on patients who have demonstrated poor or nonexistent payment histories.

Creative Advanced Technologies says that this system has proven very successful in the few months of its operation. When they initially put the system online using VoysAccess for Windows NT (OCX) they were using only two lines and call volume was about 75-100 calls per day. They then went to 4 lines and call volume increased to 200-300 calls per day. On 4/9/96 they went to 12 lines and call volume is now 400 calls per day.

Pension Plan Eligibility

Information Products, of East Granby, CT is an Expert Systems VAR. Their IVR product is called Verify. The company develops shrink-wrapped solutions for HMOs and insurance companies. The company created an interim IVR solution for the Sheetmetal Workers National Pension Plan.

It handled eligibility inquiries (for both providers and constituents). During off-hours, clients could call in and enter their account number or social security number to find out about their coverage.

Proxy Voting

Proxy voting for shareholders can also be achieved by using IVR. For example, early in 1995, Transaction Network Services, Inc. used this technology to offer registered shareholders an alternate means from the traditional written ballot.

In the traditional view, shareholders are mailed proxy materials and are asked to grant their proxy by either signing a form and then mailing it in or faxing the form in. With the new IVR system, shareholders call a toll-free telephone number. Users are then presented with automated menus that guide them through the voting process. The application allows users to vote on all issues at the same time, or they can vote on issues separately.

Access to the system is secure, since each user is asked to enter their tax I.D. number or social security number followed by a special PIN number that is provided via mail. With this information in hand, users are able to negotiate through the prompts, vote on the issues at hand, and then log-off of the system within a few minutes.

Retail Stock Brokerage

Charles Schwab & Company practices what it preaches. Just as Schwab tells its customers investing is the way to achieve their financial goals, the firm ensures its own growth by consistently spending 11 to 15 percent of annual revenues on state-of-the art technology. In 1989, well before the rest of Wall Street, Schwab began to encourage clients to place orders via automated telephone systems. Today, automated telephone and PC services account for more than a third of the trades placed through Schwab. In the past three years, technology has helped Schwab trim its cost per transaction by 50 percent.

With 20 to 30 percent annual growth of its customer base, Charles Schwab & Company needed to upgrade its Service Center technology. The company was nearing its conversion from a mainframe host environment to client-server with NT as the desktop platform.

They chose the Genesys T-Server platform along with SoftPhone (workstation software) and the Genesys Voice Response-Client.

The call centers use IVR to prompt callers for account number. The CT software routes callers, and coordinates their account data with the agent' screen (screen pop). When an agent places the client's trade, the system automatically connects to both market data and transaction service providers. Within seconds, the client receives a confirmation.

Not surprisingly, Schwab works in an extremely fast-paced environment. Seconds count. The firm tolerates zero downtime. The upgrade has produced many benefits. The exchange between Schwab customers and the voice response units eliminated 15 to 20 seconds in talk time for the customer service reps. Call center managers would kill to shave that much time off the average transaction. Time is money.

Shareholder Enrollment

The Shareholder Information System is a large automated IVR/Switching/Call Overflow application built by Shareholder Communications Corporation (SCC), a corporate information service bureau. The system was built using MasterVox Telephony Application Development software from MasterMind Technologies, and operates on Natural MicroSystems (NMS) telephony hardware.

The Shareholder Information System was developed by SCC to provide stockholder information and enrollment services for publicly traded companies. Callers dial in to the system to obtain prospectuses for stocks they are interested in buying, and enrollment materials for companies that offer direct stock purchase plans.

The MasterVox-built system uses DNIS (Dialed Number Identification Service) to identify the caller's target company, prompts for the caller's phone number, then automatically look up the caller's address from nationwide telephone directories on CD-ROM.

The system interactively verifies the caller's name and mailing address, compiles mailing lists, determines which companies the caller wants to obtain stock information from, and the nature of the desired information. Callers may transfer to an operator as well. The system can also act as a call overflow handler for SCC's large operator-manned ACD system, which employs up to several hundred live operators.

Stock Trading

Almost every stockbroker works with clients over the phone. Although many brokers work long hours, it is difficult to anticipate the odd and demanding schedule worked by their clientele. Some brokerages are associated with banks, so access to home banking services via PC data links allows 24-hour a day access for clients. Since many clients do not have PCs at home, they must wait until regular business hours to get the data they need during a live conversation with their broker.

Many times, clients want to know the balance in an account or want to make a trade after finding out what their balance is. This was the case for Charles Stanley, a private stockbroker in the United Kingdom. His company uses an IBM AS/400 to store account balance data and to generate statements. Mr. Stanley worked together with his IT manager, Stan Ede, to provide telephone access to client information. He used the Pronexus VBVoice toolkit to develop the application.

Clients call into a published number and are welcomed automatically by the system. Users are then asked to touch-tone their account number followed by their PIN number. These numbers are then automatically verified by the system to make sure that the information will only be dispensed to authorized users. The system plays-back the number that was entered in order to verify the entry with the caller. At this point, the system then verbalizes the account balance based on the field associated with the balance in the AS/400 database. Mr. Stanley plans to add a CBF capability in the future, so his clients can have statements faxed to them automatically.

Nuance Communications of Menlo Park, CA helped Charles Schwab develop the VoiceBroker (trademarked by Schwab), a phone service that uses speech recognition technology to provide stock, mutual fund and market indicator information to callers *and* lets them do business transactions over the phone.

VoiceBroker uses Nuance's conversational transaction technology, which consists of speech recognition and language understanding software that allows a caller to speak in everyday, conversational English. VoiceBroker understands and responds to inquiries about every security listed on the New York Stock Exchange, the American Stock Exchange and NASDAQ. You can just call up and say the name (or variations of the name) of more than 13,000 stocks, mutual funds or market indicators to get the latest quote information.

This job is a perfect fit for Nuance's transaction technology, since it can recognize perhaps the largest active vocabulary for phone-based services and combines this vocabulary with language understanding. This lets it recognize continuous, natural speech with meaning comprehension.

The product is completely speaker independent and no training on the part of the user is necessary. Schwab customers can create personal stock lists so customers can get multiple quotes automatically. VoiceBroker is currently available to customers in a dozen states and is being rolled out to customers on a nationwide basis.

Wire Transfer

Chief Financial Officers, bookkeepers, and everyday checking account holders need to transfer funds electronically from time to time. This usually requires a trip to the bank unless a long-standing and personal relationship exists between the requesting party and bank officials. This is often time-consuming and interrupts the overall operation of the finance department.

Voice processing can be used to automate wire transfers by allowing the transactions to be authorized with a combination of speaker verification and touch-tone input. An IVR system can be used to access account balance information in order to determine whether or not there are sufficient funds for the transfer. Once this is established, the user can speak a code word.

Pattern-matching algorithms are then used to verify the caller's identity and correlate the I.D. with the account number. The system can then ask for the bank routing number and account number where the transfer is being made. By using a system like this, IVR-based wire transfers can be made in a matter of seconds. For pre-established transfer accounts, a menu selection can be used to act as a "speed dial" list for frequent transactions.

Food Distribution

Dispatch and Delivery

DATATAL AB of Klintehamn, Sweden is an Expert Systems VAR. They sell a voice mail system called DataTal Professional and do custom IVR systems. One dairy customer uses their system to let retail outlets order new shipments of milk. It cuts down on delivery delays and saves time. Custom work like this nets 50-100% more profitability than standard offerings.

Frank Solutions, Inc. of Englewood, CO is a Parity Software VAR, Dialogic Open Solutions Developer, Northern Telecom Business Affiliate Partner, NEC Fusion Partner and a Novell Certified Reseller. The company also sells WRQ Reflection software for mainframe connectivity. The companies' product line is called TelePath. It has IVR, CTI, Host Access, ASR, TTS, fax-on-demand and credit card processing. Our focus is universities, schools, and state government.

Their most memorable sale was to the Hostess Dolly Madison (Interstate Brands Corp). They designed an IVR order drop/add system for Twinkie Delivery personnel.

The bakery requires real-time input from the drivers to make the two-day turnaround on orders. There was no quick way to enter all of this data for the local bakeries. The initial sale was one system at four ports. It's grown to a 32-system opportunity. Each system is worth about $8,000.

Order Entry and Cash Management

Voice Processing Plus, Inc. of West Bloomfield, MI is a Parity Software VAR, Nortel Business Affiliate and WilTel Business Partner. They sell custom solutions with a focus on "giving a voice to data." The company does LAN/Voice integrations, mainframe emulation and GammaLink fax stuff.

One of their biggest sales was to a Coffee Beanery near Flushing, MI. It's a unified messaging and IVR system that sits behind a Definity G3 switch. It sits on a Windows NT LAN equipped with a fax server, voice server and Internet server. It's linked to 177 retail outlets so they can trade.

Retail outlets will be able to use the system to place orders, check shipment status and trade messages. All of the EOB (end of business day) sales data will be uploaded to Flushing in order to manage cash flows and anticipate inventory levels for the week. There are 12 ports on the voice server and the same on the fax server.

Government

Auditor Voice Verification

In 1994, the Illinois Department of Revenue (IDOR) implemented a voice verification security system for off-site auditors, upper management, and programmers with special security access. The system, based on MOSCOM's voice technologies and board, is currently being updated and sold through Voice Print Security Systems.

Because of the sensitive nature of the information those employees access, IDOR was unwilling to implement keys, cards, or other security devices that might be lost, stolen or otherwise compromised. Prior to implementing the voice verification telephony application, tax auditors wishing to access information pertaining to an off-site audit were required to place a telephone call to an authorized individual at IDOR who could provide the information. Since this often resulted in telephone tag, IDOR sought a better method.

Now, authorized IDOR employees call the verification application from any telephone. Once they are cleared by the system, they are switched to a modem hookup and must pass through keyboard security for the data they wish to access. According to John Henry, the manager of IDOR's Information Systems Department, "It works very well. We are really proud of the speech system."

Government Intra-Agency Routing Platform

Treasure Coast Engineering of Gainesville, FL is a Microsoft Solutions Provider, Novell Dealer, AutoCad VAR, an HP Authorized Dealer and a TRT VP Express VAR. The company has two lines: Dial One is their own product. It has modules for outbound marketing, automated attendant, fax server, international callback and voice mail. The other is VP Express from TRT for out-of-the box shrink-wrapped solutions. One of their biggest sales was to the Island of Curacao. It's an IVR system with screen pops with ANI for the Government Resources Center, used as a router for intra-agency calls. It eliminates the need for live agents to do look-ups of phone numbers stored in multiple databases. It's connected to a timeshare Mainframe in Holland. It's 96 lines and worth $482,000. The whole platform reduced the agent requirements by 27 positions.

Motor Vehicle Registration

The motor vehicle departments in most states and provinces have become computerized. The agencies store, inspection data, vehicle identification numbers, and driving records in addition to principal garaging addresses.

When registration for vehicles become due, the computer system generates millions of forms and pre-printed labels so the solicitation for fees can be mailed to the owners. Many agencies will only accept payment for registration by mail in an attempt to reduce the staffing requirements at local offices.

Voice processing can be used to further reduce the number of clerical employees required to process the payments for registration. For example, instructions can be mailed on how to use an IVR system along with the fee mailing. Owners can answer questions on whether or not a car has been sold, destroyed, traded, or is still in service. Calculation tables can also be made available to callers in order to determine the amount owed based on depreciation, an outstanding balance, or special license plates. In fact, such a system can be used to automate the ordering of additional or replacement registration forms, custom plates, and inspection instructions. Payment for registration can be made by credit card or ATM.

Permits and Inspections

The number of phone calls that come into permit and inspection offices can be overwhelming for clerical staff and the inspectors. Calls for permit status, inspection results, and scheduling of permits get to the point where the phone can remain busy for hours at a time. Octel of Milpitas, CA worked with the Pierce County Department of Planning and Land Services in Tacoma, WA to overcome this very problem.

With as many as 600 general information calls a day with a 20% abandonment rate, the county agency was compelled to improve their service capabilities. As many as 250 technical assistance calls had as many as 60% hang-ups due to long wait times.

Octel used its Maxum voice information system with the TransAct IVR option to develop the PASS (Permit Application Status System). The IVR platform interacts with a Hewlett Packard computer, which is connected to the county's mainframe computer.

Since the volumes of data provided to callers would be difficult if not impossible to record individual prompts for, Octel equipped the PASS platform with speech synthesis technology (text-to-speech).

Callers can access the system any time before 11:00 AM to schedule an appointment for the following day. Permit application information is updated continually, and results from inspections on the previous day are posted at noon. This data is made available after the caller inputs the permit number as the identifying digits for the transaction. According to Pierce County, the PASS system handles as many as 1,300 calls per week and has extended their effective service hours by at least 20%.

Tax Office

Government agency use of voice processing technology was spearheaded by dozens of tax offices and especially the IRS over the past decade. With thousands of workers, the amount of time saved on the phone with callers is phenomenal. Of course, if a taxpayer needs to ask about a very detailed and complicated issue, nothing can replace a "live agent." In most cases, however, an IVR capability is a welcome addition to any tax office.

Take, for example, the Minnesota Department of Revenue. With only 60% touch-tone penetration in the Minnesota telephone network, a special challenge faced the agency in providing automated access to forms, refund information, and location data.

The agency installed a Wygant Scientific, Inc. (Portland, OR) Micro-ITC system to overcome the rotary dial caller problem. The system is equipped with speech recognition boards and voice processing technology from Dialogic Corporation (Parsippany, NJ). The Department of Revenue platform uses three PCs attached to a Novell LAN - totaling 72 telephone lines for the IVR portion of the platform. In addition, there is a file server, which houses the common data that the IVR system uses, and an administrative PC for remote diagnostics and application maintenance.

Minnesota taxpayers have access to refund status and forms ordering, office location information, earned income credit assistance, No-Tax filing, and taxpayer collection status. In addition, callers can transfer to the agency's ACD system during working hours to speak with a specialist.

Immedia of Montreal, Quebec, is the largest provider of Electronic Data Interchange (EDI) services in Canada. It introduced its first IVR and PC-driven payment-filing service (PFS) in Saskatchewan in 1994. Called CAN-ACT, it's a joint venture with the Royal Bank. It lets business users electronically pay their monthly national and local tax bills. Sixty percent of users pay their bills through IVR. The other 40% use their PCs, according to Bernard Dugand, technical manager at MPACT Immedia. Internet access is in the works.

CAN-ACT's payment filing service has a PC-and-IVR-driven front end. Users are set up with transaction codes to pay any of over 500 federal, provincial and municipal tax collectors and utilities. Settlement files are sent daily to the creditor institutions via EDI, e-mail or fax. The system is more than a simple monthly funds transfer of static amounts. For payments like employee payroll deductions, the IVR guides users through calculating the amount owed. Each type of payment is ordered with a separate transaction code.

Transaction types include business taxes, GST (Goods and Services Tax -- Canada's sales tax), employee payroll deductions, corporate tax installments, personal tax installments and invoiced payments to utilities.

A bilingual IVR front-end was written in MediaSoft Telecom's (Mount Royal, Quebec) IVS Builder app gen and BlaBla native language. The hardware consists of a 75 MHz Dell P75 Pentium for the IVR running SCO Unix, one Dialogic D/240SC-T1 card for 24 lines of IVR, one dedicated T-1 line, a X.25 network for PC connections, SNA links to send settlement files to financial institutions, X.400 and VANs (value-added networks) to send EDI return and payment notices to government and EDI-capable payees.

A Tandem host holding the user profile database and accumulated debit / credit information. Fax or e-mail is used to send payment notices to other payees.

Ticket Payments

No one likes to get a parking or speeding ticket, especially when you consider the time it takes to dispute the violation and pay the fine. Most municipalities offer mail-in envelopes with instructions on how to plead guilty, innocent, or waive rights to a trial. Voice processing can be used to help answer many of the questions that arise from tickets, or to automate their payment.

For example, and IVR system can be set-up to announce the meaning of certain citation codes on tickets. This can be helpful if you are cited for a violation that is not explained in sufficient detail on the paperwork. In terms of payment, an IVR system can be used to debit a credit card in order to take care of a fine associated with a citation.

This can be of great benefit to someone who waits too long to mail the payment in, because a late payment means additional fines or even an arrest warrant. Since an IVR system can man the phone lines around the clock, deadlines for fine payment can be “extended” to include weekends and evenings.

Tow Finder

It’s a frightening experience to look for your car and then not find it where you think you parked it. In many cases, the reason it’s not there is because the car has been towed. In larger cities, the hours of operation for municipal tow centers are short, and you have to wait hours to get your car back after paying your fine.

Police departments use a combination of their own tow trucks, and also contractors to haul automobiles. It's not always obvious who to call when you think your car has been towed, so a centralized voice processing system can help to streamline the process.

The tow truck operator can access a tow-finder IVR platform first. Information regarding where the car has been towed can be entered when the car is "hitched." Car owners can call a central number and input their license plate number, or search through a database of make, model, year, and color in order to find out if their car has been towed. The system can also double as a messaging system in order to inform owners of alternate towing locations when certain lots are full.

Unemployment /Job Placement

The Vermont Department of Employment and Training has successfully implemented a Periphonics (Bohemia, NY) IVR system to automate the operations of its job placement program. The agency is responsible for helping people find jobs and dispensing unemployment insurance benefits.

"The Vermont Department of Employment and Training needed a way to provide better unemployment insurance services to our clients," says Ed Orton, project administrator. "We wanted our clients to receive their payments in a more timely fashion." The department felt it was necessary to reduce phone services that required having attendant answer simple inquiries. The department also wanted to lessen the paper flow by reducing the enormous amounts of mail they receive.

When a caller reaches The Unemployment Insurance Line, they are given many options, including filing for their next weekly benefit, verifying when the last check was written and hearing general information about unemployment, district office locations and 1099 information for income tax purposes. This information is accessed with their social security number along with a personal identification number.

"The Job Line puts the public in touch with the largest employment service in Vermont. Employers throughout the state list jobs with the department," Orton says. "The Job Line lets the public learn of all available jobs, free of charge, from any telephone in Vermont. Callers are asked to select a labor market area by pressing a digit on the telephone keypad. Then they select either the job listings received within the last 24 hours or all the jobs currently in the database. Next they must choose an occupation category."

"After choosing their selections from the prompts, callers will hear job information for their field in the labor market that they have chosen. When an individual finds a job that might interest them, they are instructed to write down the job order number at the end of the job announcement. Then they may visit or call any Employment and Training office to discuss the job and a possible referral to the employer for an interview," he explains.

"The job bank voice response file is updated every ten minutes by the customer information control system," Orton continues. "The IVR system indicates all job orders that have been closed within the last ten minutes. Within ten minutes, all job openings in the state of Vermont are available.

Based in Montpelier, VT, the department headquarters acts as a central office operation, potentially serving more than 340,000 people from 12 satellite district offices throughout the state.

"The accessibility of The Job Line and The Unemployment Insurance Line has allowed the Vermont Department of Employment and Training to provide improved service to our clients, and utilize staff better in other areas," Orton says. "Our implementation has given our clients a competitive edge. Clients are now using the system, and we are receiving 700-800 calls a day. Thanks to IVR technology, we are able to reach and serve more people than ever before. Our staff immediately saw a reduction in the number of opening claims." Orton says the agency's clients have adapted to the system very well. "Everybody has seen the rewards of implementing IVR technology," he says.

"This is the first time that we have implemented a turnkey operation of new technology without having our internal staff involved in programming applications," Orton says. "The entire process went very smoothly. The system was up and running right on time. Periphonics' personnel are extremely knowledgeable and supportive, and they provided us with well-trained field support.

"The Job Line and The Unemployment Insurance Line have certainly opened the doors to the year 2000," concludes Orton. "The technology continues to meet our clients' requests and demands. Many states have already begun to take notice of the remarkable success that we have experienced."

Health Care

Bed Tracking

Concepts In Communications of Pittsburgh, PA is a Microsoft Solutions Developer and a charter Apex OmniVox for NT VAR. Concepts In Communications sells two health care applications. One is a "Bed Tracking" solution. It speeds admissions by tracking housekeeping activity and room readiness. The other is "Patient Tracking." Both are shrink-wrapped apps. We sell training and installation up-front. We also collect a monthly licensing fee.

The first sale was in 1988 for a hospital in Ohio. They needed speedier admissions. It was a one-port system. We used a one-port Talking Technologies board. It was a winner despite the busy signals. It was better than tracking housekeepers and their supervisors manually.

The biggest sale was to a hospital in Columbus that bought both solutions. The units handle between two to three thousand calls a day. It was worth about $35,000 for installation and training. The ongoing license fee is $1,500 per month.

Claim Filing

Periphonics Corporation of Bohemia, NY helped Interactive Healthcare Technologies create a claims application on behalf of Mountain States Blue Cross Blue Shield in West Virginia.

Says Mike Manco, president of Interactive Healthcare Technologies: "This was an interesting niche opportunity for us. Most doctors at hospitals, as well as a small group of healthcare providers outside of hospitals, generate the majority of their claims transactions using electronic data entry through their office PCs and have been doing so since the mid-70s. But the remaining greater majority of healthcare providers, who generate the remaining 20%-30% of claims transactions, were the most reluctant to convert to electronic data entry. They provided our motivation to develop a voice response application for claims entry."

Manco says that before the introduction of that application, less than 1% of these providers conducted electronic transactions. The voice response application was introduced in August 1995, and he reports that by the end of October that year, 25% of the physicians had already committed to using it. The reasons he gives for the success of the application are ease of claim entry and faster payment of claims.

Usage and enthusiasm continued to increase among these non-electronic providers, and administrative costs declined for the insurance company. As a result, Interactive Healthcare has developed several other IVR applications for the healthcare industry.

"We now have applications to verify eligibility, inquire on almost any transaction and automate pre-certifications, authorizations, encounter claims and dental claims," Manco says. "The basic requirements for the applications are universal for most companies. This is especially true for the claims application. The referral application usually needs more customization. We're currently running the fourth generation of this application, based on customer-requested enhancements."

Manco says many of these enhancements relate to improving the accuracy of entries. For example, the data keyed in is now reviewed by the system for plausibility whenever possible. If an error is detected, the system prompts the caller to re-enter the data, reducing the need for human intervention to resolve simple input errors. The result is highly efficient processing. Manco says that currently, 83% of the referrals the system collects go through without any human intervention. He says this is a direct result of the edits the system performs while the caller is on the phone.

"User satisfaction is of major importance to ensure continued system usage," Manco continues. "For that reason, we have added a quality control survey of our customers' providers. During the course of each month, the system randomly selects 25 providers and asks them to compare voice response to the 'old way' of using forms. Opinions are expressed on a scale of 1 to 1O, with 10 favoring voice response. So far the average is 8.5 in favor of the voice response solution."

"We have looked at other systems, but the Periphonics VPS has proved to be the most suitable for our needs," Manco says. "Our plans for the future include real-time host interaction and updates, Text-to-Speech for confirmation of physician entries on the system and Caller Message Recording so specialists can leave primary care physicians a results report after a patient's visit.

"The applications that we developed target the market that we have identified, which is the group of providers who are not submitting transactions electronically and won't computerize," Manco says. "We are marketing our services to other insurance companies and have identified about 600 healthcare companies that could use our services. We can offer several different approaches to our customers, including licensing voice response applications, integrating of applications into their environment or providing complete implementation and management, including introducing applications to potential users. We're a young company, and we plan to continue to build upon what we have accomplished thus far."

Doctor Scorecard

It's intimidating for most patients to ask their doctor for credentials, information on special schooling, and other information that can assist you in choosing a physician. This becomes especially important in the event it is recommended that you "get a second opinion."

An IVR system could be used to keep track of all kinds of statistics for doctors, dentists, and other health professionals. For example, callers could be asked by the system to input the first three digits of the doctor's last name in order to narrow the search on the professional in question. Alternately, a license number, or company name could be used.

Callers can then receive verbal prompts having to do with the type of training, years of experience, and special skill the doctor has. A medical group or insurance company could sponsor a system of this nature in order to help patients find the right doctor.

Environmental Control

MOSCOM's (Pittsford NY) Univoice voice recognition and speaker verification technologies and boards can be used to enhance existing Interactive Voice Response (IVR) Systems, or to develop new computer and telephony applications, for a variety of vertical markets. MOSCOM products are in use today in business applications spanning industries from travel to automotive, from entertainment to healthcare. In this booklet, you'll see how developers can use MOSCOM's products to design many types of systems by examining several examples of applications.

At High Tech Intelligence, Inc. a Denver-based corporation that designs and provides voice controlled computer systems, their mission is to enhance the lives of disabled people. Using MOSCOM's voice boards and technologies, High Tech Intelligence developed the Personal Attendant system to address the constantly escalating health care and rehabilitation costs of disabled peoplc.

The Personal Attendant allows people with disabilities to function independently by controlling their home or work environments with a voice-activated computer system.

"MOSCOM provided us with the technology and products we needed to develop a high quality system," says Gary Marko, president of High Tech Intelligence. "The Personal Attendant easily allows disabled people to do things like open doors, adjust beds, and control wheelchairs, all through verbal commands."

Home Care Check-In

Voice processing can be used to reduce the time and paperwork required to administer personnel forms, time sheets, and even payroll sheets. This can be especially difficult when employees work away from the office. Such is the case with home care employees or visiting nurses, for example.

The IVS company of Hamden, CT has solved this problem with a new IVR Home Care Aide Verification system. Based on the company's Vision - OS/2 platform, the IVR system accepts calls from home care workers when they arrive at the client location.

By using Caller ID and ANI technology, the IVR system is able to automatically verify that the worker is indeed calling from the place they were dispatched to. This information is matched with a system clock in order to emulate the function of a "punch clock."

The data is stored in a payroll database so time sheets and paycheck totals can be automatically generated. Of course, the system also prompts the caller for their employee I.D. number, so the correct employee is paid for the work. According to the makers of the system, "voice print" technology can also be used to absolutely verify the caller's voice as an added level of security.

Patient Charting

Medical Information Service Applications At the Penrose-St. Francis Healthcare System in Colorado Springs, and at Presbyterian Manors in Emporia, Kansas, the nursing staff employs a medical charting system that uses the spoken word as computer input and output. The Nursing VOICE System was developed and is sold by St. Louis, Missouri-based Bound Information Technology Solutions (BITS), and is designed around MOSCOM's voice boards.

The system allows eight simultaneous users located at any of sixty different locations to chart patient information hands- and eyes-free, through a single PC. The system features: wireless infrared communication devices, a decision table generator, voice training programs and a voice engine.

After users define their decision tables and train the system vocabulary, patient chart data is collected by simply responding to computer queries while service is being provided -- rather than as an extra step in the patient care process.

Mr. Thomas Vogler, CEO of BITs, says, "Due to the MOSCOM voice board design the voice recognition has a very high rate of speed and accuracy with an unlimited vocabulary. All responses are repeated back to the caregiver by the computer in a human voice. This ensures that the answers are correct."

Patient Heart Monitoring

PaceArt's (Wayne, NJ) Patient Voice System does real-time automated ECG (electro-cardiogram) testing of callers over phone lines. I was so impressed with this application, I invited the company to demonstrate the system live on stage at Computer Telephony Demo Fall 96 in Orlando. They demoed the system with my publisher, Harry Newton, as the patient. You missed a real treat seeing Harry flat on his back with his limbs outstretched.

The system was designed for physicians, hospitals and clinics. Patients can press the special phone handset to their chest (see figure 7.5), hold it in their palms, or use wrist straps coupled to the phone. Even without these trappings, the system is specifically set up so doctors / dentists can better handle their patients with automated service.

For example, it sets itself up as a mini-service bureau for medical offices. Each patient gets a mailbox where they can leave messages for their caregivers. Doctors can be paged to listen for emergency messages and dictate back their diagnoses. The return messages can be outdialed back to patients, etc.

Figure 7.5 – PaceArt Patient Voice System

And while PaceArt's system is, again, specifically crafted for this application, all the better voice-mail call-processing systems can be set up to do the same thing. Some even come with modules to just this. What none of them have, however, is the proprietary real-time ECG transmission that this system boasts. It needs a special phone. $179. Patients with serious heart conditions would replace their normal phones with this baby.

It also handles plain old telephone service. But, in case they're feeling something they shouldn't, they call into a PaceArt system, identify themselves through touch-tones or voice recognition, leave a brief message about their concerns and then grab the handset or special wrist sensors or place either of them on their chest. Bang. After the beep, an ECG test is transmitted.

From there, the system receives the info. Patient status and voice messages are stored in a box and an ECG report is printed showing the results and other pertinent info (medications, past events, etc.). The doctor is notified that a report has been prepared. After reviewing it, he calls back into the system to notify the patient what he thinks or, of course, calls the patient directly.

I am very impressed for two reasons: First, it changes our definition of computer telephony from merely "adding intelligence to making and taking phone calls" to "adding intelligence to making and taking phone calls *and* saving lives."

Second, the system proved once and for all, despite what so many have suggested, that Harry has a heart, albeit slightly irregular.

One big plus for the system: You can program it to remind patients to take their medication. Apparently taking the wrong stuff irregularly can contribute to your early demise.

Referral Form Posting

Periphonics Corporation of Bohemia, NY helped Interactive Healthcare Technologies create an automated referral form posting system. The perceived size of a task often depends on your point of view. For some physicians, completing three to four referral forms each week is a trivial task that doesn't justify the time and expense of implementing and using PC technology. From the insurance company perspective however, posting the data from three or four referral forms received each week from three or four thousand healthcare providers is not a trivial task.

Interactive Healthcare Technologies addressed this dichotomy through the implementation of an Interactive Voice Response (IVR) solution. Use of the company's services provides two benefits. First, insurance companies are able to fully automate their processes, dramatically reducing data entry costs. And second, healthcare providers receive prompt payment for services without investing their time or money.

"Managed care companies are trying to automate as many processes as possible these days to help reduce costs and improve efficiencies in this very competitive industry," explains Mike Manco, president of Interactive Healthcare Technologies. "There is, however, a relatively large group of healthcare providers who continue to submit manually prepared forms. Their transaction quantity is small, and therefore these providers find it difficult to justify the time and expense it would take to automate forms processes."

Manco says the support required to deal with the forms submitted by these non-electronic low-volume healthcare providers is a significant and disproportionate expense to insurance companies.

"Interactive Healthcare Technologies was formed two years ago in response to this problem," says Manco. "Our purpose is to provide re-engineering services to the healthcare industry, taking advantage of our staff's 40 years of combined healthcare industry experience and applying the power of interactive voice response technology."

Interactive Healthcare's first PhoneLINK product emerged when the company's first client, Blue Cross Blue Shield of Ohio, asked them to automate their referral process. This system controls the referral of patients to specialists, a procedure that is required in most managed care HMO programs.

"When Blue Cross Blue Shield of Ohio solicited our services, they were processing between 15,000 and 20,000 referrals each month, the majority of which were mailed in on paper forms," Manco says. "It was a slow and expensive process to manually enter so much data. In fact, patients sometimes received their specialist services before the referral was processed, resulting in delayed payment to the specialists."

Blue Cross Blue Shield asked how Interactive Healthcare would re-engineer their system to make it paperless and faster.

"We were faced with a network of 3,500 primary care physicians sending 350,000 patients to 14,000 specialists. It quickly became apparent to us that there was little, if any, motivation for physicians to take the time to learn how to install and use a PC-based application for only a few referrals each week, thus eliminating the PC as a solution. We investigated IVR as an alternative, because this technology is a cost-effective solution that doesn't take any special training to learn how to use.

"We developed the Electronic Referral System, a free service to providers, and in a short time we were able to support the referral entries made by all providers," Manco continues. "Providers can now call in 24 hours a day, 7 days week, and complete a referral transaction in less than two minutes using a touch-tone phone. Now specialists seldom have claims rejected because referrals have not yet been posted on the system. We fired up the system in January of 1995, and in less than a year the electronic referral rate increased to 100%."

Time and Attendance Verification

Interface Alternative, Inc. of Somerset, NJ sells three lines: Service Bureau, Custom Development and Standard Products. The standard products are based on a digital switching system. It has modules for international callback, follow-me service, dealer locator, ACD, Audiotex -- and the newest addition is pre-paid alphanumeric paging. Interface Alternative also sells software for pre-paid and 900 services.

Their biggest sale to date is a large 400-port Unix system. They used speaker verification to enable a Time and Attendance system for remote workers. A home health group bought it to confirm working hours and location for visiting nurses. These are stand-alone systems -- each box equipped with 4 T-spans. Each machine was stuffed with eight D/121s, four VR/160s and four Dianatel EA/24s.

Hospitality / Travel

Weather Forecast Trip Planner

Trip planning has been made much easier with on-line services such as CompuServe, Audiotex services, and tourism hotlines. Unfortunately, none of these methods take into account historical data, travel time, and other factors that vacationers consider when they are packing.

Voice processing can help to enhance these services with trip planning software that has real-time access to weather station data. For example, the trip planning system can ask callers to input the codes for the area they wish to visit and their proposed departure time and mode of travel. Based on the mileage, speed of travel, and weather information, the IVR system can speak instructions to the caller for what kind of clothes to pack, when to leave, or even alternate routes. A more sophisticated system would provide callers with the option of retrieving maps via fax-on-demand.

Human Resources

Absentee Call-in System

DAC Systems has created an absentee call-in system for Navistar - Columbus Plastics Operation. The system allows employees to call in and report an absence from work. By enabling employees to directly enter information about their absence via phone, the system streamlines the reporting process, saving time and money.

To build the system, DAC used Artisoft's Visual Voice telephony toolkit. Visual Voice adds telephony features to the Visual Basic development environment. With Visual Voice, any Visual Basic programmer can now build applications that place and receive phone calls, prompt for and retrieve touch-tone digits, record and play back recorded files, and much more.

The absentee reporting system can operate in two modes. For callers with touch-tone phones, the system will prompt for touch-tone input. The system allows replies to be spoken, if the caller is not using a touch-tone phone. For instance, before reporting an absence each caller must verify his identity by supplying an id and an authorization code. Touch-tone callers can use their keypad to enter the digits. Rotary phone callers can speak the numbers of their id and authorization code.

The absentee reporting system also includes Text-to-speech synthesis. When the system has verified the identity of the caller, it uses text-to-speech synthesis to speak the caller's name. Accordingly, it is not necessary to create a voice recording of each employee's name. The name is read from a database and spoken over the phone.

DAC Systems absentee reporting system runs under Windows 3.1 and communicates with a Watcom SQL database. The system uses Dialogic voice processing hardware.

Employee Training Enrollment

Siemens-Rolm Communications worked with Prodigy Concepts, Inc. of Berkeley, CA to design a special Employee Training Enrollment system based on the VoysAccess technology of Voysys Corporation (Fremont, CA).

Over the last three years, Siemens-Rolm's Human Resources department has identified and reengineered key processes. There is a need for highly trained technicians who support Siemens- Rolm products. This critical training is delivered to Siemens-Rolm customers and employees.

While there have been substantial improvements in the software used to schedule training and perform enrollments, further efficiencies were required. To provide improved service at lower cost, an Interactive Voice Response (IVR) system was created, integrating the following main functions.

Enrollment in classes via phone. Students can call in 24 hours a day, 7 days a week and determine class availability, enroll, cancel an enrollment, or transfer their enrollment to another student.

Faxback of essential documents and reports. Callers can request from over 100 documents, including a current schedule, a list of available documents, course descriptions, maps to class locations, and marketing information. Siemens-Rolm managers can also request numerous reports regarding training histories and upcoming training plans.

Online entry of computer-based training. Over two thousand Siemens-Rolm employees must complete a self-paced training course about a new product being offered. After the course is completed and a computer-based test is taken and passed, each student is provided with a "completion code." Using this code, the employee calls the IVR and enters the code. They then answer a questionnaire regarding the quality of the training course. An enrollment record is then created, indicating that the employee completed the course successfully. Statistical reports are created for management showing who has completed the course and how the course was received.

Application Description

To enroll in a class, the following steps are taken:

1. Greeting. The caller hears a greeting.

2. Request employee number and last four digits of SSN. Here, the caller provides the requested information and the system speaks back the information entered. If confirmed by the caller, the system then verifies the information matches a Siemens-Rolm employee.

3. Main Menu. This menu consists of the following choices:

A) Receive a current class schedule

B) Receive a list of documents available from the Faxback system

C) Online enrollment system

D) Faxback System

E) Speak with a live agent In this example, the caller selects the enrollment system.

4. Enrollment Options. Here, the caller can select between creating, canceling, transferring, or verifying an enrollment. In this example, the user will select the choice to enroll in a class.

5. Enter Course Code. Here, the caller enters a course code. For instance, course CU17 is entered as 2917. The system looks up the course and repeats back "CU17" for the caller to verify.

6. Enter Class Number. The caller now enters the specific class number desired. The system verifies the number and speaks back a response such as "5 seats out of 12 remaining. Press 1 to enroll, 2 to cancel." In this example, the caller presses 1 to enroll.

7. Enter Fax Number. The system requests the caller's fax number The caller enters it, and confirms it when read back.

8. Verification of Enrollment. The system tells the caller that the enrollment has been created and that a letter will be faxed momentarily.

9. Faxing Enrollment Verification. The system generates a letter containing the caller's name and enrollment, and faxes it to the caller's fax number.

Job Applicant Screening

Pronexus' (Carp, Ontario) sold their VBVoice application generator to Edge Information Management of Melbourne, FL. The company used the software to create a platform that screens prospective employees.

In order to effectively screen job applicants on a uniform basis across the country, major U.S. retailers are faced with several expensive and time-consuming tasks. Edge recognized the potential for voice processing to more efficiently and effectively deliver these screening services.

One of Edge's clients is a major auto-parts retailer with over 400 locations across the United States. Managers at each location needed a way to conveniently find out the status of background checks being performed on job applicants by Edge. Edge used VBVoice to quickly design an IVR system that lets managers from across the country use a 1-800 number to dial into the system anytime to find out the status of the background checks.

Each applicant is assigned a confirmation number that identifies him or her to the retail outlet. Using either this number, or the applicant's Social Security number, the managers of each location can check the status of background checks for applicants they have interviewed. The VBVoice system identifies several status levels so the managers can determine what stage the check is at and how soon the entire check will be completed. The system allows the managers to easily determine the status of checks 24 hours a day and eliminates the need for Edge to have an agent responding to routine telephone inquiries.

Edge also serves a leading fast-food chain in the U.S. who wanted to save money in the delivery and evaluation of psychological tests that it administers to all job applicants. Edge was able to design an IVR system using VBVoice that allows them to automate the delivery of psychological tests and provide immediate results to the hiring manager. The VBVoice application conducts a pre-employment interview with the applicant by prompting for touch-tone answers to basic interview questions.

Based an applicant's responses, the application then calculates a score. Managers can retrieve the score for a given applicant at any time by simply calling the VBVoice application. Managers can also use the system to enter answers to the same questions from personal interviews they have conducted to quickly obtain a score.

Edge runs both VBVoice applications in a service bureau setting over a 24-line T1-span, using Dialed Number Identification Service (DNIS) to route callers to the correct VBVoice application based on the 1-800 number dialed by each of their clients. Dave Bodenheimer of Edge is tremendously excited with the applications they are now using: "There is so much more to computer telephony than simple voice mail. I can see our present applications expanding to six or seven T1-spans quite easily, and we're always looking to the future for new ways to use this technology."

VBVoice's visual design interface allowed Edge to rapidly develop these custom applications. The flexibility of VBVoice and the Visual Basic platform allowed Edge to integrate the applications with existing database formats and to easily add their own Visual Basic code to calculate test scores. VBVoice's support for T-1 allows Edge and other service bureaus to deploy high-density applications using ANI and DNIS to automatically screen and route calls.

Remote Employee Timeclock Reporting

Quetzal Computers of Brooklyn, NY has created a "Timekeeper" application with Pronexus' (Carp, Ontario) VBVoice that eliminates the need for expensive time clocks and accurately tracks the hours of remote employees. As a turnkey solution for their clients, "Timekeeper" allows employees to call into a central VBVoice application and input their ID#, doing away with expensive time clocks at every site. The application logs the times for both the beginning and ending of each job, providing accurate billing and expense data.

Murray Gordon, president of Quetzal, has built many useful features into this system that can be configured by the client. Features include the use of caller-id and an "automatic call-back" feature to verify the employee is actually at their work site. It can also be set to page a supervisor if an employee does not check into the system at a designated time. Reporting features allow for all employee data to be viewed, edited or printed. Voice mail features are also available.

Timekeeper saves its users time and money by eliminating time clocks and providing timely and accurate information about remote employee hours. Using VBVoice, Quetzal developers were able to quickly develop the application and incorporate features like Caller ID, paging, automatic call-back, and voice mail. VBVoice also gave Quetzal developers the unsurpassed ability to extend their applications using the open architecture of Visual Basic. This allowed them to build in the time logging and reporting features and to integrate the application with popular payroll programs.

Insurance

Coverage and Claims Status

It is often frustrating to figure out what kind of coverage you are entitled to even when reading through a benefits booklet. There's many things to consider, including deductibles, types of procedures, special allowances, and "network provider" payment percentages. All of the rules and codes are even difficult for a doctor's office staff to understand.

IVR systems are beginning to pop-up that help doctor's offices to automate the claims reporting and authorization process. IVR can also be used to tell callers what amount of coverage will be applied to certain procedures. Many times, this kind of data is not available unless the office staff calls the insurance company and waits for a number of minutes to get help from the claims office.

If the patient is waiting to schedule the next appointment, it can seem like an eternity, and the patient may even cancel the procedure. An IVR system can ask for the procedure code numbers, and the patient's I.D. number in order to automatically report the amount of coverage that will apply. This could also include a quote based on the patient's yearly deductible and outstanding claims. Voice processing systems can also reduce the number of informational calls made to claims offices. In most cases an automated system will qualify the nature of a call, and route callers to a representative.

This type of system will ask the caller to input the date of the claim and sometimes the dates of service in question. When this data is matched-up with the policy holder's account number and identification code, an inquiry is made to the insurance company's mainframe computer. The claims status information is then downloaded onto the IVR system so it can be spoken out to the caller. Admar, a management company specializing in health insurance, has put IVR to use in this way. The company uses an IVR system developed by Centigram Corporation of San Jose, CA. The platform dispenses information on group premium statements, benefits eligibility, and of course - claims status.

Internet Service Provider

Internet Service Provider Voice Verification Gateway

Joe Baranauskas of iNTELiTRAK Technologies in Austin, Texas, has developed a voice verification gateway using VBVoice, an IVR application generator from Pronexus (Ottawa, Ontario), and VoiceBuilder for Windows, a voice verification Developer's Kit, and the model 2400 voice board from Votan, a division of Moscom Corporation (Pittsford, NY).

Citadel Gateway, the Voice Verification Gateway for Internet Access, addresses all the issues pertaining to security around the Internet by generating a volatile password for subsequent access to a secured web site on the net. The beauty of this security system is that it generates a time-sensitive password. If the caller does not use an issued password in 5 minutes he or she will have to order a new one because the old one is destroyed after that time.

The WG works like this. A caller dials into an IVR application built with VBVoice that prompts for a series of voice prints which are recorded and carried to the VoiceBuilder spectral voiceprint analysis for verification against pre-registered voice patterns. A comparison of the voice prints are made and the caller is either accepted or rejected.

If accepted, the system generates a volatile password that the caller will use to access an interactive session on the Web. As far as security systems go, this has huge appeal and potential. Current network security consists of a token card, passwords and pin numbers. For any operation with a large employee or member base with turnover implications, the cost of set up is enormous, not to mention administration and maintenance. With WG, a simple, inexpensive voice enrollment procedure is conducted on a homogenous, self-maintaining security system in minutes.

The gateway is truly cutting edge, with potential applications targeted for the Internet, Intranet, and Interactive Cable TV in both private and public sector environments.

Legal

Child Support Enforcement

The collection of money owed for the support of children by absent parents becomes a bigger challenge each day. According to state and county workers across the nation, their caseloads have easily tripled over the past decade. United Companies, Inc. (Unicom) of Anchorage, AK, has developed a highly specialized IVR application to help ease the load for county workers and custodial parents. The system is called KIDS (Key Information Delivery System) and is based on the Micro-ITC platform manufactured by Wygant Scientific, Inc., of Portland, OR. According to Kathleen Johnson, Manager of Voice Applications for Unicom, the KIDS system has become very popular with sate and local agencies in both reducing paperwork and accelerating arrearage payments for absent parents.

The KIDS system is programmed to allow access from custodial parents, each of whom are supplied with a PIN number. Users are prompted to enter information having to do with court appearances, support payments received, and overdue payment amounts. The KIDS system interacts directly with the child support agency's mainframe computer in order to update support and arrearage records automatically.

The system also doubles as a voice mail platform, so that busy caseworkers can reply to messages from custodial parents in non-real time fashion. By automating the payment input with direct caller interaction, a huge amount of telephone answering and associated note-taking is completely eliminated. One user of the KIDS system is the county of Sacramento, where officials have 36 lines installed to handle over 50,000 calls per month.

The agency has experienced over 300% growth in caseloads over the past five years, so the KIDS system is a welcome adjunct to the office's 13-person call center.

Law Office Locator

From time to time, almost everyone needs the services of a legal professional. It is confusing, however, when you consider all of the television ads, yellow page listings, and advice from well-meaning friends. The local Bar Association or social services agencies can make the search a lot simpler with voice processing.

Law practices can be divided into menus on the IVR system arranged by the type of practice. For example, family law, criminal law, corporate law, real estate, and so on. Options can be programmed into the system in terms of the size, age, and staff statistics for each firm. In fact, the fee schedule for each lawyer could also be included.

By asking the caller simple questions, such as what kind of lawyer do you kneed, where you live, how soon you need to set up an appointment, etc. - the IVR system can automatically search the locator database and come up with two or three options for the caller. If there are many options, it is possible to put the references into a "round robin" order, so that they are played-out more or less equally over time.

Manufacturing

Certification and Testing

Education is a big part of managing a manufacturing enterprise. The goal for automating testing and certification process is to eliminate paperwork, turn-around grades in an automated fashion, and build confidence with trainees. Manufacturers can develop systems to prompt dealers for their social security number followed by their dealer number at the beginning of each call. The caller then enters his or her choice for the chosen test. The options for tests are based on the dealer's record of previous tests, and the ones that he or she is qualified to take. The system can be set-up to ask questions at random and in a conversational manner. This is done to make the system sound more "life-like."

Component Availability

Shop foreman and line supervisors have a great challenge in juggling production schedules, multiple vendors, and pending orders. One missing part can mean huge delays and bring a production line to a screeching halt. The maintenance of good inventory records and access to alternate suppliers is therefore a critical function for any manufacturer.

Interactive voice response can provide an accurate and rapid means for a line supervisor or plant manager to forecast production schedules, or to accelerate production for special orders. For example, a local wholesaler can use IVR to dispense information about the parts available (by part number), alternate parts, quantities, and delivery options. This can be especially helpful if the information is delivered normal business hours.

Expert Systems, Inc. of Atlanta, GA is a charter member of Dialogic's ToolKit Developer Program. They sell the EASE Development Environment. It works on NT, Windows 3.xx and DOS. It's an award-winning apps gen. One of their greatest success stories is Drexel Heritage. They make furniture that sells in retail outlets nationwide. These had agents manning a call center fielding thousands of calls a day for inquiries on availability, colors, shipping status and manufacture runs.

Couples would come into the showroom to buy at night or on the weekends. The dealers would call central to get inventory status on the "blue leather wingback" chair only to encounter busy signals, or worse yet -- no answer because it was after hours. An Expert VARs put in an IVR system based on EASE. Now dealers can call at any time from the showroom floor on behalf of their customers.

The solution has proven to be very successful. Drexel has increased their call volume (and thus their level of customer service by 4 times). Today, they are fielding over 11,000 inquiries weekly with only 19 agents. The call center staff can now concentrate on solving real problems rather than dispensing routine information. It's a real customer-pleaser - - so much so that 7 other major NC furniture companies use similar IVR systems.

Order Entry & Order Status

As with calculating inventory levels, regular orders and the associated paperwork for distribution centers can get bogged-down due to work hour constraints. Such was the case for Domino Pizza in the United Kingdom. CT Integrators in the UK have successfully applied the Pronexus VB Voice Toolkit technology in a food products distribution application for the company. The system is used to automate bulk food orders around the clock.

Before the software was developed, the Domino staff had to place their orders the following morning, creating a delay in order processing. By using IVR technology, the users are able to record their orders at night when they have accurate information about supply usage (based on stock levels and sales of product). The suppliers now call into the IVR system and enter PIN numbers in order to obtain order information for immediate processing and dispatch. This accelerates the delivery process such that Domino employees receive shipments early in the morning. The company is also beginning to develop a means to automate pizza ordering for customers. This will allow customers to call for pizza and not have to worry about waiting in line or getting a busy signal. In this way, the company intends to improve its overall customer service capabilities.

Service Station Price Authorization

SoftAnswer of Montreal has developed for the petroleum industry a price authorization system called Servox based on the VBVoice application generator from Pronexus of Carp, Ontario. The system is designed to allow service stations to quickly get approval for a price change at the pump. Servox answers all calls from Service Stations, validates for security reasons, then personally guides callers through menu options. Users get approval to change the price or place gasoline delivery orders. The petroleum company determines the profile of each service station. This information helps them decide the level of authorization required based on competitive strategies in each territory.

Test Gear

The testing of telecommunications equipment before it is shipped is critical for any manufacturer. Switch manufacturers, for example, have developed very sophisticated programs that "exercise" switches as if many people were calling into them at the same time. This is important in order to find out if any of the new features implemented have had an adverse effect on "older features." This is called regression testing.

Most voice processing and IVR systems built to suit this testing purpose are totally customized. Some are based on technology developed by Hammer Technologies, Inc. The "Hammer" platform mimics the input made by other telephone systems or human callers. For example, the system will make (or take) calls to or from the communication device that is being tested. Hammer can be programmed to speak messages, dial digits, hang up the phone at certain times, and to do it all over again.

A system of this nature can be programmed to make thousands of calls a day into the system being tested in order to "break" it. A trail of transaction attempts are then logged by the IVR system. Based on this data, it can be determined whether or not the tested system is operating efficiently and accurately.

Marine

Charter Scheduler

A special IVR system can significantly enhance charter boat scheduling. This is due to the fact that the majority of charters are owned and operated by only one or two individuals. For hours at a time, the principals of the "firm" are unavailable to explain their services, schedules, or fees. This means lost opportunities for the business.

Voice processing can be used to greet prospects from anywhere in the world. The charter boats' normal schedule, free-range hours, and fees can be easily announced by the system. In addition, recordings made by passengers from previous voyages can be used as third person testimonials. If the system is equipped with ANI (Automatic Number Identification), it is possible to capture the caller's phone number in order to call them back to set-up an appointment. The system could also be used for callers to make tentative reservations if they are familiar with the captain and crew.

Military

Flight Simulators

Companies like CAE in Canada and Singer Link Simulation Systems have helped pilots to learn how to fly for years. Flight simulators attempt to create realistic environments that teach pilots how to react in certain flying conditions. This goes for commercial, private, and military applications of the technology. Simulators mimic the pitch, roll, and general attitude of a vessel in flight. In addition, visual indicators, backgrounds, and vibrations are used to make the simulation as life-like as possible. Voice processing is used for both pre-recorded messages and recordings of the pilot and co-pilot during the simulation. A multi-channel IVR system can play messages over a number of "speaker" channels while it is recording on the other channels.

Recording and playback are associated with timed events that are analyzed by the instructor after the flight scenario is complete. It is important to be able to accurately pinpoint what recordings were made when certain events occurred in the cockpit. Since the IVR system digitizes the human voice, it is possible to synchronize each utterance with a database of status codes for each action taken by the pilot. In this fashion, the instructor can help the trainee to analyze his or her performance during the debriefing session after the "flight."

Package Delivery

Delivery Tracking

In today's package shipping business, it's not enough to deliver on time. People want to know where their packages are. "Has my package arrived in the other city? Is it on the truck? I have to know. My most important customer is waiting." Carriers compete fiercely in this arena, offering advanced technology, live telephone representatives, and software their customers can use on their own PCs.

DHL Worldwide Express, the world's largest international air express network, had a global challenge: how could the company respond to questions from customers in 750,000 cities? The answer was a completely new, automated information system, built around Edify software.

Although DHL's competitors were offering many ways to check the status of shipments, none of them provided a truly interactive, automated information system. Most callers still ended up in live phone conversations with service representatives. DHL looked at these trends and saw an opportunity.

"Our research showed that 65 percent of customers tracking deliveries would rather use some type of automated system than deal with a live representative," says Alan Boehme, Director of Customer Access at DHL.

Boehme also knew that the telephone was the most common communications technology in traffic departments and mail rooms. He decided that an automated interactive voice response (IVR) system was the best solution to the delivery tracking problem.

DHL, however, was operating a variety of computerized telecommunications and call center systems, as well as legacy hosts and LANs. The company needed a solution more powerful than traditional IVR products, and flexible enough to bridge multiple hosts and networks. They also needed a product that would let them develop their applications quickly -- and get to market before their competitors.

DHL turned to Edify Corporation of Santa Clara, CA to help solve their problem. Edify's Electronic Workforce is a software solution for developing interactive voice response as well as Web, fax and workflow applications. The software has the ability to combine host data access with the latest client-server and PC applications. This flexibility allowed DHL to automate functions in mixed and changing system configurations. In addition, Edify's Agent Trainer object-oriented development environment helped DHL deliver GlobalTrack, a prototype automated system, in fewer than four weeks -- and become the first in the market with an interactive package tracking service.

GlobalTrack now resides on two servers at the company's Redwood City, CA headquarters, where it handles thousands of callers each day. The servers are set up to guarantee massive redundancy for back-up purposes, with each capable of handling 24 concurrent callers.

Customers dial 1-800-CALL-DHL to be connected to GlobalTrack. The system then asks the airbill of the package in question. The Electronic Workforce software agents within GlobalTrack use the airbill number to query the delivery database. The system then returns to the caller with a voice response that may even include the name of the person who signed for the shipment. If there's a problem with the delivery, the system can immediately connect the caller with a customer service representative.

Finally, if customers need hardcopy confirmation of their package's status, they have only to ask. The system will fax a copy of the delivery report within 30 seconds.

Power Utilities

Power Line Maintenance and Meter Reading

Utilities use a variety of monitoring devices to ascertain electrical load, overheating, usage and other information on power lines. Since power lines can be used to transmit data and voice in some cases, the opportunity presents itself to collect this data and make it available over regular telephone circuits for line personnel.

For example, an IVR system can be connected via serial link to a centralized computer that polls line monitoring devices on the power grid. Each device can be assigned a specific record and transaction number in a common database shared by the IVR system.

Authorized personnel can then call into the system in order to select information about the condition of any part of the power grid. This can be especially helpful in locating problems during power outages, or for emergency dispatchers at the maintenance center.

The Wisconsin Power & Light Company (WP&L) is a power utility that uses this technology to handle outage calls, IVR meter reading, construction inquiries, and transfer to agents. According to Gary Schmidt, Manager - Customer Service Center, the utility handles close to 1,000,000 calls each year. WP&L uses ANI to identify customers, so routing calls and providing a high amount of service automation is made simple. The company plans to use speech recognition and voice messaging to enhance the system. Baltimore Gas & Electric worked with Periphonics of Bohemia, NY to develop a customer meter reading platform.

Energy consumers are instructed to call a toll-free number posted on their electric bill. Users read their gas or electric meter and the use touch-tones to enter the digits. The Periphonics system automatically validates the plausibility of the digits in order to validate the reading. This helps to avoid improper billing.

In addition, the IVR system will fix transposed digits and put right digits entered from users having read the meter backwards. The system is also used for power outage reporting and power restoration status.

Service Disconnect

Customers who are moving sometimes forget to disconnect their service with the power company. The procedure is rarely complicated, but if it is not taken care of, substantial charges can be incurred, and disputes may come out of this.

Voice processing systems can be used to automate the process of service disconnection and also to extend the service hours of the utility to include nights and weekends. For example, the Sacramento Municipal Utilities District provides this service.

Of course, only authorized customers would be able to use such a system, but the account number and social security number are usually sufficient deterrents to fraudulent disconnection calls. An IVR system can be connected to a database that houses both numbers in the same record in order to allow the transaction to continue.

The main menu could be programmed to say: “Thank you for calling the automated disconnection service. Please enter the desired date of disconnection... Based on your last meter reading and pro-rated usage estimate, your final bill will be $85.66. Please press ‘1’ to confirm disconnection... At the tone, please leave a forwarding address or telephone number where you can be reached.”

Print Media

Coupons

Coupons are a form of promotion that sells newspapers. In fact, the sales pitch for subscribing to daily delivery is often the coupon section on Wednesdays and Sundays. Voice processing can act as a complimentary means to generate additional or "custom" coupons. This can attract new readers, generate subscriber loyalty, and provide an additional revenue stream to the paper in the form of paid advertisements.

A system of this nature will be promoted with a flyer. Based on the input of certain published codes, callers can qualify for "late breaking" of "non-published" store coupons by recording their name and address after the qualifying code number. In this way, the paper can sell coupon advertising space to customers even after the published deadline for printed material. As an option, coupons can be faxed to readers who have fax machines.

Entertainment / Sports Contest

Sports Buff Network (SBN) of Kenosha, WI has found a profitable way (using printed media, IVR and the Web), to actively engage the sports fan in a running test of his or her sports knowledge. They run sports contests that boost circulation and revenue for over 50 newspapers across the US.

"Fantasy" sports games are a contest, often played on a small scale in office pools. Armchair coaches invent a fantasy team made up of their favorite players from all the teams in a professional football league. Weekly performance statistics for each athlete are combined for each fantasy team to determine the "winning coach."

SBN has taken this out of the hands of the office or local tavern's official sports geek and automated it with computer telephony. They built an IVR-and-Internet-accessible application that lets participants form fantasy football teams, hear weekly results, get tips and trade players.

The computer telephony front-end was written with Expert Systems' (Atlanta, GA) EASE 3.02 app gen; computer telephony cards were Dialogic four- and 12-line boards.

The complete system provides up to 150 lines of IVR and audiotext on networked client PCs. The system accesses FoxPro databases (25 per contest) of team statistics, performance history and schedules running on a Novell server updated weekly for stats and in real-time for player trades.

Figure 7.6 – Sports Buff Network Icon

For Web access to the system, there's an Internet server (at www.sportsbuff.com) with a dedicated T-1 line. The server also houses a subset of the FoxPro data, mirrored on a SQL Server database, running on Windows NT. An Internet Data Transporter and firewall running on Windows NT is used to safely move data between the FoxPro and SQL databases. Contestants enter by filling out a form in their newspaper, establishing their passwords and picking out their teams. For a small fee, they can subscribe to Sports Buff services, which gives them five IVR- or Web-driven trades per season.

This also entitles subscribers to hear their weekly score, total score, weekly rank, and cumulative rank, current roster of players. They also get to hear the promotion's mascot, Coach B, deliver tips on hot players, weekly developments, injuries, weather and other game factors. Winners' names are printed in the paper and prizes range, depending on sponsorship, from free cellular phone minutes to Super Bowl vacation packages.

SBN downloads its raw sports statistics every Tuesday from Stats (Skokie, IL), a supplier of sports stats to the Associated Press. SBN's batch programs convert the data into contest-specific scores and rankings, which go on a Novell FoxPro server. They mirror a subset of that information on their SQL Internet server for Web-based play. Although terms vary from client to client, they pitch the whole package to newspapers as a turnkey promotion, including full-time live customer support, requiring only ad space and direction in return. They pocket the IVR and Internet service fees.

Jeff Thomas, founder and president of SBN, ran his first promotion in a specialty football newspaper in 1992. He tries to take several calls a day from contestants. "You get energized," he says. "People love the game." They also give him new ideas. "We're in the entertainment business. We're always looking for ways to increase the fun factor."

SBN also runs a more intricate promotion for 6,800 intense sports fans participating in NFL Players / Prime Sports Fantasy Football. This nation-wide contest, in concert with a weekly cable TV show, offers advanced pay-per-trade options, player profiles, weekly mailed reports and higher weekly and seasonal prizes. It also costs a lot more to play.

Temporary Delivery Stoppage

Everyone knows the tell-tale signs of a family on vacation. Lights stay on, grass grows high, and newspapers pile-up at the front door. Of course, timers can fix the light problem, and an arrangement with a neighborhood teenager can take care of the lawn. Unfortunately, last-minute packing and other arrangements get in the way of good sense sometimes, so some things are overlooked.

Newspapers can offer a special service that takes advantage of voice processing to aid vacationers. By constructing a database of routes, delivery personnel phone numbers, and subscriber addresses, an IVR system can be used to automate requests for delivery stoppage. Let's say that a family has left for vacation and discovers that they forgot to terminate delivery. Subscribers can call a toll-free number, enter their address with touch-tone digits and then enter date ranges for the temporary stoppage. Once the system updates the database, an automatic outdial to the delivery agent can be made. This phone call can be used to announce addresses and date ranges for the temporary stoppage.

Product Distribution

Dealer Locator

Product distribution companies needed to be competitive in providing services to thousands of retail outlets. But providing speedy and efficient referral services is a challenge for smaller companies, because it is expensive to staff for a large call center. This is compounded by time zone constraints that necessitate two shifts of service agents. Coast To Coast Vision of Dallas, Texas, faced these very problems in establishing a referral service for their nationwide discount eyewear program.

Craig D. Hutson, President of Tara Systems, Inc. (Irving, TX) worked with the eyewear company to develop an IVR system that could be used to direct nationwide callers to the nearest outlet for the purchase of discount eyewear. Called LOCATE24, the system is accessed by dialing an 800 number.

A series of prompts leads the caller through a simple menu which ascertains the caller's zip code via touch-tone input. An automatic database search is then executed in order to match the caller's location with a vision provider outlet closest to that "home" post office. In this fashion, prospects for eyecare products are referred to the most convenient location within 60 seconds.

Tara Systems scripted typical calls from prospects needing a locator referral. The logic flow for the new automated system was patterned after the dialog carried-out by the service agents. This included the identification of the service called, asking several questions of the caller, and ultimately retrieving the answer for referral to the caller. By scripting the new system after the manner used by live operators, the man-machine interface was designed to be friendly and easy to understand for all callers.

The software team at Tara Systems developed a zip-code search program using FoxPro. The program indexes on a zip-code field to search for participating eyewear outlets. The voice response portion of the program used TRT's Pro/Found database access module in order to interface the caller's touch-tone input with the database searches. Information in the FoxPro database then acts as a pointer to the correct message number associated with the referred outlet. The voice response system then plays-out the associated message to the caller as pictured in figure 7.7.

Figure 7.7 - Tara Systems LOCATE24

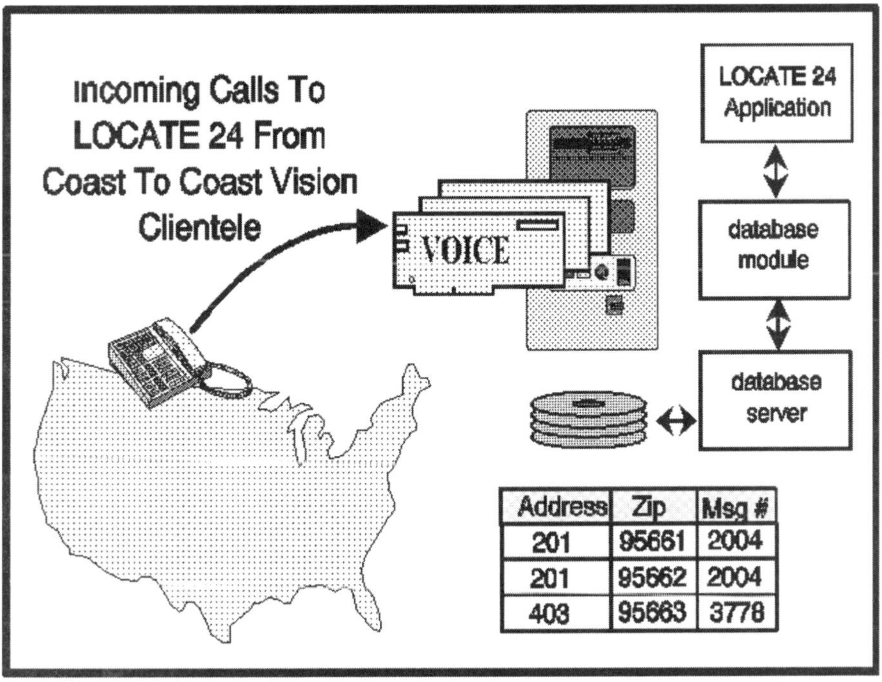

Address	Zip	Msg #
201	95661	2004
201	95662	2004
403	95663	3778

According to Joel Ray, President of Coast To Coast Vision, the LOCATE24 system was his answer to remaining competitive in the eyes of his clients. Says Ray: "We were instantly able to establish a large company image due to the fact that the service is easy to use, accurate, and provides around-the-clock service. And we did this for a fraction of the cost of a two or three-shift call center. What with over 4,000 optical providers to refer callers to, you can imagine how manually intensive the task is with live operators. Now, my staff can concentrate on other duties in order to further distinguish ourselves in the eyes of our customers."

This particular installation of LOCATE24 is abased on a 486/33 MHz PC with a single Dialogic D/21D speech card. This provides service to two callers simultaneously. The system can be upgraded to provide access to as many as 48 callers by using other Dialogic components with no changes to the system application. A 500 MB hard drive is used on the system to store the zip-code database, store location information, and the voice response software.

There are a number of enhancements that are available for the LOCATE24 system. By using T-1 access lines from the long distance carrier, it is possible to extract ANI (Automatic Number Identification) at the beginning of each call in order to further automate the process of referring customers to the nearest outlet. This can be achieved by adding an additional step to the automated database search that would include a match between the caller's telephone number and the associated zip code. Since the caller's telephone number is signaled automatically, it would no longer be necessary to prompt the caller for their zip code on each call.

Field Service Verification

Field service technicians can verify a customer's warranty, check on parts inventory, and provide invoice data via IVR. This helps to reduce paperwork, and eliminates the need for the service technician to wait for agents to answer the phone for work authorization.

The Vision OS/2 - based voice processing system developed by the IVS company of Hamden, CT provides this capability.

Service personnel can be dispatched to make a repair and then call the system to enter part numbers, serial numbers and other data. The information is matched against customer records in order to determine what parts can be replaced or repaired at no charge, versus which ones must be paid for if they are out of warranty. With the Vision system in place, invoices can be automatically mailed or faxed after the field work is done.

Inventory Inquiry and Control

Inventory control is an especially critical concept for gasoline distributors. Even the smallest miscalculation can cause a retail outlet to be closed down without product to sell. Usually, the dispatcher calls retail gas station owners and franchisees so they can report on current inventory levels and sales receipts from the previous day. These totals are compared against historical data for that outlet and then the dispatcher arranges for tank truck delivery.

The CARS (Chevron Audio Response System) solution is an interactive voice response platform developed for Chevron of Canada in Burnaby, British Columbia. The application allows gas station retailers to update sales data and gasoline inventory via touch-tone input. This input is transformed into database records associated with tanker delivery routes. Based on the daily updates made by retail owners, the Chevron dispatcher is able to partially automate the task of assigning delivery schedules and inventory for the dealers.

Some of the problems that come up have to do with the timeliness of the calls to the retail outlets. Digital Voice Response Systems of New Westminster, British Columbia, worked with Chevron to develop a means to get around these problems. By using the CARS system, retailers can update the inventory and sales database before the dispatcher arrives each morning at 7:00 AM. When the dispatcher arrives, the data is already collected and stored in the appropriate database for each station.

This eliminated the need for manual intervention, double entry of information, and significantly increased the speed in which the dispatcher could do his job.

Doug Unwin, the President of Digital Voice Response Systems approached the problem by adapting a new IVR front-end to the company's existing Clipper database server. By using the Pro/Found database module from Telephone Response Technologies, Inc., he was able to develop an application that writes records into the same inventory and sales receipts database that the on-site administrators have access to. Unwin connected the voice response unit to the other workstations with a Microsoft NetBEUI driver interface. Figure 7.8 shows how the system is put together.

Figure 7.8 - Chevron Gas Station Dispatch

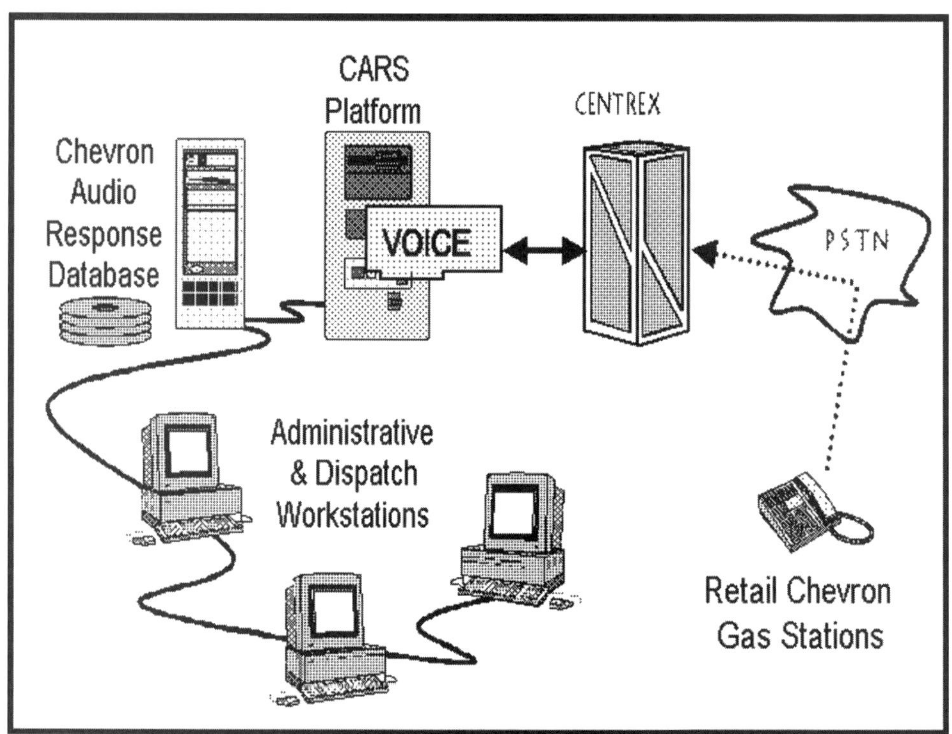

The system prompts were recorded with a high-quality microphone and then digitized for the Dialogic speech card. The development of the logic flow that "calls-up" the system prompts was mapped-out with the Chevron Marketing Department, who worked with Digital Voice Response Systems on the promotion plan for the CARS system.

According to Jim Longman, Manager of Terminal Operations for Chevron, the CARS system helps him to run an efficient operation by keep inventory levels low. Says Longman: "Besides the obvious reduction in the dispatcher's work load, we have eliminated the need to call every gas station during the day. The big return is the reduction of inventory through better control. We estimate a savings of $500,000."

The CARS system uses two Dialogic D41/D speech cards for a total of eight ports. This allows as many retail gas stations to use the system simultaneously. Digital Voice Response Systems developed and deployed the application on a 386/SX PC. In addition, Centrex lines from the local telephone company were installed so callers could be transferred to administrators and dispatch personnel when they need assistance.

Depending on system usage and traffic analysis, more telephone lines may be installed in the system. Some add-ons for the future may be a fax-on-demand capability. By adding a fax element to the CARS system, users can have confirmation receipts faxed to them.

Real Estate

Home Listing Service

Voice processing technology can be used to help potential buyers find a house. Callers dial a number and are prompted to enter information such as price range, neighborhood, number of rooms, style of house, etc. The system can then search a database of homes and respond by playing a description of any available homes that meet the specified criteria.

The number of the broker selling the home is provided in case the caller is interested. Freedom Communications of Clearwater, FL designed an IVR system for real estate and have significantly increased the overall usefulness of this solution by using Caller ID service. They developed the application for Century 21 to help the company generate prospect lists for their Clearwater operation.

The idea was to capture the phone number of IVR system users in order to build a database of interested prospects. Clients call the IVR system after reading about a property. Callers use a special IVR code number associated with the listing. The IVR system plays-out a recorded message about the property they are interested in and then writes the access code into a prospect database along with the caller's telephone number (collected automatically via Caller ID).

Each day, the system generates a report that includes telephone numbers, access codes, and time-stamp information. The Century 21 agents are then able to peruse the list and follow-up with prospects based on the associated listing. Freedom Communications built the system by adapting a Rhetorex-based Verbatim (Key Voice Technologies, Sarasota, FL) voice mail system to a Whooz Calling (Zeus Phonstuff, Atlanta, GA) Caller ID unit.

The Caller ID unit sits in-between the telephone company central office and the IVR system. The Caller ID device collects both the calling party's number and the DTMF entries. These digits are collected by the PC over the serial port and then used by a software utility that runs in the background of the IVR application.

Loan Qualification

Part of any home purchase is the sometimes painstaking procedure of qualifying the prospect's ability to buy. This is usually achieved during an interview process conducted by the real estate agent. In some sates, this interview process is governed by certain disclosure laws and requires both parties to sign a contract.

The contract indicates that the prospect understands the ramifications of principal, interest, taxes, and insurance on the hopeful mortgage. Earnings ratios applied to the prospective buyer's income and monthly living expenses are also taken into account.

The end result is a monthly dollar amount that the buyer can "afford" to pay based on the current interest rates and types of mortgages available. Both real estate companies and financial institutions can offer a special service to prospective buyers by automating this procedure with voice processing.

For example, a "Mortgage Hotline" can be accessed by dialing a toll free number and entering touch-tone digits. The main menu of such a system would guide callers to answer a number of questions. For example: "Please enter the number of income-earning members in your family... Please enter the monthly earnings for each person... Please enter the amount you spend in food, clothing, car payments, and rent..."

Based on the caller inputs, the system will be able to announce the monthly amount the caller would qualify for in making a mortgage payment. A more sophisticated system would ask for social security numbers in order to conduct an on-line search of the caller's credit history. This would enable the system to fax a qualification sheet to the agent or buyer in order to accelerate the entire process.

Retail Services

Coupon Redemption

Coupon fulfillment houses employ hundreds of workers who open envelopes, create voucher slips, fill-out forms, and then produce checks that are sent to retail consumers. The entire process is time consuming and labor intensive.

Voice processing systems can go a long way to either partially automate or fully automate coupon redemption by using a combination of ANI (Automatic Number Identification), DNIS (Dialed Number Identification Service), and caller input.

Coupons can be printed such that serialized numbers are used on each pre-printed form. This can also be achieved with point-of-purchase coupons that are created at the retail outlet. The consumer can be instructed to call a special telephone number in order to input the code number on the coupon. The IVR system can help to curb fraud by checking the ANI of the caller against a database of transactions that have already occurred. The IVR transaction can include a recording session wherein the caller leaves a message with their name and address.

This message can be linked to a report for the coupon being redeemed during the transcription process. Once the caller's address is entered into the database, the rebate can be mailed in the usual fashion. Subsequent calls to the system can be further automated by executing an ANI match-up to the caller's address.

Credit Check

There are a number of ways to get credit authorization. Some point-of-sale terminals are equipped with modems so that an automatic credit card service bureau transaction is initiated as soon as the credit card is "swiped." Other merchants call a special phone number and speak with an agent in order to get authorization. The "live" scenario is convenient for stores that do not have the automated gear, but it introduces delays at the check-out counter - especially if the call center is busy because of traffic from holiday purchases.

An IVR system can be helpful for small retail stores. Some outlets may not have the appropriate equipment to automate the process or a budget for the equipment. In addition, voice processing can be used as an alternate means to check credit if the primary system is not working.

The IVR system will simply ask the retail clerk to enter a merchant number followed by the credit card number and the amount requested. If the card is accepted, an authorization number can be played-out to the clerk. This can also be done for personal checks and ATM purchases.

Home Shopping

Shopping from home is both convenient and necessary for homebound individuals or busy families. Mail order catalogues are perhaps the most common vehicle for soliciting home-based purchases, and the use of call center personnel and overnight delivery services can put the item you want right on your doorstep the very next day. The challenge for catalogue houses is to provide quick, friendly, and efficient service in order to stay competitive. It takes hundreds of employees or contracted telemarketing agencies with call centers to do this.

One way to expand customer service is to provide a special voice processing service to frequent buyers. An IVR solution for frequent buyers could provide special incentives for using the automated method, such as discounts or coupons for future purchases. Callers can access the system with an 800 number, enter their customer ID, and then be prompted for the item number, size, quantity, and color of their selection.

By automating the order entry for catalogue purchases, some 20 to 50% of the traffic into the call center can be reduced. The IVR unit could use ANI and DNIS in order to identify the catalogue the caller is looking at (Special 800 numbers could be used for each version of catalogue), and the caller's phone number could pre-identify the default mailing address for the shipment.

Order Status

Both consumers and retail managers call wholesalers, service companies, and warehouses to check on the status of orders.

These calls require a service agent to verify the identity of the caller, and then to access the order status from a computer. In effect, the service agent acts as an extension of the company's computer system by "navigating" through the information on behalf of the caller. There are many examples of how voice processing has been put to use to overcome this labor-intensive task.

One example of how this process is automated is an IVR system for photo finishing labs. Each retail customer who drops-off film at the grocery store, for example, gets a numbered stub from the film envelope. These numbered stubs can be printed along with an 800 number for access to the IVR system. The IVR system can be linked to the photo lab's host computer system. Each time an order is completed, it is slated for delivery with the "batch" for the retail outlet the next day. This data can be transmitted to the IVR system on an as-needed basis, or the data can be housed in a common database that both the IVR system and administrative workstations have access to. This scenario is similar for other order status implementations including confirmations for appliance delivery, bulk trash pick-up, or catering outfits, to name a few.

All Things Connected, Inc. (a Voysys customer) needed an effective way to take orders over the phone when staff was not available. Customers wanted to receive product information prior to ordering items and an answering service was not an efficient way of delivering information direct to the user.

The solution was to develop an IVR application that can accept credit card orders via a touch-tone phone. The company designed the app so customers could choose from a menu to have information faxed to them. They also created a system that eliminates follow up calls by using a complete information resource database. The result is a system that produces complete phone orders 24 hours a day, 7 days a week. Mail order customers get efficient delivery of information. The system has helped with an increase in orders and a significant decrease in the outgoing phone bill.

Repossession Hotline

The prospect of having your car repossessed is not a welcome thought, however, the reality is that each day thousands of cars are "taken back" by banks, leasing companies, and car dealers. Repossessed cars are usually sold at wholesale to car auctioneers, brokers, or used car lots. In some cases, the original driver of the car can make special arrangements to refinance the car, or get additional grace periods to catch-up on payments. In most cases, the lienholder wants the driver to make payments and eventually own the car.

Voice processing can be applied to serve both repossession agents, lienholders, and those who are tardy on their payments. For example, an IVR system can be programmed to link into a payment history database for each car. The system can provide repossession status to drivers by letting them know how many more days are left before the car will be considered for repossession if no other arrangements are made. In addition, repossession agents can log on to the system, and be assigned "repo" orders automatically by the system. Assignments can be given to agents on a parity basis, or according to a contract that the agent has with the lienholder.

Tape Rental

Video rental stores purchase multiple copies of newly released videos in anticipation of strong demand. Sometimes, it takes months for customers to finally rent a popular tape. The stores are often inundated with calls each day asking if a certain tape has been returned. Some stores allow regular customers to reserve tapes, so they can be checked-out as soon as the previous customer returns them.

An IVR system can be used to help cut down on the number of calls coming into the retail outlet. In addition, voice processing can be used to augment the operation of cash registers in the store itself. For example, the IVR system can be directly connected to the rental database in order to keep track of the number of copies in use.

By using bar code scanners, keypunch entry, or magnetic strips, the tape's return can be automatically updated in the same database. Since the IVR system has access to this data, it can dispense information about a tape's availability as soon as it is returned.

Many customers have a credit card on file with rental outfits. This being the case, it would be possible for the IVR system to allow the caller to "reserve" the tape that was just returned. If the customer fails to pick-up the tape within a certain grace period, it can be billed to the credit card, or rented to another patron.

Warranty Registration

Some consumers are quick to fill-out their warranty registration cards, but most of us either throw them away or wait until something goes wrong before doing anything. In most cases, registration is as simple as filling-out a few lines on a postcard and dropping it in the mail.

Manufactures have big incentives to encourage registration. For example, manufacturers recall products, issue updates, and solicit customers to buy new items. None of these initiatives are possible unless the manufacturer knows who the buyers are. This is an issue because retail outlets usually do not provide this information to the manufacturer.

Voice processing can ease the entire process by providing a simple touch-tone registration procedure. Buyers of new merchandise can call an 800 number, enter the serial number of the product in question, and then record a message with their name and address.

This can be enhanced by providing a fax-in service, wherein the registration is scanned by OCR (Optical Character Recognition) software. In this fashion, mail handling, lost cards, and other problems associated with post cards can be totally avoided.

Security

Rounds Logger

Employee safety is a big issue no matter what size a company may be. At Bell Canada, for instance, there are hundreds of employees who need to work in the field, where safety and security issues are well outside the control of most managers. This goes for cable crews in the Outside Plant Engineering Department, Special Assignments Group, and with the installers, as well. Xenox Communications of Ontario, Canada developed an IVR system called Teletrak that maintains a telecommunications "tether" to Bell Canada's safety tracking plan.

Teletrak monitors the whereabouts and status of off-premises employees who are in potentially dangerous environments. Rapid response to emergency situations are therefore ensured, since the system uses a timing algorithm to contact employees by pager or telephone calls if they do not "log in" at preset intervals defined by their work schedule. The system matches the time of the last employee call-in with their last known location.

Workers can access the system via land lines, cellular phones or two-way pagers 24 hours a day. The system allows for both touch tone input and also uses speech recognition technology as an alternate input method. Failure of an employee to call the system triggers the Teletrak system to call the employee, and then alternate numbers if their is no answer.

According to Dave Oikawa, Manger of Special Assignments at Bell Canada: "Teletrak has met or exceeded all company health and safety requirements. The system performs its functions efficiently and accurately."

Oikawa says that the system has greatly reduced the amount of clerical time his staff spends on monitoring the whereabouts of department employees. "The system has also inspired my staff to be more accountable for their own safety."

Voice Entry

There's plenty of examples of how security gates, stands, guards, and other measures are used to provide secure access to restricted or private areas. In many communities, residents must enter a code number at the entrance in order to activate gate motors. The problem with this approach is that unauthorized people can still obtain access codes to gain entry.

An IVR system can be used to avoid this problem by using speaker verification technology. With this type of system, each resident or authorized person must record a special word or name into the IVR system. This recording is used as a basis for subsequent transactions.

Each time the authorized person wishes to enter the secure area, they use a speakerphone to gain access to the system. The IVR system will ask the resident to speak his or her name. The system will then match the recording with the "official" recording made during the sign-up procedure. If the pattern matching is positive, then the voice processing system will authorize entry. A serial link can connect the IVR system to a hard contact closure mechanism that controls the gate. Multiple ports on such a system can allow other secure accesses to occur around the perimeter.

Social Services

Child Care Hotline

Resource centers for child care have the mission of helping parents to find licensed care facilities. Non-profit organizations of this type also lend assistance in the form of parenting training, and the training of child care professionals. These organizations can help as many as 2,000 or more parents find child care facilities each month, so as you can imagine, phone calls are many and heavy. Child care centers maintain a file on each licensed child care provider. This includes their name, zip code, telephone number, and types of services offered.

Parents can call a hotline at any time to get referrals from the system. If a child care provider is not in good standing with the agency, this is also noted in their file and the information can be provided to callers.

Charity Pledges

Public television stations, school sports teams, and hospital associations are known for their frequent telethons and charity drives. It is typical for a temporary call center to be assembled so volunteers can man phones for calls from pledges. Unfortunately, it is difficult to staff these call centers around the clock and during working hours.

Interactive voice response is a viable alternative to manning a temporary call center. As many as a dozen to 64 lines can be answered by a single node simultaneously, so an IVR system could be the primary means for accepting pledges. In addition, a system of this type can be used for multiple telethons or campaigns at the same time. The campaigns can be distinguished from one another with DID (Direct Inward Dial) numbers, or by asking the caller to enter menu choices. An IVR system can be programmed to take credit card numbers for instant collection of pledges. It can also be programmed to work with Caller ID equipment, so busy volunteers can make outgoing calls to pledges who hung-up too soon.

Felon Roster

Neighborhood leaders and citizens are increasingly alarmed at the rate in which certain felons are released from prison, only to commit more crimes. There are associations in a number of states pushing for neighborhood registration of certain felons, including those involved in rape, child molestation, and other frightening crimes. An IVR system can be programmed to keep track of felons and dispatch information about where they live and what crimes they committed. In some cases, this may be mandatory depending on the state or jurisdiction in charge.

Such a system can greet callers and ask them to input their zip code. The system will then execute a "locator" function much in the same way a dealer locator does. Based on a zip-code search, the system can then provide a roster of felons living in the area.

Meals Scheduler

There are countless neighborhood programs that sponsor meals for the elderly, indigent, and homeless. These meals are either served in a kitchen, or in some cases delivered to people in their homes. The management of volunteers for the delivery of meals is difficult, because of conflicting schedules, cancellations, and shortages.

An interactive voice response system can be used to help schedule and dispatch volunteers. For example, drivers can call into the system and input digits indicating the times they are free to deliver meals, and the area they are able to work. In addition, the homebound recipients of the meals will be asked to call into the system and make requests for meals each day. Based on the availability of meals, recipient requests, and drivers, the system can automatically assign deliveries to the volunteers.

Unemployment Verification

Unemployment agencies put a lot of work into maintaining careful records on claimants and the progress of their employment searches and continuing benefits. As is the case in most government agencies, forms have to be filled-out by the client, and regular phone calls and visits have to be made to the agency in order for benefits to continue.

The State of Oregon Employment Division, for example, requires claimants to fill-out a status form every two weeks and mail it in. These cards are then read by an OCR (Optical Character Recognition) system in order to partially automate their database entry activities.

The forms and OCR process is slow and frustrating to both the clients and agency workers, so the agency worked with Wygant Scientific, Inc. of Portland, OR to develop an alternate input method. The agency not only wanted to improve service for its clients, but also wanted to find a more efficient and timely way of getting the needed information into their database.

Now callers can access the Micro-ITC IVR system 24 hours a day. Clients enter their social security number for identification, and then answer the same questions that are on the claim card. The entire interview process takes about five minutes and eliminates days of delay. There are multiple IVR systems installed that are connected via IBM 3270 emulation to a centralized IBM 3090 mainframe in Salem, OR. The IVR units automatically upload the "virtual form" to the mainframe immediately after each claimant's call. If the mainframe is off-line, the IVR units buffer the data and transmit the information when the host is available.

Telephone Companies

Cellular Dropped Call Credit

It is common for cellular phone users to "drop" a call. This occurs from time to time when the user is traveling under a bridge, tunnel, or large hill. Dropped calls are more often than not credited to the user. To receive a credit, users must call the cellular service center to make the request. This causes the caller to wait for an available service representative, sometimes resulting in frustration.

Some cellular companies are automating the way credits are given. Air Touch cellular, for example, offers an IVR capability to get around the busy call center problem. Users simply dial a toll-free number to enter the time of the call or the number of minutes they were on the phone before the disconnection. Users served by Air Touch make hundreds of these calls each day.

The IVR option actually saves the telephone company money, because the service representatives are free to help new customers and take orders for revenue generating transactions.

Cellular Feature Activation

Voice processing can be used to activate certain features such as call forwarding, voice messaging, and three-way calling. Cellular companies look for many ways to be competitive, and enhanced features are one means to do this. Enhanced features such as voice messaging and call forwarding not only add revenue on a monthly basis, but encourage network usage. The trouble is that subscribers have a difficult time understanding how to use certain features.

An IVR system can be connected directly to the telephone switch "in-line" with the caller. When the caller dials a special feature help line, the calls can be routed to the IVR system for assistance.

Let's say, for example, that the caller wants to activate call forwarding, but he or she forgets how to do it. The IVR system can prompt the caller to input the feature they want to activate based on a menu selection. The system can then confirm that the caller wants call forwarding. The system can subsequently prompt the caller for the new telephone number and then confirm the number by playing-back the digits. Once the transaction is completed with the caller, the IVR system can pulse the correct feature access digits and phone number into the switch on behalf of the caller.

Cellular Roamer

People on the move use nationwide pagers, voice messaging, and of course - cellular phones. In order to use a cellular phone away from your normal service area, special roamer service has to be activated for your unit. This usually takes a phone call with a service agent.

Voice processing systems can be used to automatically inform "roamers" of what the access codes or procedures are for each area. This is especially helpful if the user intends to roam in an area that is not serviced by the same franchise as the "home" provider.

In this case, the caller can enter a series of identifying digits which can be transmitted by the IVR system to the billing center. This will enable the use of the roaming system for long distance calls and air time that can then be charged on the "home" cellular bill. This requires reciprocity in the billing arrangements between service providers.

Telecommunications Debit Card Generation

Pegasys is a telecommunications firm that turned to Voysys Corporation of Fremont, CA to upgrade their customer service center. The company was having trouble handling the volume of incoming calls. Numerous, simple changes for address and phone numbers coupled with repetitive questions absorbed customer support time creating lags in activation and distribution of new debit cards.

By having an IVR stand alone PC it allowed more customers access through customer service, having the convenience of listening to solutions about re-occurring problems. Implementation of an IVR app to complete business transactions via touch-tone input allowed callers to update important information via a touch-tone phone.

The customer support staff is now able to spend quality time with customers. Providing technical support is now more efficient and turn around of orders through the IVR system insures happy customers.

The company is also able to generate labels on order entry of debit cards with ease. Pegasys attributes its increase in call volume by 60% (from 1,000 to 60,000) to the installation of the Voysys platform.

Line Conversion

There are literally millions of telephone systems installed all over the world. These include Key Systems, PBXs, and ACDs. Much of the equipment in use is older gear with limited features and no access to enhanced network features such as DID (Direct Inward Dial), or ANI (Automatic Number Identification). Voice processing gear can be used to enhance the functions of older telephone systems by acting as a "protocol converter."

For example, let's say that a company has an older PBX with regular DID and loop start lines for incoming and outgoing calls, respectively. If the company wants to coordinate customer record screens with incoming transactions, the service agents must ask the caller for their identification at the beginning of each phone call. This can be streamlined by placing an IVR unit in "front of" the PBX (in-between the PBX and the telephone company end office). The circuits between the IVR system and the telephone company can be equipped to handle ANI, and the circuits between the IVR system and the PBX can be configured as DID trunks.

When callers dial their service agent, the VRU can collect the calling party I.D. and then activate a program on the company's computer system to bring-up the customer record on a service agent's screen (once the screen has been presented to the agent and the agent's line is free). The voice processing system can send the call into the PBX and pulse the agent's DID (extension) digits in order to route the call. In this way, an IVR system can be used to effectively upgrade an older PBX into a modern communications system.

Line Testing

In today's highly competitive telecommunications market, service providers are aggressively applying new technology to improve customer service and save costs. A perfect example can be seen at BT - the UK's leading telco has a new way of helping its technicians to work more efficiently and it all follows a field visit on a cold, wet February day.

Aculab, Ltd. of Bedfordshire, UK worked closely with BT and used their talent and their Millennium-CT system to develop a solution for the technicians.

Two years ago, a senior BT manager noticed an engineer struggling to test lines from the top of a telegraph pole. On investigation he found that network service facilities such as ring-back and directory number identification (DNI) were inconsistent, not always available, and extremely slow. A subsequent report commissioned to investigate these problems proposed changes to BT's switching network. However, the report's proposals were rejected on the grounds that they were too costly and too time-consuming to implement.

Enter BT's FasTrak development team. Initiated to provide quick solutions that use the latest technology, FasTrak consulted David Gibson, the head of BT's Speech Platform Technology Team, about the possibility of developing a computer-telephony solution that could provide engineers with caller line information. An initial discussion was held over the telephone, with both participants on site at BT's large research and development facility at Martlesham Heath, near Ipswich in England. By the time FasTrak representatives had walked half-a-mile to meet with Gibson, the speech platform group had built a demonstration using the Millennium-CT system.

Just nine months later this demonstration - now designated Faultsman - was nationally deployed on Millennium-CT. Engineers dial Faultsman and are greeted by a recorded voice message that uses ISDN information to speak the number of the calling line. By entering DTMF digits, a ring-back test and quiet line test can be ordered.

As pictured in figure 7.9, the application sits on a LAN so lines test results can be archived in a server. Unlike a traditional switching solution, the Faultsman application harnesses the power of ISDN - meaning that it is not dependent on the vagaries of local exchange equipment. Being PC-based, the computer-telephony solution is also extremely cost-effective. (Estimates put BT's cost-savings from Faultsman at millions of pounds).

Today, Faultsman is taking over 1.8 million calls per month and has experienced no service breaks. Licensed by Aculab from BT, Millennium-CT presents an object-orientated API that shields developers from the low-level control of individual devices. With a system manager to take care of commands to speech cards and ISDN access cards, demonstration applications can be created very fast. Millennium-CT's ability to scale effortlessly from small pilot projects to national systems is also essential in the telecommunications market.

Figure 7.9 – BT Millennium-based Line Testing System

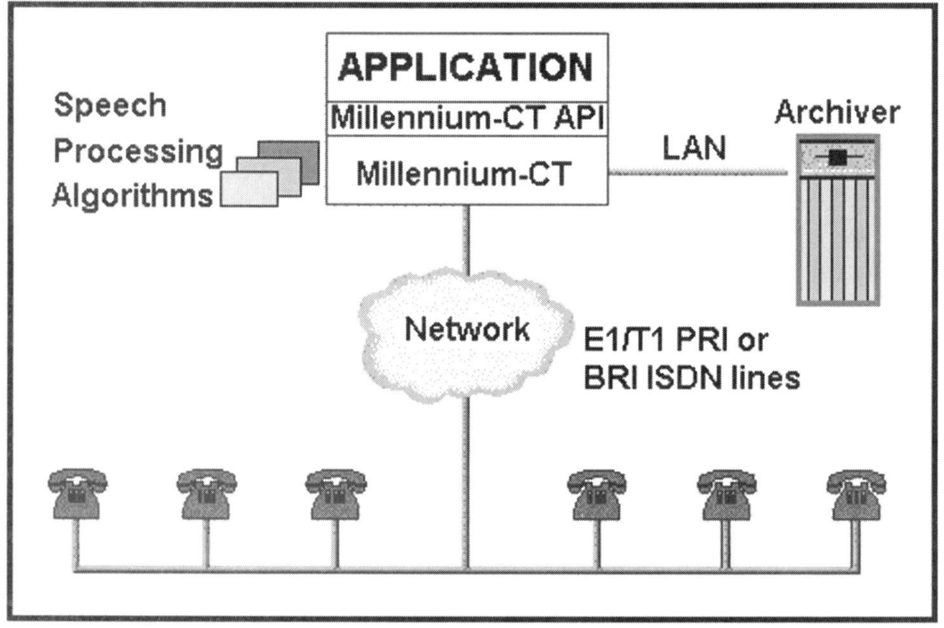

The success of Faultsman has prompted BT to trial other projects using Millennium-CT. One of the most exciting is GIRAFF - General International Recorded Announcement and Fax Facility. Callers who have dialed an incorrect international number - perhaps because of a code change in another country - may be routed to this equipment.

All callers are advised of the problem, but callers from fax machines are additionally sent a written fax message detailing the code change. Both the customer and BT benefit from this service - the fax reaches its destination and a higher proportion of international calls are successfully completed.

Millennium-CT optionally includes text-to-speech functions, so services such as remote access to email and remote advice of fax are also being developed. These initiatives confirm voice processing as a valuable tool for developing products that differentiate customer service and save costs. "Faultsman has generated a lot of interest within BT," comments Gibson. "I am frequently contacted by people from within the company who want to know if the system can be used to implement the services they have envisaged. Most often, this can be fairly easily achieved."

Pay Phone Promotion

Ameritech Pay Phone Services has used Pronexus' (Carp, Ontario) VBVoice to develop a high-density voice platform to be used for promotional activities. The Platform allowed Ameritech to run a promotion in which customers were able to call the system from Ameritech pay phones and be instantly notified if they won a prize. For the promotion, a phone number was advertised which customers had to call from an Ameritech pay phone. When callers dialed this number, the system used caller-id to determine the originating phone number then checked it against a database of eligible phone numbers.

If the call did not originate from an Ameritech pay phone, a message was played which informed the caller that the phone they had called from was not eligible for the promotion. If the call did originate from an Ameritech pay phone, the systems checked a second database to see if the caller had won a prize. Non-prize winning calls were played an informational message and told that they had not won a prize. Winning callers were informed that they had won a prize and they were prompted for information using a combination of touch-tone digits and voice recordings. In addition, major prize winners were transferred to a live operator for further verification.

The platform was developed to be capable of handling up to 100,000 calls per week and was used in each of Ameritech's major markets (Chicago, Cleveland, Columbus, Detroit, Indianapolis, and Milwaukee). Using VBVoice allowed the system to make extensive use of Visual Basic's data access capabilities. This proved crucial in the design and development of the call verification and prize distribution aspects of the system. In addition, the ease of use of the VBVoice tool allowed the system to be developed and tested in time to meet an aggressive schedule.

Phone Number Conversion

GUATEL, the national telephone company in Guatemala, is using Dialogic D/300SC-E1, D/300SC-E1-IDPD, CP-6/SC, and CP-12/SC boards to help ease the transition to a new seven-digit telephone numbering scheme. This is illustrated in figure 7.10. Designed by Kilobyte, a computer telephony solution provider, the application is working so well that GUATEL is considering adding a virtual voice mail application based on the same Dialogic network interface boards.

Figure 7.10 – Guatemalan Call Routing Platform

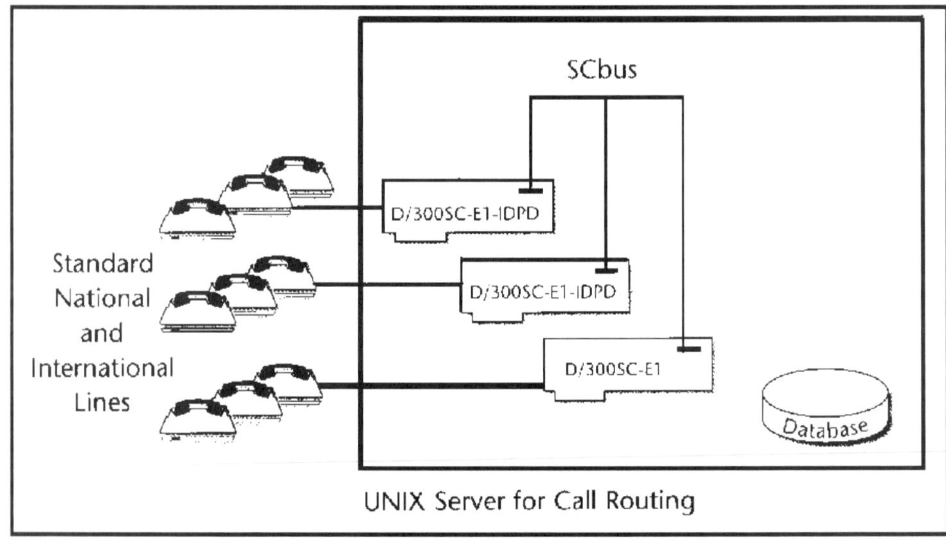

The telephone system in Guatemala has more than 200 central office switches servicing 400,000 installed telephone lines. The system was running out of phone numbers fast, and had already used two different numbering schemes (both five- and six digit numbers) to cope with the demand for service. To make more numbers available to their customers, have numbers available for other telephony service providers, and offer a unified numbering scheme, GUATEL decided to convert to a new seven-digit scheme.

GUATEL chose Kilobyte to help them implement a solution. It was a tall order. The new system not only had to accept the new numbers, but also had to inform callers of the new number and numbering scheme (for both local and international callers), accept and correctly route calls based on the old numbering schemes, and avoid interrupting service or inconveniencing customers.

In spite of the technical challenges, Kilobyte needed to have the system installed and working within a month to meet an August 15 cutover date-a national holiday for which call traffic is typically low. Kilobyte used Dialogic CT hardware for the solid customer support, as well as for the fault tolerance, EIA standard characteristics, and ISA compatibility of the boards. All went well according to Luis Maselli of Kilobyte. "I really appreciated the coordination that (the Dialogic) sales person established with every resource that we needed" during both development and testing.

To make it easy for callers to get used to the new numbering scheme, Kilobyte designed a distributed, UNIX-based system that accepts five-, six-, and seven-digit numbers. When a caller dials one of the old numbers, a message announces the correct number, and then the call is automatically transferred to the new number.

If the caller is dialing from outside the country, the application uses the language code sent from the international central office (CO) to select the language in which to play the prompt: Spanish only, or Spanish with English, French, or German. The network interface boards have dial pulse detection (DPD) to serve callers with either touch-tone or pulse dialing (rotary) telephones.

While DPD was not really necessary for handling the international calls, GUATEL wanted to be able to leverage their hardware investment for future applications.

Giving customers a second way to obtain the new number information, the GUATEL solution also incorporates an interactive voice/fax response system. By calling one of two toll-free numbers, local callers can get a new number by simply entering the old number (using either touch-tone or pulse dialing). They then hear the new number spoken in Spanish. Callers can also have faxed to them a conversion chart with both the old and new numbers. The fax can be sent during the same call or after the caller hangs up.

To handle the extraordinarily high volume of callers accessing the conversion system during the first few weeks, call centers were set up to handle overflow traffic. By 6 a.m. on the first day, 80 percent of the switches had been changed over to the new numbering scheme. By 9 a.m., all the switches had been converted and 25,000 calls per hour were going through the system-20,000 of them overflow calls. The volume of callers accessing the system temporarily dropped over the weekend, but picked back up on Monday. It was necessary to add another call center to handle all the new business traffic through the week.

Today, the volume of traffic using the conversion system has dramatically decreased, making it possible to rely solely on the automated system for new number information and routing. Emphasizing the importance of having a reliable, well designed system in place, the SubGenerente de Servicio at GUATEL, Ing., Walter Mendizabal, said they risked losing "up to $400,000 a day on international call traffic alone" if the system had failed.

Next, GUATEL plans on leveraging their investment in Dialogic hardware to offer virtual voice mail, which is CO-based voice mail for customers without personal phone lines.

Provisioning / Pager Activation System

G&H Consulting of Tolland, CT sells Technically Speaking boards and Microsoft NT and SQL Servers. Their VoiceTrac product is a full-fledged fax mail, voice mail and fax-on-demand package. It handles 48 lines. Their first sale was a pager activation system for Message Center USA (now Pactel). These guys were a LAN customer and G&H convinced them that they could do better with IVR. Instead of having agents re-key digits on keyboards for pager activation, dealers can do it themselves over the phone. The system automatically assigns phone numbers and activates beepers from the paging terminal database. In 1988, the system was four lines, but it eventually grew to 12.

Telephone Company Service Outage and Dispatch

Voice Processing Plus, Inc. of West Bloomfield, MI is a Parity Software VAR, Nortel Business Affiliate and WilTel Business Partner. They sell custom solutions with a focus on "giving a voice to data." The company does LAN/Voice Integrations, mainframe emulation and GammaLink fax stuff. One of their first sales was to Ameritech. Voice Processing Plus helped them develop a SPOC (Single Point of Contact) system. They needed a way to coordinate trouble ticket management along with MCI, Sprint, and AT&T for high capacity (HICAP, i.e. DS-1, DS-3) outages.

They developed a network - level routing system designed to handle calls during HICAP service outages. The system prompts callers to input the type of failure and then routes them to the correct HICAP service center in the five-state area. 32 ports analog. This also includes a service dispatch paging function.

Toll Calculator

Cost conscious telephone subscribers look for many ways to reduce the cost of their monthly phone bill. One way to do this is to make calls during the evening or on weekends.

Another cost-cutting measure is to limit the length of calls. A voice processing platform can be used by subscribers to find out the exact cost of telephone calls before they are placed in order to budget these transactions. Systems of this nature have been put into place on a trial basis by RBOCs (Regional Bell Operating Companies) over the past decade. One example is an IVR system designed by Gralin Associates of Fountainville, PA for Ameritech.

By using a special toll charge database, the toll calculator provides users with the ability to enter the number they wish to dial followed by the time of the call and length of the call. The system then determines the price of the call based on the subscriber's calling plan and time-of-day calculations. The system then reports to the caller the exact cost of the call. Based on that information, the caller decides if he or she wishes to make the call.

Reverse Directory

Telephone companies and phone book providers spend millions of dollars each year to maintain name and address directories in order to supply up-to-date publications. A new service that many enhanced service providers and phone companies are exploring is called "reverse directory." With this service, the name and address associated with a telephone number can be provided for a fee. Of course, there are tariff restrictions, privacy issues, and other constraints associated with this service, so it is not universal as of this date.

Voice processing can be used to collect the number in question from the user of the service. This number is then transmitted over a data link to a CNA (Caller Name and Address) database. Since the database is very large, with millions of entries, it is not feasible to record each individual name and street address. An alternate means to announce the information is to spell the names and streets with concatenated (digitized) speech, or to use speech synthesis (text-to-speech). Speech synthesis allows the ASCII text associated with a CNA record to be spoken. The resulting speech can then be played-out to the caller over the phone.

Transportation

Airline Crew Scheduling

USAir flight and work crews are now bidding for desired flights and checking on schedule adjustments with the aid of an IVR system developed by Centigram Corporation of San Jose, CA. The system is a telephone front-end to three applications that use data stored in the carrier's IBM mainframe computer. USAir's IMS (Information Management System) serves the BBA (Block Bid Award), SAP (Schedule Adjustment Period), and PBA (Permanent Bid Award) applications by keeping track of scheduled flights, employee I.D., seniority, awarded pairings, and other data having to do with bid/blockholder awards.

Since there are hundreds of employees to manage schedules for, it is easy to see how some type of automation is needed to cut down on lost information, delays, and the frustrations associated with waiting in line on the phone. With the new Centigram system, pilots and flight attendants call into the system and identify themselves with an employee ID number and a PIN number. BBA then guides them through menu choices to select and bid on the monthly schedule of trips they are interested in.

SAP provides similar access to other crew members, while the PBA application announces information about equipment type, base locations, and crew positions. The system can handle 48 users simultaneously and runs on a 486 PC equipped with a terminal emulation card for mainframe connectivity.

Gulf Air, a large regional airline carrier in the Middle East, is reaping more benefits than originally envisaged from their Periphonics Corp. (Bohemia, NY) IVR system. The airline installed a Periphonics VPS/sp system in 1994 to provide a Crew Rostering service, which enables flight crewmembers to quickly obtain their next duty roster from any telephone worldwide. The system identifies the caller and logs the time and date they called, and it helps ensure that all flight crew staff members have their new roster.

"Due to the rapid expansion of the airline, our crew scheduling staff were becoming increasingly overloaded with calls from flight crews inquiring about their new rosters," comments Zahid Mahmud, with the Management Services Division of Gulf Air. "We discovered that they were spending most of their time on the phone rather than actually optimizing crew schedules! In addition, some crewmembers told us they were experiencing a 30-minute wait on the telephone to get through to our schedulers during peak periods. This was clearly not acceptable, and we decided to use IVR to help resolve the situation."

Mr. Mahmud says the Periphonics systems were introduced to dramatically cut the number of calls to Gulf Air schedulers and to reduce the time crewmembers waited on the phone. "Both of these goals were achieved, and we have received very positive feedback from all of the users," Mr. Mahmud says. "In fact, when a new roster is released, the number of calls we receive in a single day is the same as the total number of flight crews we have." In addition, Gulf Air is realizing an unexpected benefit since the introduction of this service.

"With the system fully effective, the schedulers are now able to devote their time to planning the logistics of the rosters," Mr. Mahmud says. "This in turn results in a better match of crew skills to passenger requirements, for example scheduling Indian-speaking crews on the Indian routes. "We were delighted with the initial success of the IVR system in improving the efficiency in our crew scheduling area," Mr. Mahmud says. "But we hadn't foreseen the added benefit of improving the quality of customer service in the air by simply reducing the time the schedulers are on the phone."

Airline Frequent Flyer

Before IVR systems became popular, less than 50% of the calls coming into airline reservation centers were revenue-producing. Requests for frequent flyer mileage credits, general information, or even flight status are still common. Frequent flyers expect instant access to their account information so they can plan for free trips and other program benefits.

The airlines spend millions of dollars each year not only in awarding the mileage credits, but simply managing the data and distributing the information to its customers. The service centers for frequent flyers are open during business hours, thus putting the strain of this traffic on 24-hour reservation centers.

Fortunately, IVR platforms can completely automate these calls. These systems are programmed to prompt callers for a PIN code and frequent flyer number I.D. code. Once this information is entered, callers are asked to choose from a menu selection of accrued points, flight segment verification, and options to hear about point redemption. USAir uses this technology for their frequent flier program and has recently installed a speech recognition upgrade to their system. The IVR unit allows callers to enter menu choice selections by speaking the digits in addition to using touch-tone entry.

Airline Flight Status

Although flight status information is computerized for scheduling and safety reasons, most airlines lack the ability to provide this information to friends, relatives, and business associates of the party who is traveling. Limo services, tour managers, and travel agents are forever calling airlines to access flight status information in order to coordinate customer pick-ups. Airlines prototyped a number of IVR systems in the mid eighties to overcome this problem.

Voice processing systems of this type are directly connected to the airline flight status computer. These computers are the same ones that are used to control the television monitors in the airports that provide arrival and departure information. Users of the system are asked to enter the flight number of the plane in question. The IVR platform then sends a request to the flight status system to extract the appropriate information. The IVR system then associates the resulting information with a series of pre-recorded prompts: "Flight number 457 traveling from Philadelphia to Atlanta is on time. The scheduled arrival at gate 44 is 8:50 PM."

Invotec of Marietta, GA is an Expert Systems VAR. The company also sells Informix and Sun Microsystems products. The company developed an IVR system for Carnival Airlines. The solution is a front-end to Carnival's reservation system (A Mitel 2000 PBX). It provides flight status including ETA, so callers don't swamp the agents with non-revenue calls. The system was connected to the old Eastern Airlines flight reservation mainframe with terminal emulation software for the IVR system. The system handles 15,000 calls each day with 36 ports.

Cab Company Dispatch

American Taxi of Mt. Prospect, IL is has a voice processing system that automates 350+ cabs in Chicago and surrounding suburbs. Customers avoid busy lines and long waits by entering their requests from a touch-tone phone.

The system was integrated into the Cab Company's dispatch system by Interactive Communication Systems (Colorado Springs, CO). They built it with VOS from Parity Software of Saulsalito, CA and Dialogic cards. A Btrieve database on a Novell network provides the interprocess communication between the IVR and automated dispatch applications.

Future enhancements are planned for online credit card validation and payment processing, order cancellation and status features, package delivery and Global Positioning Satellite (GPS) to track the exact location of each cab.

Handicapped Transit

Toronto Transit Commission's new computer and phone system, called Ride-Line is used to book trips for "para-transit" (read: handicapped) customers on the 340 vehicles running in the Toronto area. The system has made major improvements, reducing the advanced booking times from four days to one day ahead.

They used Computer Talk Technology's (Richmond Hill, Ontario) Intelligent Call Exchange (ICE) PC server-based system to enable "para-trans" folks to touch-tone in Wheel-Trans registration numbers and then PIN numbers to find out the night before when their exact pick-up times will be for the next day.

The Ride-Line system runs on a 486 PC server. Information retrieved is stored on an HP 9000 host housing an INGRIS database. Access to data is done over an Ethernet LAN running TCP/IP protocols. Ride-Line handles up to 3,000 call a day.

The ICE/IVR system used by Wheel-Trans is only one component of CTT's Intelligent Call Exchange suite of products. The ICE family provides IVR, a standalone Automatic Call Distributor (ACD), voice recognition, and call/data integration and fax services with CT integration built-in, all under the control of a GUI. It's for call centers and is scaleable from four agents to 50 and more.

ICE uses Natural Microsystems AG cards. The master system console runs on OS/2 but will run on NT as soon as drivers are available. The "IcePhone" client runs on the agent's PCs - it's Windows 3.1 and Windows 95-based.

Pilot Flight Planner

The Federal Aviation Administration (FAA) requires pilots to file flight plans in order to help air traffic personnel to control traffic and ensure safe air travel. These flight plans can be hand delivered, faxed, or communicated by computer. A number of alternate approaches for filing these plans has been explored over the past twenty years.

IVR systems can be used to help pilots with this task by automating the input and confirmation of flight plans over the phone. A system like this was designed by Input Output Computer Services of Waltham, MA for the FAA. The company designed a telephone response system that connected directly to a series of DEC VAX computers.

Pilots are greeted with a menu and are asked to enter identification information including the craft, departure airports, arrival airports, and time of travel. The system also asks pilots to input the coordinates of their travel. Once the data is entered, any authorized air traffic employee or FAA official has access to the information via computer terminal.

Postal Package Tracking

Millions of parcels are delivered each day by the postal service, trucking lines and overnight delivery providers. Critical packages with medical information, contracts, and valuable instructions are often delivered to the address in question, but sometimes overlooked by the appropriate individual on the other end. Because of this, customers who inquire whether or not their package has been received make thousands of calls.

Federal Express has a way to automate these inquiries by providing customers with direct telephone access to their delivery transaction database. With the system in place, callers simply enter the airbill number of the package in question. The system then speaks prompts to the caller indicating if the package is en route, or whether it has been delivered. In the case of a successful delivery being made, the IVR system announces the time of delivery and spells-out the first initial and last name of the person who signed for the delivery.

Rail Car Weight and Locator System

Invotec of Marietta, GA is an Expert Systems VAR. The company also sells Informix and Sun Microsystems products. They developed an eight-port IVR system with 3270 terminal emulation purchased by Norfolk Southern Railroad. Users key-in an eight digit railroad car number. This is matched-up to a database of location posts that are logged with optical scanners -- 5,000 of them. The system tells the callers what railroad the car is on, when the last scan was and what the ETA of the car is.

Shippers want to know the car's empty weight so they can tell exactly what the transportation cost will be for the car. It's important for the customer to know this so they don't end up paying for a ton of snow, dirt or mud the cars can pick up along the way.

Shipping Line Global Tracking

Shipping lines receive many telephone calls each day regarding the progress and whereabouts of containers and other cargo. It is difficult to handle these calls, considering the time it takes to research and report the tracking information to each caller. Since the containers on board ships are assigned special numbers, this data can be stored on a centralized computer system.

In addition, geo-tracking devices can be used to pinpoint the exact location of vessels, no matter where in the world they are. Ship captains also radio the line's central dispatch in order to report on delays, repairs, and weather conditions. Mearsk Lines, for example, manages vessel data like this for thousands of ports worldwide on a 24-hour basis.

Interactive voice response systems can be used to greatly simplify the inquiries associated with container delivery. Mearsk uses a PC-based system that does this by asking callers to enter shipping manifest and container numbers. These queries are transmitted to the main computer with terminal emulation technology. The retrieved data is then automatically spoken to the caller. A typical response from an inquiry would be: "Your cargo is scheduled to arrive in the port of Seattle on September 16, 1997. You can call at any time to receive an update on this shipment."

Train Trouble Ticket Reporting

When you're running a five star rolling hotel on the rails, customer satisfaction is paramount. "In the old days, when we had a problem on a passenger train, on-board personnel would write it up in the Maintenance Analysis Program (MAP) book."

"After this, the problem would be fixed at the train's final destination," relates Ed Adelman, Director lndustrial & Systems Engineering at Amtrak's largest maintenance facility in Beech Grove, Indiana. "But, as more people began riding Amtrak trains nationwide, there was less and less time to mobilize repair facilities, crews, and materials. We decided we wanted to go totally paperless by automating as much as possible how a problem was reported and solved. Our voice mail system vendor recommended Show N Tel." The product is a PC-based application generator sold by a BrookTrout of Southborogh, MA.

Show N Tel works for Amtrak - and many other corporations nationwide by giving callers anywhere anytime access to corporate information systems via the telephone. Show N Tel includes all the tools for building enterprise-strength voice, fax, messaging, and telephony. Experienced developers and even non-programmers can design an application quickly and easily using Show N Tel's breakthrough object oriented development environment.

The end result? The right people get brought into a situation and can deal with it expeditiously. For example, at Amtrak, Show N Tel provides a fully automated trouble-ticketing system. "We call it IDRS the Interactive Defect Relay System," says Mr. Adelman. If a passenger train has a problem, a train crewmember calls a special 800 number on a cellular phone, often as the problem is happening. The call is always answered, thanks to the Amtrak PBX's four-line circular hunt group. Then, the Show N Tel system takes over. Here's the sequence of events:

First:, the employee enters two numbers via the keypad - the last four digits of his Social Security number and the Unit number of the car in which he's traveling. He then has three minutes to record details of each situation in each car affected. Upon report completion, the worker hangs up.

Second, Show N Tel works through Amtrak's dedicated workstation in Beech Grove to find train or unit information. Amtrak uses a 486 DX/4 133 MHz machine with 32 megabytes of RAM, a one-gigabyte hard drive, a four-line Rhetorex board and a four-port Brooktrout card.

Third, Show N Tel identifies the closest downline maintenance facility. Lastly, Show N Tel automatically generates a trouble ticket number and faxes the trouble ticket to repair personnel. They call the train and/or nearest maintenance facility and fix the problem "on-the-fly. "

For Amtrak, these capabilities are the true power of Show N Tel. Amtrak personnel can now manage repair resources more efficiently and cost effectively. "With Show N Tel," says Mr. Adelman, "downline maintenance facilities often get hours of lead time so they can prepare both maintenance forces and material to resolve the problem without undue passenger inconvenience. The solution often arrives before the train does." Here's an example of an actual Show N Telgenerated Description of Defect and Description of Repair (since the IDRS program's inception, average problem resolution time has been cut to less than 24 hours). Note the short elapsed time between notification and remedy:

2/10/96 10:54AM: "I had a passenger complain about no hot water in the sleeper car. Also the PA is not operating between the cars going out of the dining car which is the next car adjacent to the sleeping car. It won't go out to the rest of the train. That should be it. Thank you."

2/1O/96 5:04 PM "Car 32021 came in on Train 59 Outbound 58 today, Feb. 10, 1996. All repairs were completed. We repaired the hot water and the PA is OK. Thank you very much. Have a good day!"

Notice the kinds of problems reported. Amtrak customers expect the best and Amtrak is determined to give it to them. "When we put IDRS into operation across our entire network-22,000 miles of track, 600 cities, 1,000 cars, and 200 locomotives (rolling stock value of $1 billion)-we expect the Show N Tel/lDRS system to help us maintain the trains and our riders' satisfaction to the highest level humanly possible," says Mr. Adelman.

Amtrak has been using the Show N Tel/lDRS system since 10/95 on two of its premier runs:

The City of New Orleans (running from New Orleans to Chicago) and The Lake Shore Limited (going from Chicago to New York City). "Thanks to the excellent support from Technically Speaking, only three months elapsed from the time we reviewed Show N Tel until the start of our pilot program," reports Mr. Adelman.

Travel Agency Travel Reporting

Voysys Corporation of Fremont, CA solved a big problem for Associated Travel Int'l, a California based travel agency. The agency was experiencing excessive faxing of incomplete travel reports to travel coordinators. The faxing was extremely costly and their service was available during business hours only. Voysys helped the company develop an IVR application to enable travel coordinators to request specific travel reports. Now, travel coordinators receive detailed reports on flights, cost, rental cars, etc. This cut the company's fax expense by 50%. Travel coordinators now have the freedom to make requests at their own convenience

8

Messaging

Perhaps no other technology has done more to transform corporate communications than voice messaging. Information overload is a problem in today's competitive marketplace. Busy workers get dozens of mailed communications, faxes, e-mail, and voice mail messages to sift through every day. What makes the process more taxing is the fact that disparate terminals and telephones control the management of this media. Recall the days of pink slips, busy phones and overloaded reception desks. If this conjures images of the way you work, then it's time to seriously consider an investment in this technology. With voice messaging, you can:

- Avoid jammed switchboards
- Increase the accuracy and privacy of messages
- Eliminate "phone tag"
- Distribute messages to many recipients in one call
- Retain the true meaning of a message

Most voice mail packages for business come with an automated attendant. These systems will route callers using touch-tone directories, automated screening, and coordinated transfers with on-site telephone systems. In addition, most voice messaging systems provide message waiting indications so users know when they have "mail" to pick up. Capabilities of modern voice mail systems thus allow the switchboard operator to concentrate on helping visitors or callers that require a human operator for assistance.

Depending on the nature of an employee's duties, voice messaging can increase the efficiency of a person's day-to-day chores. Take, for example, a salesperson or field technician. Customers can leave messages regarding appointments, reminders, or directions. This can be handy when salespeople are about to board an airplane, since they can be advised of an alternate flight or a meeting that is pushed-back. Service technicians use the technology to re-schedule appointments and load their vehicles with the appropriate material for the next service call.

Since voice messages can be retrieved from any phone, and do not require operator intervention, messages can be picked-up and sent from anywhere in the world. This helps to overcome the constraints of time-zone communication lag and after-hours communication. A new development is the use of the Internet for transmission digitized voice messages for retrieval from your PC. As you can see, there are numerous ways messaging helps you communicate, and we've only scratched the surface.

Perhaps the biggest trend in messaging is *unified* or *universal* messaging. Unified messaging allows both keyboard/screen access and telephone access to your messages. Many systems now give you a visual representation of your messages. You can view the time, date and subject on your voice mail as if it were e-mail. You can double-click on an icon representing a voice message and you will hear it over the phone or on your PC's speakers. Figure 8.1 is a screen shot of CallWare's ViewPoint environment. This unified messaging program communicates over a LAN to coordinate message elements from a voice and fax message database, as well as interfacing with the enterprise e-mail system.

The display allows users to visually select messages they wish to review. For example, by clicking on a voice message element in the list box, the user activates an automatic outdial to his or her telephone extension from the voice mail platform. Alternately, the digitized messages can be sent over the LAN for playback on the PC's multimedia sound system.

Figure 8.1 – CallWare ViewPoint Universal InBox

CallWare ViewPoint

File Messages Help

Record Play Reply Forward Route Note Lists Delete Save Help Fax

From	Subject	Number	Length	Date	Time
Craig Hansen	The future of CTI	3219	13s	Mon 12/16/96	8:41 AM
Lori Gauthier	Fax integration	3207	2p	Mon 12/16/96	8:48 AM
Robert Buitron	Telecommuting from Chicago	3202	22s	Mon 12/16/96	9:11 AM
"Ivan"	Psuedonymns	3203	29s	Mon 12/16/96	1:22 PM
Bob Johns	$tock option$!	3202	18s	Mon 12/16/96	1:31 PM
Harold Toomey	Technical reports	3222	21s	Mon 12/16/96	1:33 PM
Jenni Lofgren	Changes to documentation	3270	15s	Mon 12/16/96	1:35 PM
Outside Caller	No Subject	Outside	20s	Mon 12/16/96	1:36 PM
Jay Wilson	Training schedule	3211	14s	Mon 12/16/96	1:37 PM
Mike Sawyer	Return Receipt	3273	0s	Mon 12/16/96	1:37 PM
Reino Kerttula	Staff meeting notes	3200	30s	Mon 12/16/96	1:38 PM

11 messages | 3273 | 2:16PM | Mon 12/16/96 | CAPS | NUM | SCRL | OVR

The user's telephone extension number is programmed into the voice mail software. When the phone is answered, the message associated with the screen selection is played-out. The same procedure is true of e-mail messages and fax messages (these are managed in the classic sense of an enterprise e-mail and fax server system).

There are many types of messaging systems to choose from. There's central office based messaging, micro-messaging systems for small offices, and large voice mail systems that can be accessed by hundreds or thousands of users. There are a variety of PC-based systems that you can design on your own.

This chapter highlights popular uses of the technology, so you can develop your own sense of how it can be put to work. And you'll see voice logging, dictation and other voice store and forward applications not usually associated with straight messaging.

Automotive

Dealership Automated Attendant

Voice Processing Plus, Inc. of West Bloomfield, MI is a Parity Software VAR, Nortel Business Affiliate and WilTel Business Partner. They sell custom solutions with a focus on "giving a voice to data." The company does LAN/Voice Integrations, mainframe emulation and GammaLink fax stuff. The local Ford auto dealer was having difficulty handling inbound traffic for service, personal calls, and sales. They had an AT&T system 25 and an Audix voice mail system with no automated attendant.

They were using 17 Centrex lines as trunks on the switch. Voice Processing Plus replaced the Audix system with their own VOS-based one. They mimicked the user interface at the customer's demand and added an auto attendant and ACD function. The company backward engineered a proprietary AT&T feature phone. We monitored the LED pin outs on a big feature phone that had the ACD group agent status on it. In effect, this computer-enabled a busy lamp field in order to route calls to non-busy extensions.

After the installation 41% of the callers knew the extension of the party they were trying to reach, increasing the volume of answered (no voice mail jail) calls by 30%. Customer service improved instantly and the dealer was (and is) very happy.

Trade or Sell Hotline

Anyone who has placed an ad in the classifieds knows what it's like to sell a car. There are crank calls, answering machine messages and unqualified prospects to wade through before your vehicle is sold.

This problem stems from the fact that there is limited space to describe your auto in the ad itself. In addition, you can't be home all the time, so you miss opportunities to speak with prospects.

A special voice messaging system can be used to augment the capabilities of a classified ad. In fact there's plenty of newspapers offering this type of service for cars, personals, and even real estate. The concept is to associate a voice mailbox number with each ad. Ad buyers can then create a customized greeting that describes the car in greater detail. If the prospect is interested, they can leave a message for the owner of the car. The messaging system can be accessed from anywhere. Some systems offer a paging option for users with beepers. An added benefit with this type of service is that you aren't required to publish your home phone number.

Construction

Messaging & Paging Service

Construction foremen and contractors are a busy group. Besides managing workers on the job site, these individuals have to maintain an open line of communication with the home office, customers, vendors, inspectors, and prospects. Because construction people are not always available to take a phone call, voice messaging can play a special roll in helping them to stay in touch.

For example, a plumbing inspector can leave a message in a contractor's mailbox indicating that an appointment has to be rescheduled. The use of a pager can be especially helpful when workers are on their way to a construction site. Take, for example a foreman who wants a worker to pick up some materials on the way to the job site. Since messaging systems can be programmed for "urgent" delivery, these types of messages can activate beepers. Anyone getting "beeped" knows to immediately make a call into the voice mail system for an important message.

Corporate

Automated Attendant

Besides the messaging element itself, voice mail systems usually have the ability to route and screen calls. This is referred to as an automated attendant feature. Automated attendants are helpful in streamlining the traffic flow for any company. In fact, the use of an automated attendant can sometimes eliminate the need for a full-time switchboard operator. With this option, callers are presented with alternatives to route their calls.

For example, most systems provide callers with the ability to press a one-digit code for access to a certain department. The menu will say: "Thank you for calling Beauty Industries. Press 1 for sales, 2 for service, 3 for a directory, or your party's extension number. Press 0 or wait on the line if you want to speak with an operator." If callers don't know who to speak with, but they are interested in sales, the automated attendant will transfer the call to any available line in the sales department. Callers who know the person's name they wish to speak with (but not their extension number) can use the automated directory.

The directory search allows callers to enter the last three digits of the party's name. This causes the messaging system to search through a list of users and then announce the extension number associated with the user's record. In some cases, the directory function will either return the caller to the main menu or allow the caller to transfer to the chosen directory name at the push of a button.

Many voice messaging systems offer a special screening capability as part of the automated attendant function. This screening can mimic a secretary who asks for the caller's name and business before connecting the call. If users do not want calls automatically sent to their extension, they can access the messaging system and type a code that indicates they want their calls screened.

When the system is instructed to do so, it will ask callers to speak their name. This spoken name is then temporarily stored so that it can be played-out to the called party. The system will put the caller on hold and then ring the called party's extension. The system announces the call like this: "You have a call from Bill Smith. To accept this call, press 1, or press 2 to have the caller leave a message." If the called party accepts the call, the transfer is made; otherwise the caller is asked to leave a message.

Figure 8.2 - CallWare Viewpoint Topology

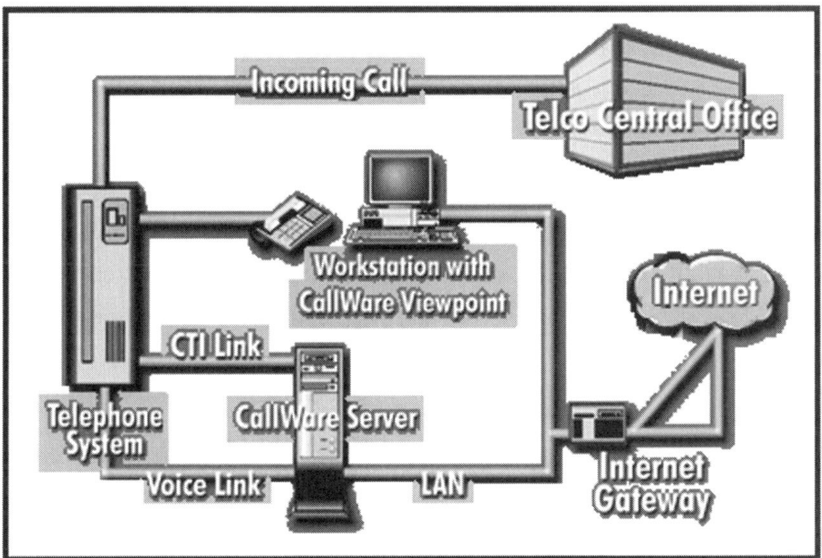

This message taking and transfer capability can further be enhanced by using software that integrates these functions. This makes the status of transfers and messages directly available on a computer screen for each internal user. Such is the case with products like the CallWare Viewpoint system pictured in figure 8.2. Software packages that integrate message-taking and other communications activities at the desktop are becoming more popular all the time. By borrowing elements from call center applications such as PBX integration and CTI links, voice messaging takes on a whole new meaning.

International Employee Broadcast

It's difficult to communicate with a large base of employees, especially when they work all over the world. Global companies get around this problem with private television networks, e-mail, recorded tapes, and fax broadcasts. Unfortunately, each of these methods require the creation and production of materials and often require employees to gather in centralized meeting places at a specific "broadcast' time. Voice messaging has none of these constraints, because any employee can access a voice mailbox from any phone 24 hours a day.

For example, let's say that a company president wants to announce a new quality program or record a monthly "fireside chat" for employees. The executive could rehearse the script for the announcement and then record it as a voice message. By directing the voice message to a pre-defined "distribution list," the message can be accessed by every employee. In fact, the message could be recorded in several languages. Employees can be instructed to pick-up messages like these on certain times of the month, or a message waiting light could alert them.

Sales Force Reporting

Trip reports are submitted as a means to communicate the outcome of meetings, customer requests, forecasting information, and action items for other employees. The filing of these reports has never been a popular task. This is due to the fact their preparation is tedious.

Trip reports are important, however, because delays in distributing this knowledge can degrade customer service. A "trip report mailbox" is the voice messaging equivalent of written trip reports. Many companies now instruct their field sales staff to record messages in a pre-defined format. In addition, sales managers can review these messages and offer advice to the entire field sales force in one phone call. Companies like Dialogic Corporation of Parsippany, NJ have been using voice messaging to do just this for years. At Dialogic, each field sales employee calls a general mailbox after every customer visit.

The sales team identifies the people in the meeting, customer concerns, and forecast information. By using a security access, managers log-on to the mailbox as the owner in order to review and save trip reports. The messages are saved for about a week and then they are transcribed to tape for archival purposes.

Telephony Server / UN-PBX

I enlisted the help of Rick Luhmann, editor and Harry Newton, publisher of Computer Telephony magazine to explain how Java means portability for computer telephony. Without hesitation, they say the NTS Telephony Server from NexPath (Santa Clara, CA) is a perfect example. Our friend Brian McConnell, president of Pacific Telephony Design (San Francisco, CA) couldn't wait to give us the scoop on NexPath. Here, he unveils some promising news for Java developers looking to bolt telephony onto their apps.

But first, consider the world NexPath is from. The company manufactures a new breed of communications server I like to call the "UN-PBX." It connects regular telephones on the outside world to regular phones in your office. But it does this in a PC without the use of a stand-alone PBX. This alone is impressive. Add the notion of picking up voice messages from either your phone or from your PC, and you've got the makings of an UN-PBX. Users can answer and control the phones like they do today with any dumb analog phone -- punching obscure buttons on the phone's touch-tone pad. Or they can log their desktop and control the movement of calls over the LAN via a GUI interface. All this while speaking over the phone.

At last count, there were at least six (shrink-wrapped) systems fitting this model: AltiGen's AltiServ, Interactive Intelligence's Enterprise Interaction Center (EIC), NetPhone's NetPhone, PhoNet's EtherPhone, Sphere's Sphericall and now NexPath. You can achieve similar results by *building* your own ATM-based "UN-PBX." This with hardware and middleware from Madge and Mitel, respectively. Ditto for the Artemis boards from InnoMedia Logic (IML).

Until now, when people talked about computer telephony, it was understood this meant "CT for Windows" at the client level. There were exceptions -- some TSAPI implementations handle multiple clients. But generally, Mac users (and people using anything *except* Windows) are plain out of luck.

Here's where NexPath distinguishes itself. The NTS Telephony Server supports clients on every major desktop and workstation OS except MS-DOS. They brought this off by developing software that uses the Web and Java to run client-side CT *applets*. Rick calls NTS a communications server. Harry calls it a PBX in a LANned PC with a twist. Whatever the name, the essence is the same: A client-controlled communications controller with a universal interface.

NexPath functions as a PBX, voice-mail system, auto attendant and small ACD. The PBX part works with ordinary analog handsets and loop-start phone lines and can expand up to 48 ports.

The PBX also includes just about everything you'll find in a traditional system except digital display phones (the display phone functions are mimicked by the Web / Java applets over the LAN to the PC at the desktop). Basic PBX features include:

- Caller ID on trunk lines and pass-through Caller ID to analog phones;
- Call waiting, call forwarding and do not disturb;
- Message waiting indication via lamp or stutter dial tone;
- Virtual extensions (i.e. multiple users can share a single physical phone);
- Night and day modes with automatic changeover;
- Integrated voice mail and automated attendant;

- Simple ACD features (i.e. dynamic hunt groups with round-robin call distribution);
- Barge into voice mail (i.e. intercept a call while caller is leaving a message);
- System activity log via SMDR (Station Message Detail Recording);
- Feature programming on a user-by-user basis;
- and Support for TAPI on Win 95 and NT workstations.

As a combination switching / messaging platform, the NTS Telephony Server offers most of the features you'd expect to find in a similarly sized PBX with a third-party voice-mail system attached. But what really sets NTS apart is the degree to which it's integrated with the Web and Java applets.

Figure 8.3 – TelOper Control for Mac Clients

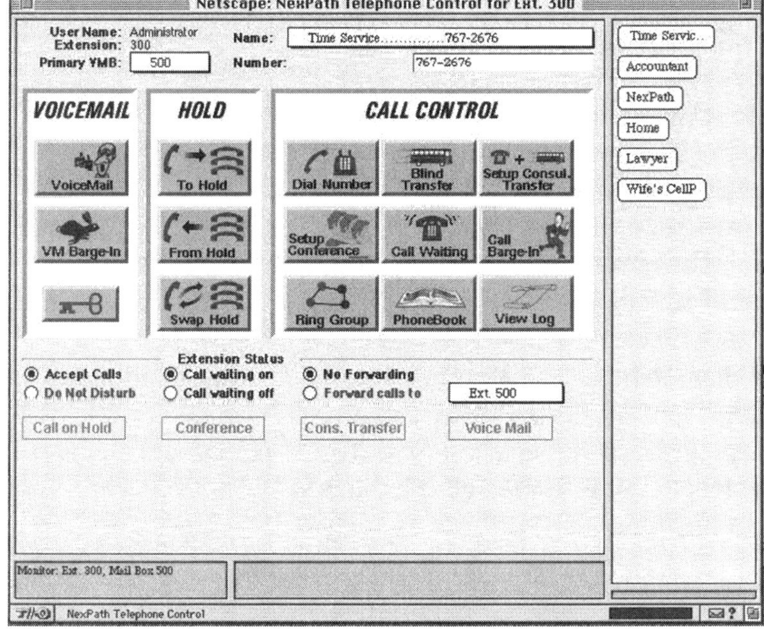

To gain access to the computer telephony integration features of the NexPath system, all one needs to do is launch a Web browser and aim it at the NexPath server (i.e. *pbx.phonezone.com*). After logging in with a user ID and password, the system displays the main screen. This Web / Java computer telephony interface provides call screening, call control and messaging services on Mac, Windows, UNIX and other Web-capable platforms. This is the Windows 95 version. Figure 8.3 shows the kind of control you get:

- Asynchronous display of incoming call information (Caller ID and name -- if present in the phone book);

- Ability to modify station settings (i.e. call waiting, do not disturb, call forwarding);

- Ability to place calls on hold, retrieve calls from hold, set up conference calls, transfer calls, etc.;

- Speed dial frequently called phone numbers;

- and Check voice mail messages.

NexPath has taken unified messaging a step further by integrating voice mail into the Web, making it easy for users, regardless of operating system, to retrieve voice messages via the Internet. This interface works equally well on the road or in the office. The NTS Telephony Server doubles as an in-house Intranet Web Server. It's equipped with an HTTP daemon, so you can point to it with your browser as if your were cruising the Web.

Rather than dumping voice-mail messages with their large audio file attachments into an e-mail inbox (as most vendors do), NexPath makes voice messages accessible via a Web forms interface as shown in figure 8.4. This lets the user selectively play messages, thus minimizing network usage and download times.

Figure 8.4 – Netscape Basic Audio Player for Voice Mail

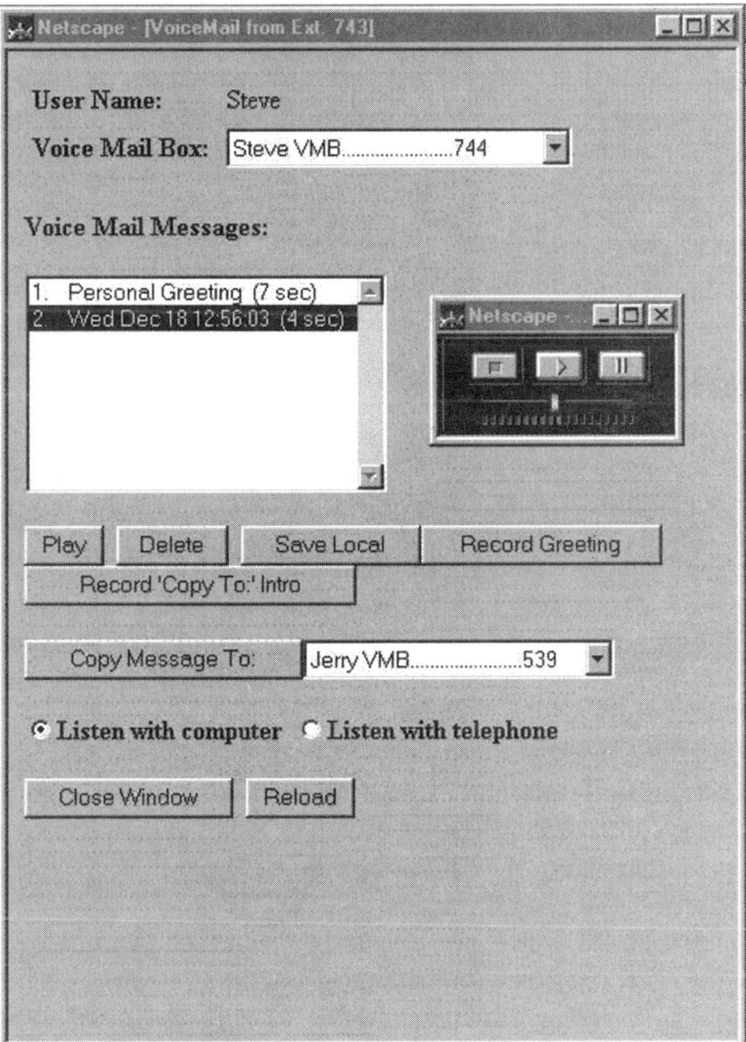

The beauty of NexPath's approach is that client side software has already been developed for virtually every platform imaginable in the form of Java-capable Web browsers. Here's a quick run down on where things stand in relationship with OSs and available client software support.

DOS. Again, this is the one place where there's no support.

Windows 3.x. Use Netscape Navigator. Netscape is currently beta-testing Java on Win3.x, but does not expect a product release Java-enabled browser until the release of Netscape Navigator 4.0.

Windows 95. Use Netscape Navigator or Internet Explorer or Sun Microsystem's HotJava. Netscape Navigator provides excellent Java support; so I recommend that browser over Internet Explorer.

Windows NT. Use Netscape Navigator or Internet Explorer or Sun Microsystem's HotJava; see previous recommendation.

MacIntosh. Use Netscape Navigator. It rules on the Mac.

NeXT. Use NextStep Browser. NeXT is heavily into the Internet and Java application development through their Web Objects initiative.

UNIX. Use Netscape Navigator or Sun Microsystems HotJava or OS-specific browsers. Depending on which version of UNIX you're running, you may want to run either Netscape Navigator or Sun's HotJava browser. As a general rule of thumb, use Sun's browser on Solaris platforms and use Netscape's browser on other UNIX variants.

Web Terminals. These may become popular as low-cost data entry terminals (i.e. in call centers). Ask your Web terminal vendor if their products support Java. If they do, they'll support Java CT.

This ability to handle different clients is a neat idea from NexPath. With the Web and Java, you write your app once and it runs on just about everything. Talk about reducing the cost of software development. One program, dozens of platforms, no installation headaches.

NexPath is sold as a pre-assembled unit (8 to 48 ports) which is ready to attach to outside phone lines and ordinary analog phones. The server is configured entirely via a LAN or WAN using a browser. The only configuration done at the server itself is to tell the server what IP address (Internet equivalent of a telephone number) to use.

This differs substantially from other PBX platforms (both PC-based and traditional PBX systems) in that the customer uses a familiar World Wide Web interface to configure the system instead of the vendor's own proprietary configuration programs. This is a relief for system administrators, who tire of learning new interfaces for the variety of platforms they must support on behalf of their users. Even the Phone Log (Station Message Detail Recording – SMDR) applet is accessible via the Browser interface. Figure 8.5 is a Web form display showing recent activity on an extension. Better than hunching-over a character-based console in a crowded equipment room.

Figure 8.5 – NTS Telephony Server Administrator Interface

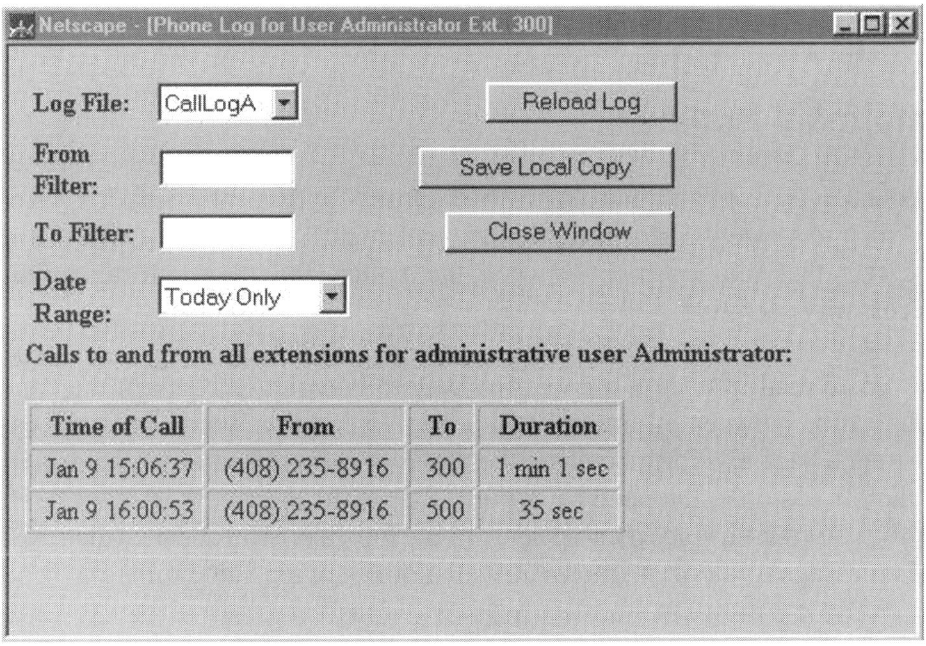

Netscape - [Phone Log for User Administrator Ext. 300]

Log File: CallLogA

Reload Log

From Filter:

Save Local Copy

To Filter:

Close Window

Date Range: Today Only

Calls to and from all extensions for administrative user Administrator:

Time of Call	From	To	Duration
Jan 9 15:06:37	(408) 235-8916	300	1 min 1 sec
Jan 9 16:00:53	(408) 235-8916	500	35 sec

NexPath's server is UNIX. Since the customer configures the system via the Web, there's no need for the customer to know anything about UNIX. What about client software?; that's going to cost a couple hundred bucks a seat, right? Nope. It's free, or almost free, depending on whose Web browser you use. Your browser is already paid for. $3,895 for a 16-port system; that includes the PC. My thanks to Brian McConnell for this report.

Corporate Web-Based Messaging

Enhanced Systems, Inc. of Norcross, GA sells unified messaging products such as Hello!, Hello!NT. These handle voice mail and auto attendant. Enhanced also sells complex IVR systems and Web-based unified messaging. The companies' latest technology allows unified messaging over the Internet. The product includes global unified messaging. You can attach voice messaging and other forms of messaging to the Internet including faxes. Instead of having a client solution, it runs on a server. As long as you have a browser, you can access your messages.

Crime Prevention

Informant Hotline

Police detectives and patrolmen work closely with informants. It is often difficult to locate informants to arrange meetings. This difficulty arises from the fact that relationships between the police and their informants are necessarily secretive.

A voice mail system can go a long way to ensuring privacy in the communication between informants and police. In addition, voice messages can be traded back and forth between the parties in a rapid and accurate fashion. Take for example, the need for a policeman to locate a certain suspect. He or she can create a voice message describing the individual and then distribute this message to a dozen informants' mailboxes at the same time.

If informants have a beeper, the system can be programmed to automatically page them. The users of the system can then retrieve the message and instantly reply to the request within minutes. It is therefore possible for police to retrieve tips and other information from dozens of informants within minutes of requesting the information. This could not be achieved otherwise, unless entire teams were dispatched to individually contact the informants on a one-by-one basis.

Parole Reporting

A special relationship exists between parole officers and their charges. Parolees must meet with their sponsors at regular intervals and are also required to "check-in" from time to time so their activities can be monitored. Voice messaging offers a modern and efficient means to achieve this.

For example, each parolee can be assigned a mailbox on a special parole reporting system. The voice messages between the parole officer and a parolee can be used to convey information about whereabouts, job searches, or any problems encountered. If a parolee misses a scheduled appointment, the voice messaging system can be programmed to call the individual at certain telephone numbers or to page the parolee.

In addition, Caller ID or ANI technology can be used to confirm the whereabouts of parolees when they call into the system. If the number they are calling from is not the one indicated in their personal profile database, it may signal that a breech of parole limitations has been made.

Education

Homework Assignments on Voice / LAN

Combining voice and data traffic in any organization is difficult, but it is especially tough in a place where there haven't been any telephones or computers to begin with: the public school classroom.

Yet, at the Stoughton Area School District in Stoughton, Wisconsin, voice and data traffic already are running over a common fiber optic network that interconnects classrooms in multiple buildings over a wide area. And a raft of new educational applications and opportunities has been born. As part of a referendum to overhaul and expand its communications systems, Stoughton has deployed a fiber-based wide area network that serves the classrooms and offices of its six schools, administration building and the local public library.

The network will support voice, data, video and Internet applications, giving students and teachers access to a wide array of information that was never available to them before.

"We're going to be able to do things that most people in the school district didn't even know could be done," said Bob Smiley, district technology director for the Stoughton Area School district. "And yet our telephone bill will actually drop by 30 to 35 percent."

The central switching element in the Stoughton network is NeVaDa (Networked Voice and Data), a system that combines today's typically separate voice and data infrastructures locally and across metropolitan areas using asynchronous transfer mode (ATM) technology. An essential component of the product is an ATM/SONET module that was jointly developed by Mitel Corporation, a maker of computer telephony integration solutions, and Madge Networks, a supplier of end-to-end switched networking systems.

Using ATM technology, the Mitel/Madge module provides the capacity to carry multimedia traffic across an enterprise via a single 155 Mbps fiber backbone. By combining "voiceLAN" and data switching technologies, Mitel and Madge have produced a high-speed network "pipe" that gives users access to both voice and data technologies at the desktop.

More than two years ago, Stoughton Schools began seeking an upgrade solution for a mixed bag of non-interconnected token ring and Ethernet LANs that had been installed separately at the schools and at the district's administrative building. The goal of the new solution was to interconnect the buildings and to expand the applications available to students in computer labs.

At the same time, Stoughton Schools was evaluating replacement solutions for its older telephone system and centrex service provided by Ameritech, the regional Bell operating company in Wisconsin. Capacity had already been reached at two of the district's buildings, so an upgrade of the telephone system-as well as a data network overhaul-was critically needed.

Initially, Stoughton Schools pursued separate approaches for its voice and data requirements. “We weren’t initially looking for a voice solution when we talked to Ameritech about fiber," said Smiley. "They had promised to extend fiber to the schools to support data, so we pursued it from that direction. But as we looked at the possibilities, it was clear that combining the traffic opened up a lot of new ideas."

Stoughton Schools considered a few different voice/data alternatives, including Integrated Services Digital Network services from Ameritech. But ISDN was quickly dismissed, because basic rate interface services were too slow to support the applications and were not available in the school district’s service area. High-speed local data services - such as Tl and T3 services - created too many recurring costs for the public school system.

NeVaDa turned out to be the right solution for Stoughton’s needs. "NeVaDa provided us with the opportunity to leapfrog forward with the technology we needed to support our voice and data applications plans, all within a single infrastructure. By combining what were two distinctly separate operations, we were able to simplify our network and realize cost savings in the management of our overall network."

Stoughton has installed an 18-strand hybrid fiber (12 single-mode and six multi-mode) to each site in the school district. Operating at SONET speeds of 155 Mbps between sites, the network provides very high bandwidth capabilities that enable the schools to transmit a large amount of data while maintaining a measure of guaranteed bandwidth (16 Mbps) to support high-quality voice traffic. And because they run over the same network, Stoughton can carry both voice and data traffic from building to building and then split them into separate lines for service to each classroom.

At the corc of thc combined network is the Madge Multinet switching hub (LET-36 model), a chassis product that can support both hub and switch functions, at the Stoughton Area School District’s central administrative building.

Smaller Madge hub models (LET-20s) reside at each school and at the local public library, enabling LAN- attached end stations at each site to communicate with each other across LAN boundaries. A Cisco router provides routing services between LANs.

However, most of the "heavy lifting" is handled by the LBT-155, a NeVaDa interface card that resides in the Madge chassis and provides ATM switching between sites at SONET speeds. With NeVaDa, the schools can exchange data and files with the library, the central administration building, and with each other - all at speeds more than ten times as fast as typical WAN services such as frame relay.

Stoughton also has attached its telephone switch - the Mitel SX-2000 LIGHT distributed platform - to the NeVaDa system, enabling it to handle voice calls over the same high-speed fiber that it uses for data. Calls to and from classrooms and administrative offices are routed across the fiber to a common switch, eliminating the need for separate telephone switches in each building and enabling Stoughton to reduce its telephone service costs.

"The new phone system also is giving us very high call quality," Smiley said. "We're still testing the lines, but so far our decibel loss level is about 0.5. And the phone system is much easier to manage - we think we can manage it in about one hour a week, once the initial setup is worked through."

The NeVaDa solution was implemented by the Wisconsin office of Mitel Telecommunications Systems, Inc., the direct sales and service arm of Mitel Inc.

While Stoughton Schools is pleased with its ability to reduce service, equipment and management costs with the new network, officials are more excited by its ability to support new applications for students.

In its five-year information and telecommunications strategic plan, Stoughton Schools has outlined an ambitious plan for voice, data, video and Internet applications. An entire curriculum is being developed to take advantage of the new infrastructure.

On the voice side, classrooms will be able to send and receive voice mail via a common system. An Octel voice processing unit will enable students and parents to call into the voice network for homework assignments on any given day. Caller identification capabilities will be provided for incoming calls, and outgoing call activity will be monitored to prevent any possible abuse of the system. By 1998, the system will even automatically call the parents of absent students to confirm that students are safe at home.

Data applications offer even more potential. Students will be able to exchange electronic mail from building to building, and exchange files with the library or other school district resources. The network also will provide access to CD-ROM towers in several of the buildings, enabling students and teachers to share information resources across town. Students will be able to directly access the public library's card catalog or resources available in the central Career Center. Direct desktop access to the student database and the district's financial database is scheduled for the end of 1998.

Naturally, Stoughton Schools is anxious to give its students access to the many information resources available on the Internet. This year, all of the Stoughton District's students and staff will be given Internet e-mail and access to the World Wide Web. Next year, file transfers will be possible via the Internet File Transfer Protocol (FTP). Security will be provided through Novell 4.1's Encrypted Bindery, an Internet firewall, and virus control, which will take place at each server.

Video applications also are in the works for the Stoughton network. By next year, distributed video capabilities will be provided between buildings and distributed satellite feeds will be available over the same network. By the end of 1998, the school district hopes to provide live video feeds to the local cable network, and perhaps even broadcast- quality, two-way interactive video to support distance learning and home education services.

"It was important to us that this network be able to support not only new applications that we can envision, but new applications that we haven't even thought of yet," Smiley said.

"We believe that what we're doing here will not only give us new services and capabilities for students in the short run, but it will provide a strong platform for implementing new applications even beyond that five-year window. That's important not only for the students we have now, but for the ones who will be coming into our schools in the future."

School Board Messaging

Some school boards are geographically scattered, and it's difficult for meetings, votes, and other information to be communicated easily. For example, the Dufferin-Peel School Board in Ontario, Canada, serves 2,000 French schools. The district needed a way to trade messages between school board members, teachers, and parents without having to hire additional administrative workers.

The Dufferin-Peel School Board worked with Info Systems, of Downsview, Ontario, to develop a custom voice processing system to solve the problem. Info Systems' Talkie is a voice processing platform with many modules. With it, you can do IVR, fax, international call back and messaging. Parents, board members, and teachers are now able to broadcast messages, respond to messages, and answer questions in both French and English. Additionally, the system provides live conferencing and allows for the casting of votes. Parents can also request fax documents including registration forms from various colleges and universities.

Electronic Media

News Tipsters

Television stations, newspapers, and radio stations encourage viewers and listeners to "donate" their time by "tipping" reporters on neighborhood news and other events. Everyday citizens bring much of what is reported by newscasters to their attention.

It's no wonder that many stations are now offering voice messaging as a means to relay late-breaking news stories and other information.

Reporters and copy writers can pick up mail in a series of mailboxes according to their "beat." These mailboxes are assigned by subject matter, making it easier to report information. For example, a news tip hotline will answer the phone and say to callers: "Thank you for calling the KYW-TV tipster hotline. To let us know about a fire, press 1. Reports on neighborhood events, burglary, and theft can be left by pressing 2. Press 3 to let us now about emergencies. By pressing 4, you can tell us about human interest stories, community announcements, or other newsworthy information."

Electronic Media Radio Broadcast Retrieval

A Web site located in France (www.francelink.com) collects unedited recordings of radio broadcasts from the leading programs in that country, allowing fans the opportunity to listen in remotely. But what if your home communications system is limited to a .7-watt phone speaker? Jim Chiu of Catalyst Group in La Mirada, CA has an answer. Catalyst downloads these recordings and distributes them via the Internet to various Pronexus (Carp, Ontario) VBVoice servers around the U.S. Catalyst has provided radio enthusiasts a way to dial in to their favorite shows.

Traffic Reporting

Commuters count on timely reporting of traffic conditions in order to choose alternate routes or modes of transportation. Of course, the use of planes and helicopters for this function is unbeatable, but even aircraft can't be at all places at once. Voice messaging systems can provide an effective means for gathering and distributing information about traffic conditions. A news station, telephone company, or cellular service provider can easily employ voice messaging to provide traffic reports.

By using a network of "spotters," the voice messaging system gathers information from each area and then provides access to any caller. In the case of a news station, the reporters can call into the system and check on the status of certain mailboxes in order to piece together a traffic report. Users can dial individual access codes for the area they are interested in. This type of system was first successfully deployed by Cellular One for commuters on Long Island, New York. Systems like this receive thousands of calls each morning and afternoon during rush hour.

Entertainment & Sports

Conference Bridge

The 976 and 900 service provider industry gave rise to the popularity of conference bridges. Examples include conference bridges for teenage gossip, adult-oriented subjects, and "dial-a-friend" hotlines. With a conference bridge system, users call a telephone number and enter codes to "jump" into a live conference call with other people. An operator who has the ability to limit the number of conferees usually monitors these conversations. He or she can "expel" people from the conference who are too loud or obnoxious.

Voice messaging can be used to enhance the usability of these conference bridges by allowing callers to leave non-real time messages for one another. Callers can dial a code to jump out of a conference and then be prompted to dial a mailbox number to leave a message for someone who may not be actively speaking. In this fashion, users of the system can log-on to the conference bridge in order to speak with each other "live."

Dating Service

There's many ways for single people to meet one another. For example, singles clubs, dances, and social events managed by church groups, community organizations, and schools. In addition, there are numerous dating services that use video tapes, catalogs and newsletters to distribute information about "unattached" people.

Special voice messaging systems have been developed to help singles "meet" one another in a convenient, low-cost, and safe manner. One of the first voice messaging services for singles was developed by Gralin Associates of Fountainville, PA for a local information provider.

The "Dial-A-Date" system allowed callers to choose from a variety of public and private messaging bins. If a caller wanted to review public messages, he or she would select from a menu of male-oriented versus female-oriented general announcements. These announcements either included the person's phone number or a mailbox number. Systems like this automatically assign a temporary mailbox number and security code for users who don't want their home phone number publicized. In this fashion, messages reviewed by other callers can be responded to and a non-real time dialogue can continue between the parties.

Psychic Hotline

A look into the future, advice on life, love, and business are among the many topics brought up with psychics. Millions of people worldwide seek advice on these topics. Psychics are sought at an increasing rate due to easy access afforded by conference bridges and messaging systems.Psychic hotlines allow advisors to record messages on a variety of subjects. These subjects can be general in nature, or they can pertain to individual constituents of the service. For example, one psychic can record messages having to do with a subscriber's profile and leave messages on a daily basis, or on demand. Requests for advice can be left in the morning by a caller and then picked-up and responded to by the psychic during the day. In this fashion, an advisor can listen to and respond to hundreds of messages and provide "answers" for many subscribers in a single day.

Rodeo Events

It's difficult to handle information requests for attendees, volunteers, and participants both before and during these sporting events. In a sense, rodeos are like a trade show or convention.

It's typical for only a handful of workers to manage the communications for these events. A voice mail system has a number of applications for rodeos including messaging, Audiotex-based information delivery, and reservation confirmations, to name a few.

Pacific States Communication (PSC) has experience with installing systems just like this. The company is a reseller for the Smooth Operator product manufactured by Octel of Milpitas, CA. At one customer site, a Smooth Operator system handled over 40,000 calls in a 75-day period. The system is used to coordinate daily event information, provide information on tickets, and distribute reservation information on some 720 horse stalls for the participants' use.

About half of the calls were received in a three week period around the time of the rodeo itself. During one eleven hour period, the system logged over 1,600 calls. Anyone can image the relief of the rodeo staff after having the system installed. Before using voice mail, many calls would simply go unanswered.

Financial

Cash and Treasury Automation

Louba Rapoport is a principal with Alterna Technologies Group, a treasury automation consulting and software development company in Calgary, Alberta. Alterna developed a wonderful cash management system called Midas. It uses voice processing technology manufactured by PIKA Technologies, Inc. of Kanata, Ontario.

I cajoled the company into a live demonstration on stage at Computer Telephony Demo Fall 96 in Orlando. Midas was a hit. I asked Louba to explain how voice processing can help to protect corporate liability and facilitate deal making:

"The modern corporate Treasury is facing information capture, management, and delivery challenges second to none. Volatile international money markets, geographically and technologically diverse business units, and highly mobile decision makers make it exceedingly difficult to develop, authorize, monitor and attune a cohesive corporate cash and risk management strategy."

According to Alterna, the collapse of a well-known British bank was attributed to the lack of automated tools and controls. Other well-publicized incidents have brought renewed scrutiny of risk management strategies from boards of directors, shareholders and executives alike.

The trading environment is especially susceptible to costly errors due to multiple layers of staff who are involved in the decision making, authorizing, making and processing of "deals." There is a real danger of missing critical information in the avalanche of data that is constantly flowing in and out of treasuries. This problem is further compounded by the highly distributed and mobile nature of the work force in this business and the fluidity and volatility of the markets.

Business realities today demand systems that are accessible from anywhere in the world. Alterna's products feature location independence. This means access to critical information to those who need it, from anywhere in the world. This can be done at any time using the interface of their choice. Alterna's suite of products offer alternative independent means of accessing and delivering information.

Midas is based on Alterna's Electra Computer Telephony tools. The system integrates Intranet, and database technologies to effectively manage the complexities of the treasury business. As pictured in figure 8.6, Midas captures and delivers critical financial information to decision makers responsible for corporate cash assets and successful execution of corporate risk management strategies.

Figure 8.6 – Midas Treasury Management System

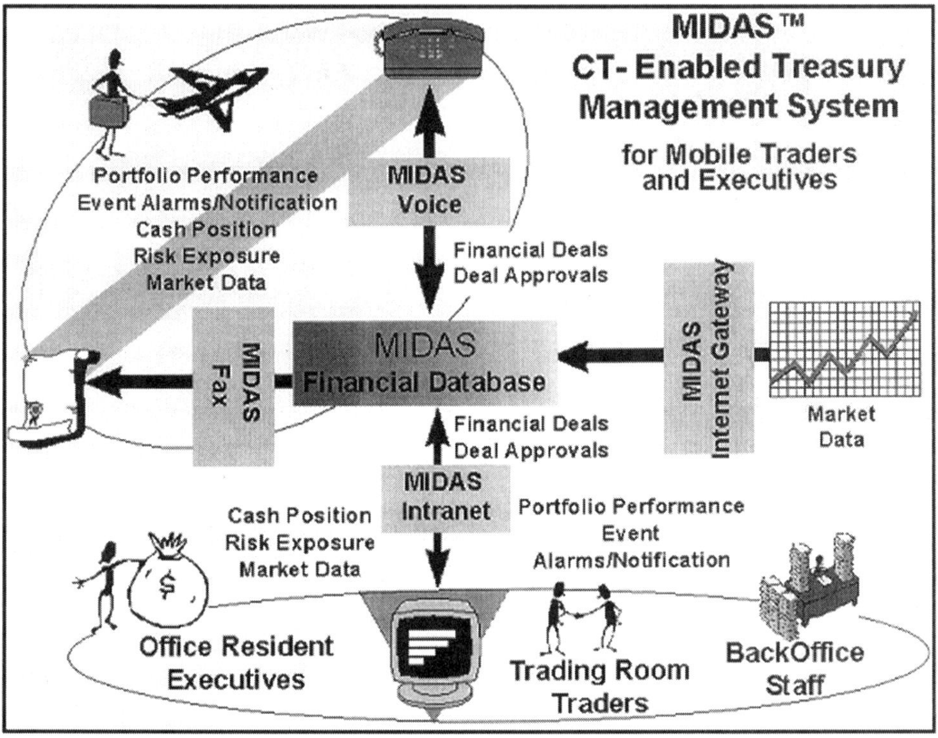

Midas captures and integrates financial data from internal and external networks through the telephone and desktop computers, then uses computer screens, printed or faxed reports, and automated voice delivery over the telephone as mechanisms to capture and deliver critical financial information.

Here's the sample scenario Alterna demonstrated. It is based on the premise that corporations want to combat the "rogue trader syndrome" in the high pressure, high stakes and high-risk environment of cash management & trading. The demo focused on a "slice" of the deal-making process in the trading room of a large corporation. Additionally, the demo showed how Midas can provide mobile executives access to crucial information while away from the office.

1. First thing in the morning (i.e. 7:00 AM) deliver information to staff:
A) Markets have opened in London - lots of volatility.
B) Read some prices and foreign exchange rates.
C) Lots of activity expected due to revenue inflow the day before.

2. In the trading room:
A) Deliver voice message from head trader advising of strategy and activities for the day.
B) Cash position notification / alert executive / inform cash manager / notify trader.
C) Deal making / call up counterparties-obtain quotes / capture information during conversation - during deal inform trader that he is at point of breaching exposure with one of the counterparties.
D) Settle deal with counterparty / inform banks of payments, and send executives summary of executed deals.
E) Broadcast risk exposure / portfolio details to traders and management

3. Executive (CFO) stuck at the airport:
A) Had to adjust agenda and fly to New York to meet with important potential (but jittery) investors.
B) Needs to provide reassurances that company XYZ has a firm handle on trading activity.
C) Calls in to obtain activity and exposure assessment.
D) Schedules a callback into the meeting with investors for delivery of up to date activity and exposure information.
E) While on the phone, Electra reminds the CFO that his sons birthday is the following day.
F) Electra asks whether he would like to initiate any other calls or schedule other activities during this session.

4. Finale:

A) Trader needs VP approval for large deal, calls VP but reaches secretary instead who asks the trader to call Electra.

B) Electra informs trader that VP is at a client site, and proceeds to track VP down.
C) Electra reaches VP who is just lining up a putt on a golf course.
D) VP answers cell phone and approves deal.

Residential Mortgage Messaging

Mortgage originators experience the same communication delays and frustrations associated with a busy switchboard. One of the unique problems mortgage companies deal with, however, is the “peak interest” period of mortgage prospects. Most home buyers meet with their real estate agents after work. By the time their interest in a home is peaked and they are ready to talk about financing - the mortgage company is closed for business. Without the prospect being “booked,” the chances of them losing interest is very high.

Voice processing can help solve this problem by providing a “quick response” to financial applications.Business Communications Resources (a reseller for Octel of Milpitas, CA) worked closely with mortgage professionals in New England to devise a solution based on the Smooth Operator voice mail system.

At one mortgage company, they installed a 12-port voice mail system alongside an AT&T System 25 switch. The standard logic flow of the system was modified in order to assign territorial mailboxes for each mortgage agent who was “on call.”

Any real estate agent that calls the system after hours can indicate where they are calling from, and the system will automatically page or call the on-duty agent and deliver the message. This solution extends the service hours of the mortgage company and helps to increase the closing ratios for the “busiest” part of the day.

Government

Assessor's Office

County assessors spend most of their time dealing with the review of property values, making site visits, and working closely with homeowners and other government offices. Voice messaging can make the assessor's office run more smoothly by relating messages and appointment schedules while the assessor is in the field.

For example, let's say that the assessor has four appointments scheduled in the afternoon. If these appointments are in different areas of the county, the drive from one place to another could take as few as several minutes to more than an hour. If appointments need to be rescheduled, either the clerks at the office or the homeowners themselves can record and post messages directly to the assessor's mailbox.

By using a cellular phone or payphone, the assessor can check his or her voice mail at the end of each appointment to confirm the next visit. This can also be helpful when coordinating site visits while permits are in process.

Business Violations

Unfair trade practices and consumer fraud often result in the filling-out of forms and making a variety of phone calls to speak with government officials. This requires a lot of legwork on the part of consumers and careful follow-up by investigators. A special voice processing system can help to ease the flow of communication for business violation complaints.A special 800 number can be published to inform companies and individual consumers of a "Business Report Card Hotline."

This hotline can provide a menu for callers to select the type of complaint they have. Complaints can be recorded in a conversational fashion, so that the system asks for specific questions regarding the violation, making recordings easier to transcribe.

For example, the system might say: "Please state your full name and then spell it at the tone... Now please state your address and phone number at the tone... Please state the name of the business in question... Please record your complaint...." Administrators of the system will then be able to transcribe the messages they are assigned to follow-up on. As an option, callers with complaints can be offered temporary mailboxes on the system, so they can call to hear about the disposition of a complaint they lodged.

Child Support Hotline

Enforcement agencies that handle child support cases are under an ever-increasing strain to handle delinquent payments, disputes, and pending court actions. With budget cuts and downsizing, these agencies can barely keep up with parent and court communications. Voice processing systems are being installed to solve many of these problems with mainframe-connected IVR, fax-on-demand, and voice mail capabilities. For example, United Companies, Inc. (UNICOM) of Anchorage, Alaska has designed a sophisticated telephone answering system called KIDS (Key Information Delivery System). Based on the Wygant Scientific (Portland, OR) Micro-ITC platform, KIDS enables custodial parents to report on late child support payments with PIN numbers.

Parents can also leave detailed messages for follow-up by agency employees. Sacramento, California's child support agency worked with UNICOM to develop a system to help answer the calls workers were receiving. According to agency chief Dick Williams, the callers would experience busy signals for as many 6 of the eight hours the center is in operation during the day. After the installation of the KIDS system, more than 50,000 calls each month are handled efficiently alongside the 13-person call center staff.The Sacramento system is programmed so that the most frequently asked questions are answered with a series of scripts. This cuts down the live handling of "information only" calls and enables the staff to return messages within 24 hours of their receipt. As a result of this faster response, the number of face-to-face visits made to the office has dropped by two-thirds.

Health Care

Health Care Medical Dictation

Video Voice & Data, Inc. of New York, NY is a Parity Software VAR. The company has also used Technically Speaking's Show N Tel (apps gen) and the Demosource Vicki (voice mail) products.

On of their biggest sales was in Beverly Hills, CA to a Medical Dictation service. $150,000. The owner wanted to turn around the Doctor's transcriptions fast and cheap. He was paying $15 an hour to local transcriptionists. He was not competitive.

Video Voice & Data developed a centralized medical dictation and transcription platform based on Show N Tel. They used a Novell File server, a Micom Marathon LAN/WAN bridge, the IVR system and Delrina's Win-Fax-Pro for networks with four ports of fax as the fax server. They installed DID trunks so doctors could call into their own (personalized) dictation "inbox."

Doctors input the patient number, status, and then the verbal dictation. Both the digitized voice files and status codes are then transmitted over a 56K data link to a transcription house in the Philippines. The transcriptionists use the PC Voice (Paoli, PA) foot switch to pause the recordings and type-up the text from the doctors' messages.

All of the text and associated patient codes are then sent back over the 56K line to the main system. The system automatically fulfills the completed documents by faxing the transcriptions to the doctors with the fax server.

Medical Messaging

Doctors and other health professionals have been long-time users of live answering services and special paging systems.

The ability to have a real person intervene on behalf of a patient can be critical, especially for emergency situations. Adding an adjunct voice messaging and paging capability to any live answering service can significantly enhance the screening and relaying of messages.

An operator can access each doctor's profile. He or she can then determine how to treat each call. For example, emergency messages require the operator to place outgoing calls manually, on behalf of the caller. Personal calls can be automatically transferred to a voice mailbox. Non-emergency callers can do likewise, but with the option of message or paging delivery.

When a doctor gets "beeped," he or she knows the nature of a call by viewing the display on the pager. If it is a less important message, no visual indication will be required on the pager. Another popular use of messaging for doctors is dictation. With a voice processing-based dictation system, health workers can access house phones in order to record patient status information.

Transcriptionists, who type-up the messages into medical reports on behalf of the doctor, then access these special mailboxes. Many of these systems are installed worldwide by companies like Dictaphone, and Harris Lanier, to name a few.

Medical Paging

Medical technicians, nurses, and doctors can use a combination of voice messaging and paging devices to ensure the accurate and timely delivery of emergency messages. In this case emergencies are categorized by different codes, which are then linked to alphanumeric dialing sequences in a special voice messaging system.

Dispatchers, secretaries, and messaging operators can record emergency instructions for an individual and then enter a series of codes that identify the emergency. Immediately upon posting this message an attempt is madc to activate the recipient's beeper.

The codes appearing on the beeper can be automatically transmitted by the voice processing system. Once the recipient has access to a phone, more detailed instructions can be provided with a recorded message. This is helpful if emergency lines are busy.

Hospitality & Travel

Convention Message Board

Trade shows and conventions attract thousands of visitors and exhibitors. The message board at these events is a busy hub of communication. The types of messages posted include notes to call home, contact customers, or get a hold of someone at "the office."

Unfortunately, many of these messages are lost or never picked-up. These problems usually occur because recipients of these messages have to physically visit the message board to see if they have any communications from the outside world. A voice mail system can be used to assign temporary mailboxes to both attendees and exhibitors to avoid these problems.

As part of the pre-registration or site registration procedure, each conventioneer can be assigned a mailbox number. The instructions for using the system can be put inside the seminar pamphlets or other "care package" material. In fact, if a voice mail provider is sponsoring the system, the instructions for using the system can be printed on that vendor's collateral.

It is typical for telephones to be hooked-up at every trade show booth, since it is difficult for exhibitors to get away from the booth during show hours. These phones can be used to access the messaging system by the folks running the booth, and for visitors in the booth as well. If the phones are equipped with message waiting lights, it is feasible to illuminate them when messages come in. In addition, the phones can be automatically called if the messaging system is programmed for outdial notification.

Voice mail on the trade show floor can also be used to streamline the communication between trade show coordinators, plant providers, and shippers. For example, a broadcast message can be sent to a group of mailboxes associated with an area of booths that will be serviced next during set-up or tear-down. This eliminates the need for many exhibition employees to run around the show floor to coordinate these activities.

Travel-Buddy Hotline

Travel agencies can sponsor a special hotline for use in arranging special tours and "buddy packages." From time to time, special deals become available for travelers who can pack up and go at a moment's notice. More often than not, the travel agent will call his or her customer to let them know about these deals, but contacting everyone in time is difficult.

To alleviate this problem, customers who are interested in special travel packages can be assigned a voice mailbox on the travel buddy system. Agents can record a message about a certain trip, and ask customers to respond if they want to go on the trip. The administrator of the system can create a list of travelers who are interested so they can pair-up couples for room and seating assignments. These mailboxes can make arrangements for the trip, discuss preferences, and get acquainted.

Human Resources

Human Resources Executive Recruiting

AltiGen Communications of Fremont, CA is proud of their AltiServ. It's an "All-in-One" multi-application platform environment. With AltiServ, connecting multiple computers and telephone systems together with communication links is a thing of the past. AltiServ platforms do basic PBX switching, automated attendant, voice messaging and automatic call distribution. You can forward urgent messages to a user-specificd pager.

You also get "Dynamic Messaging" to do "special delivery" messages on an ad hoc basis for individual callers. One useful goodie is the "Boomerang," allowing users to automatically call the person who left the message. The system can even initiate calls to multiple numbers so important callers can get through a la "One-Number-Access." Under $4000 for a 4-trunk-by-8-station starter kit gets you going.

AltiGen's bundled AltiWare and TAPI apps make all this possible. At the hardware level, you need an NT-based server and AltiGen's Quantum boards. The boards are full-length ISA components. They handle all the basic PBX and voice-processing duties. According to Gary Andresen, AltiGen's VP of Marketing, one executive recruiting firm can't live without voice mail. They're constantly on the go. They can automatically return voice-mail messages via Boomerang and have important callers forwarded to them by AltiServ automatically anywhere in the world.

Insurance

Insurance Transaction Archives

Insurance companies use a variety of technologies to help run their businesses with efficiency. For example, automated claims service, premium quotations, and forms by fax to name a few. Voice mail can also be used to develop an audio archiving system for insured parties and personnel who provide detailed interviews over the phone.Claims adjusters and investigators, for example, depend on messaging systems to record interviews, conduct investigations, and forward their findings to other employees.

When a claims adjuster or investigator wishes to make a recording with a customer, he or she can "conference" the call into the voice messaging system and follow the prompts to leave a message in a claim mailbox. With three-way calling enabled, conversations can be recorded. Due to the fact that the conversation can be stored on a computer, it is easy to forward recordings to other personnel to get advice on claims.

Legal

Legal Expert Conferencing

Judges and other court officials sometimes use the telephone to get testimonials and other information from subject matter experts and witnesses. The use of telephones can cut down on rescheduling of court dates, unnecessary travel, and improve on the availability of expert witnesses. By using a conferencing system, along with a voice mail system, the coordination and scheduling of expert testimony can be made simple.For example, a court clerk, attorney, or judge can provide written notice to a group of experts (and alternate experts) they are needed to give a deposition or participate in a conference call as part of a court proceeding. This notice will include instructions for these people to "log on" to the voice mail system at a pre-determined date and time.

A general message can be broadcast with further instructions on how to join the conference call, or messages regarding the delay of it can be picked-up. Once the dialogue has been established, the court agents can use the messaging system to inform alternate witnesses that they will be needed due to a schedule conflict or some other problem. If the conferencing system is integrated with the messaging system, conferees can be added-on to a scheduled conference when they enter their mailbox number.

Manufacturing

Subcontractor Scheduling

The management of subcontractors is a major job for most companies. Take an automobile industry subcontractor who assembles engines, stereo equipment, and seats. In order to avoid cost overruns and delays, voice processing can be used to coordinate delivery schedules, locate subcontractors in case of emergency, and to forecast orders for the next production run

.Subcontractors can use general mailboxes to receive messages from the production supervisor. Mailboxes can be set up by company, person or on a project basis. In addition, subcontractors can use project codes associated with certain mailboxes to report on the status of jobs.

Special mailboxes can be programmed for forecast and status, another one for cost adjustments, and one for personnel-related issues.Another example is offered by Creative Voice Systems (CVS), a reseller of the Smooth Operator system from Octel (Milpitas, CA). According to the company, one of their manufacturing clients has a 40,000 square foot facility. The client receives over 900 calls a day, most of which for managers walking the plant floor.

Creative Voice Systems installed the messaging system alongside an Inter-Tel GMX-48 switch and achieved a tight coupling of the systems through intercom paths in the PBX. By using the "Intercom Redirect" feature, callers can be automatically placed on hold, while the automated attendant pages the called party over the intercom. With the Inter-Tel switch, the called party can redirect the phone to the nearest extension, so no calls are missed. Considering that this does not require an operator, CVS clients are thrilled with the efficiency gains.

Marine

Marine Vessel Trader

Voice messaging can be used to provide hundreds of boat owners, buyers, brokers, and sellers to communicate with one another and to bargain for trades. One of the difficulties in buying or selling boats is first locating them, and making arrangements to meet with the owner or broker. To make vessel location easier, a radio paging system can be used so automatic outdials occur when inquiries are made on a certain boat. With each vessel having its own mailbox, a broker can trade messages with dozens of prospects in one hour - much more than would be possible by one person over the phone.

A system used for this purpose would have a main menu that sounds like this: "To get more detail on a vessel you are interested in, please type-in the vessel code number. Press '1' if you know the owner or broker's name." If the caller wishes to search by broker name, another menu associated with that broker may offer several choices.

For example, the broker's name could be announced followed by a list of available vessels along with their codes. By dialing any of the codes, the caller can then hear a full description of that vessel and have the option of recording a message. Each vessel's mailbox can be programmed to page or outdial the broker, or alternately, may be programmed to call the owner.

Military

Military Base Messaging

In many ways, military bases are run like a large business with many offices. Due to the fact that there are dozens of outbuildings on a base, it is difficult to distribute messages quickly. The campus environment of most bases places physical constraints on those who do not have instant access to a phone.

Such was the case for the Elgin Air Force base in Fort Walton Beach, Florida. Lt. Phil Block, Senior Systems Administrator at the base, wanted to find a way to help 1,200 military and civilian personnel to communicate base-wide. Block installed an Active Voice Corp. (Seattle, WA) voice mail system.

Block used it to integrate desktop messaging service along with voice mail features. With the system in place, any user can view messages that are pending on their workstations, or call into the system from anywhere on the base. The system is used to communicate duties, arrange for leave, coordinate projects, and receive messages from family and friends.

Pet Industry

Lost Pet Animal Locator

Stellar Communications of Calabasas, CA has developed a unique (patent pending) lost pet locator system called the *Pet Lost and Finder*^{sm} *System.* Stellar says it's a product that could end the heartbreak of a lost pet.

The system includes two pet ID tags, and an identification and instruction card. Following purchase, customers attach the tag to their pet's collar. Users dial the 800 number on the instruction card, and follow the simple package instructions to securely record pet information and return instructions into a personal voice mailbox. Pet owners can record the pet's name and description, owner's name, contact information and data on the pet's health and vaccinations.

When a pet is found, the finder dials the 800 number from the pet tag and enters the 5-digit code number on the tag. The finder hears specific instructions about how to care for and return the pet. Pet finders can also leave voice messages in the pet owner's voice mailbox.

"Owners of lost pets know the agony of waiting for someone to phone when a pet is found," says Stellar Communications president Cheryl Elliott. "We weren't satisfied with any of the pet location or ID products on the market because they're too complicated, expensive, or they're useless when you travel with your pet. We developed the Pet Lost and Finder System as a better solution to the problem."

The Pet Lost and Finder System requires no mail-in registration forms, monthly or annual fees, so it can be activated immediately. It also links pet finders directly with pet owners via a 24-hour voice mail system, rather than relying on third party pet-finding services or animal control agencies. Finally, it allows pet owners to update information in their voice mailbox at anytime, so return instructions can easily be changed whenever the owner travels or relocates. All you need is a touch-tone phone.

A pet retailer can use the system for voice broadcasting sales and special events to customers of his/her store. Stellar can download the phone numbers and have the system call the customers automatically.

Stellar uses the Telephony Experts (Los Angeles, CA) Telepro 2.0 voice processing system, with a Dialogic card and battery back-up. The software runs on a 486/133 Pentium, with 32 MB of RAM.

Product Distribution

Home Party Sales

Many companies distribute through retail chains, wholesalers, catalogues and factory outlets. Perhaps the most communication intensive type of distribution is that of independent salespeople, or consultants. This is especially true of fast-growth companies like Artistic Impressions (AI) of Chicago, Illinois.

AI sells paintings and frames through a channel of consultants who arrange for neighborhood home party sales. In eight years, the company has grown from only eight consultants to a nationwide organization of over 1,0000 consultants.AI salespeople show the actual paintings at the home parties that are hosted by friends and family.

Once a client has chosen a painting, they can choose what type of frame they desire. The price range for the pieces is between $64 and $300. In the early years of AI sales, it was easy for Chicago-based consultants to drive over to the office to pick-up frames and make their deliveries.

This was not possible as the sales force grew into a nationwide organization. Eric Brackett, President of BTI Communications, helped AI solve the problem. BTI is a reseller for the Smooth Operator product manufactured by the Compass division of Octel in Milpitas, California.

BTI installed the voice mail system along with a Prostar switch. The system helps the majority of the nationwide consultants deal with the home office in ordering supplies, asking questions about inventory, and sharing messages on sales reports.

According to Brackett, the AI system makes excellent use of the auto-forward mechanism for subscriber messages. He says that if messages left for a customer service representative are not picked-up and listened to within 25 minutes, the system automatically alerts a manager.

This is a bonus for the field consultants, because they are assured that their requests are taken care of on a first-in/first-out basis.Bob Stodola, Director of Operations at AI summed-up his feelings on taking advantage of voice processing technology: "Around here, customer service is an income opportunity, not a cost center... the enthusiasm and the compliments from the field (consultants) tell me the whole story... With the Smooth Operator system, our expectations have been exceeded."

Test Marketing Feedback

Managers are forever looking for efficient ways to get feedback on newly launched or tested products. This is especially true of personal products distribution. Managers use demographic surveys, census reports, point of sale reports, and a variety of other intelligence to make decisions on product launches. On sure way to get un-filtered input from these test marketing initiatives is to offer a voice mail feedback system to consumers.

Users can call an 800 number to report on their likes and dislikes of certain products. The voice messaging system can be set-up to take messages on a variety of products. Each product manager would be in charge of maintaining the messages for all of his or her products on the system.

The marketing manager could also be equipped with special software to route calls. It also makes sense to use DID (Direct Inward Dial) or DNIS (Dialed Number Identification Service) in order to pre-identify the market that the customer is calling from.

It is then possible to assemble considerable feedback from several markets at once without having to worry about the information being "filtered" by an intermediary agent.

All of the messages can be transcribed in order to tally specific points of data. For example, a product manager may solicit callers to indicate their feelings on price, color, and smell of a certain product. By adding an IVR capability to such a system, the results of the test market survey could be automatically compiled by the system.

Wholesale Dealerships

Without tools such as hand-held terminals for delivery personnel, voice messaging, or pagers, the fulfillment of wholesale orders can take as long as two or three days. This was true of Doosan Beverage Company in Seoul, Korea. The company used a combination of hand-held terminals, a PBX, IVR, and voice mail to shave their delivery turn-around to less than one day. The owners of the company are very enthusiastic about the technology.

Locus Corporation of Seoul, Korea developed the system. As explained by Hyewon Lee, Associate Marketing Manager for Locus: "It was a critical point for Doosan Beverage to be able to immediately commit to orders from dealers. The TeleOrdering system we developed uses voice mail to make this possible."

Locus programmed the system so that each field salesperson's mailbox was set-up to activate pagers. In addition, employees in the field can use hand-held terminals in order to automate DTMF dial strings. These digits are captured by the IVR system for placing orders. These transactions use touch-tones to coordinate screen pops from the VAX computer over to a group of customer service agent terminals. The TeleOrdering system is also attached to workstations to relay voice messages and order instructions from the hand-held units.

According to K.S. Lee, Manager of Sales for Doosan Beverage, the Locus system has significantly increased the productivity of his staff.

Says Lee, "The new TeleOrdering system not only shortens the delivery times, but also saves on sales call time, and the cost associated with each customer visit."

Real Estate

Real Estate Maintenance Dispatch

Property management companies have a real challenge in scheduling maintenance and repairs for condominiums and apartments. The difficulty stems from the fact that special permission has to be sought in entering homes if the tenant is not present. In addition, it is not possible for maintenance personnel and contractors to always be available to the phone - especially if they are right in the middle of a repair.

The management company for the Sundance Apartment Complex (Citrus Heights, CA) has discovered a simple way to solve this problem. They use voice messaging to field maintenance calls for the Sundance complex.Tenants are asked to call a special number if they have a plumbing, electrical, or other problem with their unit. A simple menu allows the caller to choose the correct complex. Users are then asked to leave a detailed description of the nature of the problem, the unit number, and permission to enter the apartment.

The voice mail software is programmed to immediately alert maintenance personnel, who then retrieve the messages and schedule a maintenance visit. The system provides the property management employees a powerful means to not only log the times of a request, but also have recorded proof of authorization to do the work. By using this system, the maintenance personnel are able to handle more than one complex, and they are not required to publish their home telephone numbers.

Retail Services

Complaint Hotline

Fielding complaints is a special problem for many retail outlets, especially nationwide chains. This is true because local managers are often young and inexperienced, thus they do not always have the maturity or authority to handle every complaint. One way to solve this problem is by centralizing a nationwide complaint line that is manned by several experienced managers in the corporate office.

This type of hotline can be promoted with an 800 number that is printed directly on receipts, bags, and signs in the store. Callers are asked to select from a menu of mailboxes according to the store location. Dozens of complaints can be taken in the form of a voice message, and then listened to by management personnel.

By appending comments to the complaint message, the home office staff can then forward the complaint to the store manager's mailbox for quick resolution. The turn-around time on complaint resolution can be shortened significantly this way because paperwork and time zone problems are eliminated.

Computer Chain Message Access

AltiGen Communications of Fremont, CA has a computer chain store customer. This customer has numerous employees sharing the same desktops. They use the log-in/out and virtual mailbox. It lets you have as many mailboxes as you wish. The boxes don't have to be associated with station lines on the system.

With the AltiServ (PC-based PBX and Messaging System), employees receive messages whether they are at work or not. Allowing them to forward passwords. When users arrive at work they can log-in at any telephone. They use the same phone until they log-out.

All services associated with the user "follow" the person to the location of the logged-in phone. In the early eighties, Northern telecom called this automatic set relocation. It was part of the business generic for the SL-1 (now Meridian) PBX.

Social Services

Animal Control

County animal control agencies can benefit greatly from the use of voice messaging. Due to the fact that calls come into these agencies for a number of reasons, the treatment of each call can be prioritized.

For example, informational calls regarding animals available for adoption can be serviced with a separate "adoption menu." Regular business calls into the facility can be processed with an automated attendant, and agency workers can file incident reports for transcription by other workers. In addition, a "Lost Pet" section can provide individual descriptions of animals that are rescued or captured by the workers.

A system similar to this is in use by the Sacramento Area SPCA. DaviSoft of Citrus Heights, CA installed the system. Ken Davis, President of the company worked closely with Telephone Response Technologies, Inc. (TRT) of Roseville, CA and one of their developers (IPEX Software Services, Victoria, Australia) to design the PC-based system. Based on IPEX VOCAL voice messaging software, the system was assembled using Dialogic speech processing cards and a 386 PC. Davis customized the system to provide message waiting integration for the SPCA's Panasonic PBX.

Food Distribution

Social service agencies can also benefit from the use of voice processing to distribute food to needy citizens. The coordination of food donations, deliveries, and events is difficult for these agencies.

They are often short-staffed and volunteers are enlisted to help. By giving volunteers and local food donors their own mailboxes, these agencies can increase the efficiency of their communications. For example, a general "call to action" message can be recorded by an administrator and then sent to a distribution list of contributors. The messages can contain information on the needs for a certain food distribution drive, a time table, and a confirmation request. In addition, the same type of broadcast can be used to solicit assistance from volunteers (on the day of a food drive or distribution event). This helps the administrators to avoid having to call all of these individuals one at a time. The system can be programmed to call people at their place of business or at home to deliver the messages.

Neighborhood Watch

Local neighborhood watch programs are popular nowadays in helping to curb crime and to improve the relationship between local authorities and the citizenry. One critical aspect of managing a neighborhood watch is the idea of communicating to a large number of citizens in a short period of time. Some neighborhood watch volunteers achieve this by going door-to-door or by manually calling from a list of names in their assigned "phone-chain."

A voice mail system can be used to automate much of the communication between volunteers, citizens, and local police. For example, police officials can record and distribute messages in the form of warnings of certain criminal activity. Any citizen can call into the system at any time to learn what they should be on the look-out for. The "phone chain" calls made by volunteers can be replaced by one broadcast message.

Village Voice Mail

The communications infrastructure of many countries is such that only a fraction of the population has a private telephone. This is true of India, Peru, Brazil, and China to name a few.

A large voice messaging system can offer a "virtual" phone service to residents by allowing them to trade messages in non-real time fashion in lieu of having their own phone service.

The way village voice mail works is simple. Someone who wishes to make a "call" to a friend or relative accesses the system from a public phone. Public phones are either payphones or "phone parlor" terminals in this case. The system will greet the caller and ask the user to identify him or herself by entering a user ID and passcode.

Once this is done the "virtual phone number" of the called party is entered. The caller can leave a simple message to arrange a meeting or a message to simply say hello. The recipients access the system in a similar fashion to pick up messages in non-real time. For those who don't have their own phone, this offers a significant improvement in lifestyle. Citizens living in rural areas do not have to travel to communicate with someone, nor do they have to communicate by mail for simple arrangements such as doctor appointments or business meetings. Boston Technology (Wakefield, MA) has designed systems for this purpose.

Telephone Companies

Cellular Messaging

One thing many cellular phone users have in common is that they want to be in touch at all times. Users want messages to be relayed immediately, and they want total control over their communications. The use of voice messaging along with mobile communications is a powerful combination. This is true because of the options on cellular messaging systems.

Users can program a cellular mailbox, so it's always activated when calls are made to the cellular number. This allows the user to screen calls so that the most important ones can be called-back first.

Messaging is also critical in case the user has traveled outside of the transmission radius of the cellular service provider. When this occurs, no calls are lost since the messaging system can answer every call.

Users also prefer cellular messaging as a means to avoid peak traffic phone call rates. For example, some subscribers will purposely forward all calls to their voice mailbox until after the peak calling rates go down. At that time, they call into their mailbox and make the return calls. Systems manufactured by Centigram (San Jose, California), Octel (Milpitas, California), and Comverse Technologies (Long Island, New York), are among the most popular used by cellular providers. AirTouch, Cellular One, L.A. Cellular and virtually every other provider offer some kind of messaging service to their base of subscribers.

Paging Services

Radio paging providers were among the fist group of telephone service vendors to pioneer the use of voice messaging with paging. Today, paging companies like SkyTel offer e-mail links, fax forwarding, alphanumeric pagers, and a nationwide network to keep their subscribers in touch with virtually every aspect of their communication "linkage." Users of a paging service can use voice mail and their beeper as a way to react quickly to caller transactions.

SkyTel users, for example, simply call 1-800-SkyTalk to pick-up messages from callers who opted to leave a recording. Regular callers can be instructed to use codes that are displayed on the subscriber's alphanumeric beeper. For example, two business associates could use code "1" to mean "meet me at the office," or another code to mean "meet me at the customer site."

If a more descriptive instruction is required, then the caller can leave a voice message. This form of message waiting indication is useful when subscribers are not convenient to a phone. For example, hotel repair maintenance people, corporate executives, and sales agents can be sent important messages and tracked-down via paging.

The means to prioritize messages can be controlled by the user. For example, a series of special codes can be used with the messaging system in order to activate the paging device. If these codes are only given to special customers or co-workers, the user can be sure that when the beeper is activated, it's an important call.

Payphone Message Forwarding

Since voice store-and-forward is the very basis for voice mail systems, it makes sense that the technology can be used in public communications. Some payphone providers and long distance companies are now providing a special forwarding service for users at airports, convention centers, and other high-traffic areas. For example, some AT&T payphones now have this feature.

These payphones are routed through a special messaging system that acts as a store-and-forward concentrator. When you make a phone call and encounter a busy signal or no answer, a voice prompt indicates that you can leave a message that can be delivered later to the same number you dialed. The application then attempts to deliver your recorded message over a period of time. This can be especially helpful for travelers about to board a train or plane. A typical message would be: "Hello, Frank? I'm taking an earlier flight. It's 809 out of Philadelphia and should be arriving at 2:00 PM. Can you make arrangements to have someone pick me up at the gate?"

Residential Messaging

Every major RBOC and large independent telephone company offers voice mail service for their subscribers. These systems are based on technology from Unisys (Blue Bell, PA), Boston Technology (Wakefield, MA), and Octel (Milpitas, CA) to name a few. Promoted as a modern replacement for answering machines, telephone companies offer the service for as little as $3 per month depending on the options provided.

These systems are centralized, so calls forwarded to the platform are directed over the local exchange network to the "toll tandem" or "operator center" level of the network. As many as 50,000 subscribers or more can be handled with some of the larger configurations.Subscribers activate their voice mail by dialing a sequence of codes on their regular phone.

Once activated, the voice mail system will intercept messages and answer on behalf of the subscriber, just like an answering machine. The system can be programmed to do the same thing for ring-no-answer conditions as well. Users are alerted of messages waiting with a stutter or "gallop" dial tone. There are many twists on service offerings of this type, and the telcos are testing new offerings and features constantly.

For example, BellSouth is offering Knoxville, TN subscribers a service to store messages for delayed delivery. The MemoryCall feature allows users to program telephone numbers and delivery times for messages using their existing voice mailbox service. Users can activate this feature in order to send themselves messages (wake-up calls, anniversaries), or as a means to communicate with family members before a long commute home.

Voice mail manufacturers go to great pains to ensure that these systems will operate in a resilient fashion, providing uninterrupted service. It is a challenge to put these systems to the test before they are installed, because doing a "load test" manually is not feasible.

The voice prompts on residential voice mail systems can be complex, supporting many prompt trees, menu options, and features. For example, Boston Technology's C.O. ACCESS system has over 1,400 different prompts in the English version of their system alone.

One company that specializes in testing large-scale messaging systems is Hammer Technologies, Inc. (HT) of Wilmington, MA. According to Steve Gladstone, general manager: "The testing software must consider a comprehensive study of the user interface. This is also required whenever there are software patches, new software releases, or configuration changes.

"Manually testing each path of a prompt tree with hundreds of speech files can be an extremely time consuming and tedious task."

The approach was to design a test system addressing the specific problems of regression testing on the voice/user interface. Every time a new feature is added to a voice mail system, the user interface usually changes and must be regression tested. As these systems have become increasingly complex, the skill level required of manual testers has increased to a prohibitive level. The goal was to design an automated system that could be operated by anyone.

The Hammer system keyboard is transformed into a "virtual telephone" with the help of SoundSeer software, also provided by HT. SoundSeer eliminates the need for the testing lab to purchase expensive channel banks or other telephone equipment. For example, SoundSeer allows testing personnel to execute valid telephone transactions on the Hammer keyboard itself, as if input by a telephone caller on his or her keypad.

Like a word-processing macro, SoundSeer "captures" the keyboard transaction steps as well as the responses to those keyboard entries. The output of SoundSeer is an actual test script that can be run as-is or enhanced. In this way, SoundSeer is a critical facilitating testing tool because it automates an otherwise laborious process.

A special feature called HammerGen automates the creation of tagged voice prompts and their associated test routines, further automating the process of testing. The Hammer performs regression tests of user interfaces in a 100% automated fashion. Testing personnel can schedule a test suite to run, say, overnight and come in the morning to a report verifying that all voice prompts are still functioning properly. Boston Technology says the testing system helps them deliver reliable systems to their telco clients.

Transportation

Car Pooling

High-occupancy vehicles (HOVs) are the rage in most metropolitan areas. The use of traffic lanes for HOVs is taken very seriously. In fact, if you're caught in an HOV fast lane as a single occupant of a vehicle, you can be heavily fined. There are a number of approaches to encourage commuters to use alternate forms of transportation.

Besides taking the bus, subway, or train, many people have turned to car pooling as a way to cooperate with local HOV initiatives. One of the problems associated with car pools, however, is the manning of telephone lines by sponsor agencies or volunteer organizations. The people manning the phones have to be half matchmaker and half dispatcher in order to get the job done.One way to streamline the car pool matching procedure is to employ a voice mail system.

With such a system, people wishing to be in a car pool are given a personal voice mailbox. These voice mailboxes can be associated with a series of origination and destination codes. Anyone wishing to "hook-up" with a car pool can be directed to the mailboxes associated with that route.

System administrators can send broadcast messages to certain geographically oriented groups to help speed the process of a car pool search. An alternate method would be to use IVR, so users with the same zip code can be automatically grouped. It would be therefore possible to direct new users automatically to the mailbox of someone who needs another member in their established car pool.

Postal Package Re-Delivery

Some packages require a signature and confirmed delivery in order to be dropped off by couricrs. When no one is at home, delivery persons often leave a note regarding the delivery attempt.

A special voice mail system can give customers 24-hour access to a re-delivery appointment database. This requires a combination of voice mail and IVR to coordinate the delivery route with each customer's schedule. With such a service, the customer will access the system and enter the package tracking number indicated on the note left by the courier. The system will then execute a search and prompt the caller to choose from a selection of proposed delivery times over the next several days. If none of these suggestions is workable, the caller can leave a detailed message for an alternate delivery address, or an indication that he or she will come into a regional office to pick-up the package.

Trucking Dispatch

Perhaps one of the busiest types of call centers is that of a trucking dispatch office. Employees at these centers are deluged with calls from drivers, who are anxious to get instructions for the next pick-up or drop-off point on their route. Drivers often make repeated attempts to get through to the dispatcher and are put on hold for minutes at a time.

These delays are more than an annoyance, especially if the driver is on a rigid schedule or carrying perishable cargo.Voice messaging can solve most of these problems and significantly reduce dispatch switchboard traffic. Drivers get a personal mailbox number.

The dispatcher then updates employees on schedule changes, re-routes, or cancellations. Drivers can also use the mailbox to post delivery status. This can be especially helpful if the dispatcher has to coordinate the delivery of packages on behalf of a sick employee. The dispatcher can assign extra drop-offs for drivers who are available.

Truck Driver Messaging

Long-haul truckers are on the road for huge spans of time, and are therefore out of touch with family and friends. Due to the fact that time zones interrupt the regular flow of phone calls, it is difficult to relay messages from the East Coast to the West Coast, for example.

Driver messaging systems can do much to overcome these problems. Since voice mail can be picked-up and responded to at any time, family members can record messages for drivers according to their own schedule and time zone.

Take, for example, children who are at school during the day and have early bedtimes. The window of opportunity to call home and speak with the children is very limited. In this case, children can call the messaging system, enter the access code for their parent and record a message about school that day, a baseball game, or a simple greeting. The driver can then respond to the message and have it delivered to the "guest mailbox" associated with the driver's mailbox. This allows family and friends to communicate back and forth regardless of the time zone, traffic, or delivery schedule.

9

Outbound Systems

Voice processing systems with outbound call haven gotten smarter over the years. But when most people think of outbound dialing, they conjure images of interrupted dinners, unwanted wake-up calls and long sales pitches. We've all heard recordings for timeshares or newspapers over the phone. Don't think these are the only uses of the technology.

Take, for example, emergency notification systems, appointment reminders, and service dispatch systems. One not-so-obvious scenario for the use of outbound dialing systems is that of adjunct to an operator-aided call. As much as 30 percent of a service agent's time is used-up by placing calls and waiting for the called party to answer. Without voice processing technology to lend a hand, even these operator-aided calls are truly inefficient. With all of these apparent benefits, it is still important to consider laws that govern telemarketing.

In some areas, it's O.K. to have machines call people from a list - if you already do business with the called party. For instance, this would apply to collections systems, or service reminders. In some states, no outbound systems are allowed unless an operator comes on the line immediately upon connection. There are statutes specifying the legal calling periods for making outbound automated calls. Most of these guidelines apply to the use of outbound dialers for the purposes of a commercial solicitation. Some laws don't apply if the system is being used for security, safety, or telephone network long distance dialing. This chapter represents the best examples I've found.

Automotive

Maintenance Reminder

Car dealerships derive much of their revenue from maintenance contracts and post-sales service-related business. This is necessary, because the margins derived from the sale of the cars are thin due to fierce competition. This makes the follow-up on service appointments, recalls, and inspections a critical aspect of the dealer's marketing efforts. Voice processing technology can be used to increase the number of contacts made with customers each day for this purpose.

An outbound system can be programmed to dial telephone numbers associated with a database of service customers who are due for a maintenance visit. The system will dial the telephone numbers of these customers and deliver a friendly reminder message indicating that it's time to schedule an appointment for the car. A more sophisticated approach would allow some interaction with the customer. For example, the system may ask the customer to select from several appointment slots that are available, or to transfer to a service agent to set-up an appointment: This is Peter Sullivan Autos calling. You are scheduled for a maintenance and oil change for your car. Press 1 if you would like to select from a list of appointment times, or press 0 to be connected to a service agent.

Corporate

Network Alarms and Security

There are literally millions of local area networks installed worldwide, and countless wide-area bridges, routers, and packet switched networks. Many software products enable network alarms and security. Alarm packages can notify a technician too many collisions are on the network, or if a node is not operating properly.

In addition, there are many ways to add security to LANs and wide area networks with password protection, encryption, and call-back schemes. Voice processing technology can be used to act as a compliment to these packages.

Take, for example, a university with a variety of bridges, routers, and cables between each building. Information Services employees are often away from their computer because they are stringing cable, making connections, or training users. With an outbound dialing system attached to the network, these service people can be contacted when network alarms occur. For instance, network notifications can be captured by the voice processing system and linked to discrete voice prompts.

These prompts might say: "Network Alarm 2 - drive F 90% full, or; Network alarm 11 - southwest fiber loop is down." Service personnel can input their beeper number and program the system to outdial and deliver the message from their voice mailbox. The system could also be programmed to call the technician at a certain telephone extension or beeper number that was input before leaving his or her desk.

Remote log-ons can also be made more secure with the use of voice processing. For example, remote terminal users can use a modem to access a network remotely, and then enter the telephone number where they can be called on the phone.

The network security software can send this information via data link to the voice processing system, or write the information into a database that is shared with the outbound system. The outbound system can then call the remote user on the phone and ask for positive voice verification. Once confirmation of the user is established, he or she can then be allowed to access certain network privileges.

Correctional Institutions

Collect and Debit Calls

Voice processing systems can be used to concentrate privately managed payphones at prisons and other correctional institutions. Due to the fact that prison administrators discourage inmate transactions involving hard cash, the use of debit vouchers and collect calls is a favored method for placing calls to friends and family. An outbound dialing system is used in conjunction with an on-site PBX or a direct connection to a long distance provider's switch. All calls made from payphones are routed to the voice processing system first, so the inmate can be prompted to either record his or her voice, or enter a series of digits that reference a long-distance credit allowance.

Home Arrest

Due to overcrowding in prisons and the resources required to run the court system, many states and municipalities are beginning to use home arrest as an alternative to incarceration. A person under home arrest is limited to activities in or around the home during certain days or certain times of the day. The monitoring of these individuals can be very expensive.

For example, special radio transmitters, on-site dialing gear, and face-to-face visits are used to ensure that the individuals are indeed where they are supposed to be. A special centralized outbound dialing system can be used to cut-down on the complexity of this arrangement, and also reduce the cost incurred.

Using MOSCOM's (Pittsford, NY) voice board and speaker verification technologies, a security company can develop a home incarceration system to monitor prisoners' activities. With the application, law enforcement agencies train the system to recognize the prisoner's voice print. Each prisoner is required to train several passwords, which are chosen by whomever is administering the system.

At random times during the day or night, the system will call the prisoner's home and ask the person to speak their name, password, and/or pass phrase (s) into the phone. The prisoner's voice print is compared against the previously recorded words, and the application automatically checks it for accuracy. If the voice prints match up, then it is guaranteed that the person speaking is the prisoner in question. If the voice print does not match up, the security forces or police are immediately notified to take action.

Crime Prevention

Curfew Conformance

In 1990, Michael Lindsay, then-director of special projects for the Los Angeles County Probation Department, used a MOSCOM voice verification system as a deterrent to teen gang violence. Lindsay, currently a training specialist at the Central District of California US Probation Office, trialed voice verification as a way to keep teenage gang members home during a rumble.

The application, based on MOSCOM's (Pittsford, NY) voice board and speaker verification technologies, was supplied for six months. One hundred juveniles had his general data, previous arrest history, special notes, financial requirements and curfew schedule entered into the computer database. Officers had the teens train the system to verify their identity through passphrases, and a voice prompt was made for each person so that the program could ask for them by name.

When probation officers became aware of a rumble, they had the system call each teen's home and verified that they were home through voice prints. If the person called didn't pass the test after several chances, the machine printed out the name and address information and notified the probation officers via a pager.

While the system was discontinued when the program's funding ran out, Lindsay considers it to have been cost-effective as a control mechanism. "The system required juveniles, and their parents, to take responsibility for their actions. It had great potential."

Today, Lindsay points out that in Los Angeles, with as many as 5,000 adult offenders on bank case loads per probation officer, a voice verification system could greatly reduce the administrative costs involved in keeping track of them, and could increase the productivity of the probation department.

Education

Overdue Book Reminder

Librarians face a burdensome task in contacting readers who have failed to return a book on time. Often, this task is given to less senior staff members or volunteer workers. The cost to recover missing books or to replace them can run into the thousands of dollars each month. Voice processing can be used to automatically call library clients both before and after the book is due as a friendly reminder to make a trip back to the library.

This can be achieved by linking the library member card database directly to the voice response unit. During the book check-out procedure, the librarian will enter the book ID number into the client's record and enter the number of days the book is being checked-out for.

The outbound dialing system can be programmed to call the client one day before the book is due as a reminder. In addition, the system can be programmed to make repeated calls after the book is overdue. In making the "overdue" calls, the system can optionally announce the amount of late fees that have been incurred.

Parent-Teacher Conference Scheduler

Whether a student is doing well in school or whether he or she is having trouble, teachers are encouraged to keep an open dialog with parents. When a child needs help, a parent-teacher conference can be scheduled quickly. These conferences often coincide with PTA meetings, school night, or they can be scheduled at any time.

Teachers often call parents at home after hours to schedule these appointments, because they are busy during the day. Letter writing is often ineffective, because conflicts and confirmations are usually discussed over the phone. Voice processing can help simplify this procedure by extending calling hours to the entire day, and by automating the task so many calls can be made simultaneously.

For example, an outbound dialing system can be programmed with parents' home and work telephone numbers. In addition, the teacher or an administrator can program a range of available conference slots into the system. The system can then be made to call every parent until each one has chosen a conference slot with a touch-tone entry.

A more sophisticated outbound dialing system will generate a report showing who was contacted, what time slot they chose, and a list of the unsuccessful attempts. The teacher can then limit his or her personal calls to the uncontacted parents. If the system is equipped with four or more lines, it is possible to automate the entire procedure within a 24-hour period.

Truancy Reporting

With class sizes in the hundreds, many school districts have literally thousands of students to keep track of during the day. Even if truancy percentage rates are low, as many as several hundred parents may need to be notified of a child's absence. There are a number of challenges school administrators have to overcome in dealing with truancy. First, sending a truancy notice by mail does not enable instant, corrective action. Secondly, the labor involved in compiling truancy lists, printing-out forms, stuffing envelopes, and then mailing them can be very costly. Lastly, the mail can be easily intercepted by children or misplaced.

One popular means to solve this problem is the use of a specialized outbound dialing system. These systems call parents to let them know that their child was absent or tardy either the entire day, or for an individual class. These systems often keep track of the number of times this has occurred in order to automatically suggest a conference with the principal, or other administrative review meeting.

For example, the San Juan Unified School District in Sacramento, CA uses a system like this, as do hundreds of other school districts. A typical outbound message to a parent is: "This is the Casa Roble High School Attendance Office calling. Your 10th grade student has missed one or more periods during the previous school day. Please call the attendance office at your earliest convenience to discuss this. Thank you, good bye."

Financial

Client Contact Management

There are a variety of personal productivity tools that can be used to make client contact management an easier task. In a typical brokerage house, for example, each representative is required to call customers and advise them of new products.

In addition, margin calls, portfolio advice, and other information can be conveyed at the same time. Depending on the length of each call, as many as 100 of these calls can be made by each representative.

The means to partially automate this procedure includes the use of sales contact management software like TeleMagic, GoldMine, and ACT!, for example. These PC-based software packages allow brokers to scroll through customer records automatically. For example, you can filter a list of clients based on their portfolio, scheduled contact date, or any other pre-defined field.

Collections

The collection of past due accounts is one of the most important functions a call center can perform for a company. Some of the largest collection efforts are those of telephone companies and power companies. A company well known for its innovations in the automation of collections is Melita Labs International of Atlanta, GA. Melita was one of the first companies to develop sophisticated outbound dialing and call progress detection software. In fact, board suppliers like Dialogic Corporation incorporated some of the call progress technology.

There are two types of approaches to automated collections. The first approach is to program a power dialer that simply calls clients to deliver a canned message: This is the power company calling about your overdue account. You must call us to arrange for payment or your service will be disconnected. Reconnection fees apply if you do not call right away.

The second way to streamline collections is to automate every part of the transaction up to the point when the called party answers the phone. At that point, a customer service representative comes on the line to speak with the customer live. This is necessary, because many people will simply hang-up on a machine that's announcing an overdue bill.

Government

Jury Duty Scheduling

The procedure in selecting a jury depends on where you live. In some places, a random selection of licensed drivers is used. In other areas, registered voters are the list that's used to select jurors. If you are selected to participate in a court proceeding, you will receive a written notice.

The notice identifies a time and place for you to appear. Sometimes court delays, schedule conflicts, and other circumstances render the initial date and time invalid. Voice processing can be used to help automate the procedure of letting jurors know about the status of a case, and whether or not they need to reschedule a trip to the courthouse. This can save time, court costs, and a lot of effort for both the jurors and officers of the court.

For example, the jury duty notice can include instruction for each juror to call a special jury duty VRU. The voice processing system can then ask a series of questions, such as confirmation of the person's obligation to appear, their case number, and the location of the trial. In addition, the system can ask the jurors to enter their daytime, nighttime, car phone, and beeper numbers. With this data captured, the system can then instantly call dozens of jurors to inform them of rescheduled court dates or other instructions.

Polling

Politicians need to know what their constituents are thinking. This goes for referendum votes, the stand on taxes, crime, and a variety of topics that are distinct for their area. Politicians use various communication vehicles to get feedback. For example, opinion polls sent through the mail, radio talk shows with live call-ins, and door-to-door surveys. Voice processing can also be used to collect thousands of opinions in a very short time period.

An outbound dialing system can be programmed to call numbers sequentially, or numbers pertaining only to a certain ward or district. Constituents can be contacted at home and asked for their opinion on a number of issues. Recorded messages can be solicited from each constituent, or simple yes or no answers can be asked in combination with touch-tone input.

For example: Say *yes* or *no* to this question after the beep, or Press 1 if you agree, 2 if you disagree, or 3 if you are undecided. In this fashion, politicians can get quick feedback on a bill, a speech, or any other issue.

Voter Registration

There are a number of influential groups that can have an effect on voter turn-out. These include unions; voter leagues and church groups, to name a few. Each of these organizations has a vested interest in supporting their favored political platform, and therefore, want their members to be registered to vote.

Voice processing systems can be used to encourage voter turn-out. In addition, the same equipment can be used to educate voters on the issues of most concern for that group.

An outbound system can be programmed to automate a voter registration drive. Such a system will dial each member of the group and deliver a special message that could include, for example, the endorsement of a candidate, instructions on how to vote, and the time and place to register or vote. In fact, a system equipped with as many as 48 or several hundred ports can easily do the work of dozens of volunteers.

For those members of the group who want to know more, they can be prompted to enter a digit that transfers them to a volunteer's phone. The volunteers could be located in a call center, or they can be contacted at their homes and connected via conference circuit with the called party.

Health Care

Appointment Scheduling

Employees at medical centers and private offices spend a lot of time on the phone scheduling appointments with patients. This is necessary for follow-up care and also for routine check-ups. These phone calls can be made in addition to the mailing of reminder postcards, a practice commonly used by many dentists. Unfortunately, the time it takes to schedule, confirm, and re-schedule an appointment is often too much to handle, so many appointments go unconfirmed.

Outbound dialing technology can be used to augment the efforts of the medical office staff by automatically dialing patients with reminder messages. A system like this will say to the patient: This is Dr. Brown's office calling. You have a scheduled appointment this Wednesday at 3:00. If you would like to confirm this appointment, press 1 or press 2 if you would like to re-schedule. Subsequent menu choices may allow the patient to pick from a list of alternate dates for the appointment. By providing the option for live connection to the doctor's office, each patient will have the ability to talk about any conflicts or other problems that require personal attention.

Home-Bound Monitor

There are millions of homebound individuals who live alone, don't drive, or are otherwise restricted to their residence for medical reasons. Many of these individuals are elderly, and family members call them frequently to make sure everything is O.K. Visiting nurses, volunteers, and neighbors also make these check-ins.

Not everyone has someone to take care of them, however, and the cost to monitor someone can be very high. In these cases, an outbound dialing system can be very useful, and perhaps offer life-saving capabilities to the homebound.

A system of this nature will be linked to a database that includes information on the home-bound individual's address, phone number, doctor's phone number, and emergency contact information. The software governing the outdial procedure can be set-up so that it expects a verbal or touch-tone response from the called party to indicate that he or she is well and not in need of assistance.

If the system does not get a response, an event trigger will cause a series of stand-by phone calls to be made. For instance, the system will call a neighbor or friend and announce that there was no response at that phone number. These individuals can enter codes to indicate that the person in question is either with them and O.K., or is known to be away. The system may also be programmed to call an ambulance or an on-duty agent who will then make phone calls manually.

Prescription Reminder

Pharmacists and doctors work very closely with one another to ensure the correct prescription instructions are given to patients. Pill bottles, for example, now have very complete instructions on side effects and other warnings. Despite these measures, many people will skip a dose or completely discontinue a course of medicine due to forgetfulness. The effect of discontinuing medicinal prescription can cause an illness to be aggravated, or sometimes even lead to a more serious condition. Unfortunately, the time required to continually contact patients and remind them about their treatment is prohibitive. Very few pharmacies or doctors have the manpower to make these contacts regularly.

This is where a voice processing platform can be of great assistance. Nationwide drug store chains, medical centers, or insurance companies can sponsor the use of an outbound dialing system that reminds patients to take their medicine. For example, a list of patients can be called with a general message: Hello, this is the prescription reminder system calling.

You should have two day's worth of your prescription left. Please call us to have your prescription refilled. It is also possible for the system to use Audiotex for providing detailed information on the side effects, warnings, and limitations of the prescription.

Hospitality & Travel

Time & Charges Call-Back

Small motels often get time and charges information from the local telephone company. This ensures proper billing of their guests for telephone calls. Since the telephone company has centralized automatic message accounting capability, it is not difficult for a TSPS (Toll Service) operator to call with this information. Unfortunately, the cost to deliver this information manually makes the service unprofitable for the phone company. There are many ways around this for the hospitality industry.

For example, the use of SMDR (Station Message Detail Accounting) software on-site, or use of a long distance provider that handles billing automatically. In addition, HOBIS (Hotel Billing Information Service) data links can be provided by the phone company to deliver this information to a Teletype terminal or the hotel's Property Management system.

The challenge for smaller properties is that all of these terminals, data links, and specialized services are expensive, so they often simply forgo offering direct dialing service to their guests.

Several telephone companies have worked with vendors like Voicetek Corporation (Chelmsford, MA), DATAP Telecom (Atlanta, GA), and Summa Four (Manchester, NH) to develop a low-cost alternative to CPE-based SMDR and HOBIS. Called *time and charges call-back*, the solution automatically collects dialed number information from each room, rates the calls, and then stores the data so a voice processing subsystem can announce it.

The system will automatically notify properties after calls have been made. The information is delivered to the hotel operator over the phone. In addition, these systems can be called by the property to confirm time and charges. Some of the first telephone companies to use these methods were Bell Atlantic and Southwestern Bell.

Legal

Deposition Scheduler

Depending on the law firm, it is possible for dozens of depositions to be conducted each day. In addition, there are client conferences, interviews with experts, and other matters that require a lot of phone work to arrange. A legal secretary or assistant usually handles the scheduling.

A computerized schedule could be used, so it can be shared by an outbound dialing VRU. The VRU simply dials from a list of phone numbers associated with the people who are scheduled to be at the conference.

If these individuals have to break the appointment, the system can transfer them to the legal secretary, or the system can suggest and alternate time based on the attorney's schedule. Considering the value of a lawyer's time, the outbound dialer can save a lot of money for the firm.

Manufacturing

Subcontractor Scheduling

Just-In-Time inventory is a concept now widely embraced by suppliers and manufactures. The idea is to pull together all of the needed parts and assemblies as close to the time of shipping as possible. This requires subcontractors and other suppliers to deliver the material on demand. The benefits of last minute fulfillment are varied.

First and foremost, the cost to maintain a large inventory is virtually eliminated. These costs include overhead, carrying costs, and real estate. In addition, since material is turned-over with each order, the worry of dealing with outdated or useless inventory is done away with.

A voice processing platform can use the database associated with material requisition and planning to aid in this process. For example, the telephone numbers for each vendor's dispatch center can be linked directly to the ordering system and the VRU.

When inventory is required to build a certain order, the VRU automatically dials that vendor and announces the part needed, the quantity, and the delivery requirements. If the vendor uses an IVR system to automate their own dispatch, the outbound dialing system can be used to mimic the touch-tone sequences that a person would have input in communicating with the vendor's VRU.

Military

Mobilization Orders

Outbound dialing can be used to automate the mobilization or formation of any military unit. For a system like this to be effective, it is important for each enlisted person to submit a schedule of phone numbers, beeper numbers, or other contact information to his or her commander. This information is input into a computer database so these numbers can be located instantly.

The use of speaker verification can be especially useful in order to confirm that the called party is indeed the enlisted party. This is helpful not only for confirmation of the receipt of orders, but also to ensure that the person receiving the content of the orders is authorized to hear them. Each person can be asked to speak selected words or to type digits indicating their receipt of the message.

Power Utilities

After-Hours Dispatch

Supervisors depend heavily on service dispatch operators and standby management employees to call workers into the field after-hours. This is a very time consuming process that keeps the operator from attending to the water and electric systems. A PC-based IVR system can keep a more accurate record of call-out acceptance and declines, save money on the labor involved in making the call-outs, and to allow the staff to concentrate on more productive tasks.

Systems like this are introduced to employees with a special pamphlet or Audiotex prompts to guide the users and their families. For example, an IVR system can call and announce the employee's name. Anyone answering the phone can press digits to indicate that they are the employee, or that that person is available and is coming to the phone, or that he or she is not available. The system operator can initiate a call-out procedure from the system's keyboard or by phone.

The system could be programmed to sequentially call people from a job class list until it gets a *yes* to the call-out request. Once a yes is received, the system can make a phone call the system operator announcing the name of the person reporting to work. Employees who desire overtime work must call the system themselves in order to enroll for overtime. These are the only employees who will be called-out for that type of work.

Print Media

Subscriptions

Telemarketing call centers are famous for outbound newspaper solicitations. Most of these calls are made manually, but the use of predictive dialers and fully automated systems is becoming more popular.

The chances of closing a deal, however, are much higher when the prospect can have a live dialog with the solicitor. In either case, voice processing can be used to place each call, determine if a person has answered the phone, and complete the transaction.

PAGE TeleCOMPUTING, Inc. (Delray Beach, FL) helps telephone solicitors “go fishing” legally with their live calling LINK and Sales Call Software. LINK and Sales Call Software integrate with any TAPI or TSAPI compliant telephony device. This goes for PBX’s from Comdial, Inter-Tel, Lucent, NorTel, Mitel, Panasonic, Siemens and Toshiba. Ditto for Caller ID modems, NetPhone’s NetWare Server TSAPI PBX boards and TAPI devices like Comdial’s PATI 3000, etc. LINK does screen pops with sales contact management software, too. Sales Call Software’s power dial and restrict features are helpful for outbound telemarketing apps. You can maintain an aggressive marketing program in the face of increasingly restrictive state and federal legislation. Automatic restriction of "Don’t Call Me" telephone numbers is accomplished with this software as illustrated in figure 9.1.

Figure 9.1 - PAGE TeleCOMPUTING Call Restriction

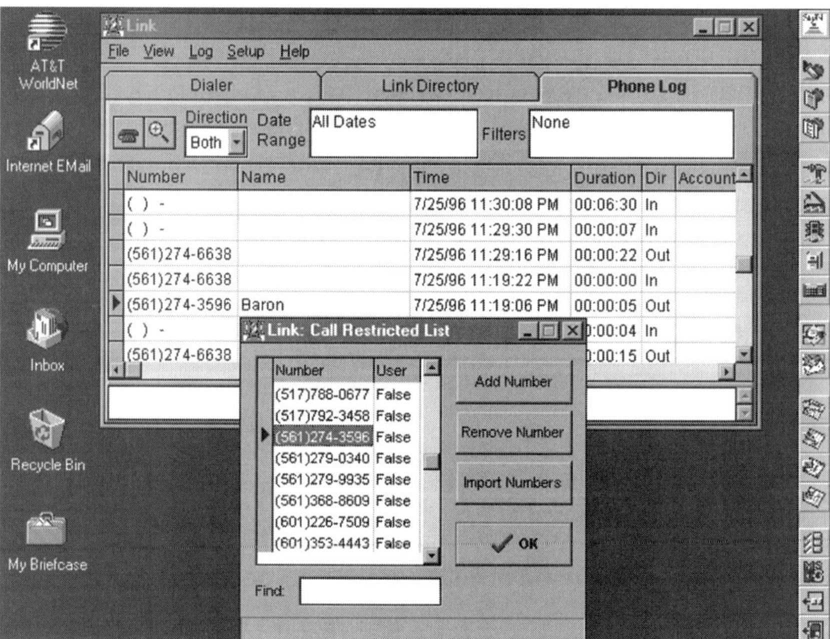

Retail Services

Birthday Reminder

American Express is offering reminder services to their customers. Reminder services collect important dates from their clients (birthdays, anniversaries, review dates, and vacation schedules).

These dates are used to create a profile for that client. The profile identifies what kind of gifts, flowers, or other presents that person thinks are appropriate for each person on his or her reminder list.

These services are tightly coupled with the merchandising of gifts, so the cost associated with providing this type of service is typically absorbed as a cost of sales. The idea for this service is to establish a lasting relationship with the client, who represents continual purchases over a long period of time.

The use of outbound dialers can help the enterprise to grow in service volume very quickly. Since a VRU can be programmed to make calls at any time (and do it constantly over many lines), the amount of labor required to notify clients can be greatly reduced.

In fact, a large VRU can be used to augment the efforts of a three-person call center, producing the results of a center with dozens of agents. The system will speak the client's name and ask for a password before announcing the purpose of the call. This helps to avoid spilling the beans about a gift to a spouse or business associate who may answer the phone.

Based on the client's profile, the system might say: "This is the gift-minder service calling. It's your spouse's birthday in one week. May we suggest one dozen roses? If this is O.K. with you, press 1 to record a message for the note. If you would like to hear about other choices, press 2 or dial 0 to speak to a service representative.

Rain Check

Sometimes, even a plentiful inventory of sale items can be depleted during special promotions. This happens as a result of miscalculation, or in some cases, a back-ordered shipment. Most store managers will offer shoppers a rain check, thus allowing the purchase to be made at a later date at the sale price. To add to the frustration of getting a rain check, the product may be discontinued, or it may sell-out again before the coupon can be redeemed.

One way to solve this problem is to link the issuance of a rain check with a customer service database. When the item is delivered to the store, the store manager can update the rain check list to look for anyone who may be waiting for the item.

In some cases, the store manager will set-aside the item for some grace period before selling it to another customer. A voice processing system can then be programmed to call rain check holders to announce the arrival of the sale item. Customers can then be prompted to indicate if they still want the item. In the case a customer declines to pick-up the item because they either lost interest or bought elsewhere, the system can then log this information so a report can be brought to the attention of the store manager.

Security

Alarm Monitoring

Security companies can use voice processing as a means to dispatch field personnel or police on the receipt of alarms. Most alarm systems use a regular phone line to outdial to a central monitoring station. The monitoring station is able to determine what alarm has been triggered by detecting touch-tone digits or a string of data characters. An outbound dialing system can be used to concentrate many of these alarm circuits or to tap in on t data links. The system can be programmed to make telephone calls from a pre-defined database in the case of an alarm.

For example, the system can call the building where the alarm occurred and announce to the person who answers: This is the alarm monitoring station calling. Please enter your secret code to indicate that all is well. If you do not enter your code, we will dispatch the police. In addition, alarms can be used to trigger action when water levels get too high in pump stations, or when electrical circuits are out of order.

Telemarketing

Home Agent Routing

Compro Technologies, Inc. of Barnegat, NJ specializes in service bureau telemarketing systems. Summa Four and Excel switches are used to connect callers to the system via the PSTN. These calls are handled by local agents or re-switched to PSTN-connected home phones or remote agents behind a channel bank. The remote centers can be connected via OPX (Off Premises Extensions) or with regular dial-up telephone lines. The company has designed the system for a "home agent" telemarketing company in Mexico among others.

The ACD service bureau application supports local, remote, and home based agents. The idea was to provide service bureau customers with unlimited access to an agent labor pool without the overhead of large facilities traditionally associated with big call centers. In two months from start to finish, one service bureau was up and running using a programmable switch driven by a C application running on a Unix workstation.

Through approximately 20 E1 interfaces, the programmable switch accepts premium, toll, and local calls from a variety of carriers. It then guides callers through a series of prompts and pre-ambles and ultimately selects an agent using a variety of standard and customized ACD decision criteria. Once the agent is selected, the application determines the best route based on least cost route and other factors, places the outbound call (or rings a telephone extension) and monitors the progress.

Upon answer, a prompt is played to the agent informing them of the nature of the call. The agent then greets the caller accordingly. When the call is complete, the agent and caller hang up and the appropriate billing records are generated. This application provides a great deal of flexibility to the service bureau and their customers. Very often, a service bureau may experience a sudden increase in call volume. Without the at-home or remote agents, the call center would not be able to add more agents than they have physical seats. With this application, they can add as many at-home agents as needed in very short time.

The service bureau charges pennies per minute for switching time depending upon the value of the service provided. The bureau keeps costs down by minimizing physical real estate, selecting the least cost routes, and the availability of a virtually unlimited labor pool. It's ideal for entertainment services such has horoscopes and psychic advisors, information services such as doctor and lawyer advice, and even technical support using programmers and engineers that work at home or at a remote office.

Telephone Companies

Bulk Call Generator

Telephone companies specialize in communication, so it's no wonder that they use voice processing and outbound dialing systems. One of the most critical aspects of managing the telephone network is the ability to accurately route calls. The fees to do this are based on tariffs approved by state government. For example, if a subscriber makes a call to a neighboring town and is charged for two minutes of usage when the call only lasted for one, it is cause for great concern.

The problem not only has to be fixed, but there are costly billing adjustments, labor, and investigations that might ensue. In fact, when a telephone company petitions to the governing authorities for a rate increase, an over-billing incident can seriously harm that effort.

Unisys of Blue Bell, PA developed a special UNIX-based system called Sentry that monitors the routing, billing, and completion of calls controlled by telephone company switches. The system simulates real telephone calls in order to find out if any of the switches are programmed improperly, and therefore billing incorrectly.

Voice processing systems can mimic this same function to create chain call scenarios. For example, VRUs can be placed at either end of a switched network connection. These units will place calls and then connect them together to complete a physical loop including switch connections.

Since the VRUs are looped together, they can log time-stamp data into a common database from which reports of the transactions can be run. This same data can be automatically compared to similar reports generated by the message accounting links on the switches. This is one of a number of ways that outbound systems can be used to automate testing on the network.

Collections

Apex Voice Communications, Inc. of Sherman Oaks, CA is a Dialogic ToolKit Developer. Apex sells the OmniVox line. It handles Windows NT, UNIX, X Windows and some host stuff.

One of their largest sales was a 1,320 line system for Argentine Telecom. Apex worked closely with Ameritel and Virtual Phone (Apex VARs in Argentina). The system acts as 1,320 agents calling customers on past due notices and billing inquires. It handles both inbound and outbound calling and uses R2 signaling on E-1 trunks.

It's connected to SQL databases on the back-end part of the telco's provisioning system. The database holds customer records on balances, past due invoices, etc. On the telco end, the system is connected to a central office switch several Meridian SL-1 PBXs. The whole thing sits on a TCP/IP UNIX network.

Service Verification

Utilities are beginning to use voice processing technology to conduct follow-up service calls after the activation of phone lines. This is done in order to decrease the holding time for inbound calls to the service center. By automating this service verification process, service agents are available for more critical transactions.

The primary function of these systems is to contact new subscribers and confirm that their phone service has been activated as promised. They can also collect feedback from new subscribers regarding customer satisfaction. A simple questionnaire can be answered with touch-tone input or with speech recognition to achieve this.

Voice Dialing

Hands-free use of the phone is commonplace nowadays. For instance, many residential, car, and PBXs have speakerphone capability. Users of these devices appreciate the freedom of being able to work with both hands while talking on the phone. This is the same benefit derived from the use of voice dialing. With this type of service, subscribers simply go off-hook and speak in order to place a phone call. The applications for this service feature are immediately apparent: Assistive technology for the physically impaired, emergency calls, and a more intuitive interface.

Voice dialing uses speaker-dependent words, numbers, or phrases to trigger phone calls. These systems can also use speaker-independent technology to recognize a limited vocabulary of numbers or letters. Subscribers in West Palm Beach, FL for example are now using this technology on a trial basis. BellSouth Telecommunications developed its voice dialing solution based on Texas Instruments software with 30-number memory.

When callers go off-hook, the telephone circuit is immediately connected to a recognizer channel. Once the recognizer confirms the appropriate number to dial, the local switch places the call, and the recognizer channel drops-off so someone else can use it.

Transportation

Courier Scheduling

Same-day local couriers provide local package delivery within minutes as an alternative to postal delivery. These services are very competitive, so the communication link between dispatchers and couriers is critical. Perhaps one of the most important aspects of the job is getting to the pick-up point as quickly as possible. Dispatchers arrange for this by placing couriers strategically, so customers are never more than a few minutes away. Depending on the town in question, the mode of transportation could be bicycle, van, or motorcycle.

VRUs can aid in the dispatch of couriers by answering the phone at the same moment the dispatcher does. By using Caller ID or ANI, the caller's phone number can be matched-up with a customer name and address database. Once the pick-up is confirmed with the dispatcher, the VRU can page or call the closet courier and play-out prompts with the address of the client. If the courier cannot make the pick-up, he or she will so indicate with the push of a touch-tone sequence.

In this case, the system will call the next closest courier and try again. If no one can take the job, the dispatcher can handle the matter manually. During the time the outbound system was making these calls, however, the dispatcher was able to take incoming calls, thus improving on customer service.

Utility Companies

Remote Telemetry Reporting

Treasure Coast Engineering of Gainesville, FL is a Microsoft Solutions Provider, Novell Dealer, AutoCad VAR, an HP Authorized Dealer and a TRT VP Express VAR.

The company has two lines: Dial One is their own product. It has modules for outbound marketing, automated attendant, fax server, international call-back and voice mail. The other is VP Express from TRT for out-of-the box shrink-wrapped solutions.

The company first developed a remote sensing application. As an engineering firm with four EPA labs, they wanted to use the phone to find out about hydrocarbon, ground water and other alarm conditions. Treasure Coast could not find a company who knew how to both sense remote alarms and deliver progressive, time-sensitive voice information.

They rigged some hard contact closures to an autodialer. They used the prototype for engine family certification. If a remote power supply dips below a 12V output, the dialer calls folks on the phone to report the specifics and ready them for dispatch. It reports progressively at different voltages. A local power company also adopted this solution.

Utility Emergency Outcall

Technically Speaking (a Brooktrout Company) of Southborough, MA is an Oracle Business Partner, an IBM Business Partner and member of the Microsoft Early Adopters Program. The company started out selling solutions in the service bureau business, and got into tool development.

The company helped one of their VARs with an emergency notification system for a nuclear power plant. The system would call people after tracking them down and get them to report to work in the case of an emergency. The system would take a call from an administrator who set the whole process of outdialing in motion with the entry of some secret digits. They test the system monthly as part of a government regulation on emergency drills.

Appendix A

Recommended Resources and Bibliography

The publications cited here offer a comprehensive and detailed understanding of the subjects covered in this book. The publisher would like to acknowledge the authors of copyrighted material for the permissions granted to use portions of their work as solicited by the research staff.

Great Books on Computer Telephony:

1001 Computer Telephony Tips, Secrets and Shortcuts,
Edwin Margulies, Flatiron Publishing, Inc., 1996.

Basic Book of Information Networking,
Motorola Codex, Motorola University Press, 1992.

Basic Book of ISDN,
Motorola Codex, Motorola University Press, 1992.

Client Server Computer Telephony – 2nd Edition,
Edwin Margulies, Flatiron Publishing, Inc., 1997.

Complete Traffic Engineering Handbook,
Jerry Harder, Flatiron Publishing, Inc., 1992.

Computer Based Fax Processing,
Maury Kauffman, Flatiron Publishing, Inc., 1994.

Computer Telephony on the Sun Platform,
Patrick Kane, Flatiron Publishing, Inc., 1996.

Customers: Arriving With a History and Leaving With an Experience,
Andrew Waite, Flatiron Publishing, Inc., 1996.

Customer Service Over the Phone,
Stephen Coscia, Flatiron Publishing, Inc., 1995.

Fault Resilient Computers,
Richard Grigonis, Flatiron Publishing, Inc., 1996.

Handbook of Telecommunications,
James Harry Green, Flatiron Publishing, Inc., 1993.

International CallBack Book,
Gene Retske, Flatiron Publishing, Inc., 1995.

Local & Long Distance Billing Practices,
M. Brosnan & J. Messina, Flatiron Publishing, Inc., 1993.

Newton's Telecom Dictionary - 12th Edition,
Harry Newton, Flatiron Publishing, Inc., 1997.

PC Telephony,
Bob Edgar, Flatiron Publishing, Inc., 1995.

Predictive Dialing Fundamentals,
Aleksander Szlam, Flatiron Publishing, Inc., 1996

Reference Manual for Telecommunications Engineering,
Roger L. Freeman, John Wiley & Sons, publisher.

SCSA - Signal Computing System Architecure - 2nd Edition
Edwin Margulies, Flatiron Publishing, Inc., 1996.

Secrets of Windows Telephony,
Edwin Margulies, Flatiron Publishing, Inc., 1997.

Speech Recognition,
Pete Foster, Flatiron Publishing, Inc., 1993.

Telecommunications Management,
Harry Green, Flatiron Publishing, Inc., 1993.

The MVIP Book,
GO-MVIP, Flatiron Publishing, Inc., 1995.

Telephony For Computer Professionals,
Jane Laino, Flatiron Publishing, Inc., 1994.

Testing Computer Telephony Systems and Networks- 2nd Edition,
Steve Gladstone, Flatiron Publishing, Inc., 1994.

The Guide to T-1 Networking,
Bill Flanagan, Flatiron Publishing, Inc., 1990.

Understanding Computer Telephony – 2nd Edition,
Carlton Carden, Flatiron Publishing, Inc., 1997.

Understanding Data Communications,
G. Friend & J. Fike, Flatiron Publishing, Inc., 1993.

Understanding Java Telephony,
Edwin Margulies, Flatiron Publishing, Inc., 1997.

Understanding Telephone Electronics,
J. Fike, Flatiron Publishing, Inc., 1993.

Understanding The Voice-Enabled Internet,
Edwin Margulies, Flatiron Publishing, Inc., 1996.

Visual Basic Telephony,
Krisztina Holly & Chris Brookins, Flatiron Publishing, Inc., 1995.

Voice Processing,
Walt Tetschner, Artech House, 1991.

Windows Telephony,
Jeffrey Shapiro, Flatiron Publishing, Inc., 1996.

Glossary

10Base2
IEEE standard for baseband Ethernet at 10 Mbps over coaxial cable to a maximum distance of 185 meters. Also known as Thin Ethernet.

10Base-T
IEEE standard for operating Ethernet LANs on twisted pair wiring that appears like telephone cabling. Sometimes old cabling will not work.

A & B Bits
Bits used in digital environments to convey signaling information. A bit value of one generally corresponds to loop current flowing in an analog environment. A bit zero corresponds to no loop current, i.e. to no connection. Other signals are made by changing bit values. For example, a flash-hook is set by briefly setting the A bit to zero.

A & B Leads
Additional leads used typically with a channel bank two-wire E&M interface to connect certain types of PBXs (also used to return talk battery to the PBX).

A & B Signaling
Procedure used in most T-1 transmission links where one bit, robbed from each of the 24 subchannels in every sixth frame, is used for carrying dialing and control information. A type of in-band signaling used in T-1 transmission, A and B signaling reduces the available user bandwidth from 1.544 to 1.536 Mbps. See also PRI.

A4

Basic Group 3 standard defined for the scanning and printing of a page 215 mm (8.5 in) wide. An A5 page is 151 mm (5.9 in) wide, and the A6 is 107 mm (4.2 in) wide.

A LAW

The PCM coding and companding standard used in Europe. See A LAW ENCODING and E-1.

A LAW Encoding

The method of encoding sampled audio waveforms used in the 2048K bit 30 channel PCM primary system, widely used outside North America. See E-1.

Access Tandem

A special type of local phone company central office. It's designed to provide equal access for all the long distance carriers in that area.

Active Controls (ActiveX)

A set of technologies that enable software components to interoperate in a networked environment, providing end users with a richer, more interactive experience. These are interactive objects created by programmers. They can be embedded in documents, programs and on Web pages. ActiveX video control could be used to enhance a Web page with real-time video sequences. The controls are language-independent and can be programmed using C++ Visual Basic or Java.

Over 1,000 ActiveX controls are available today. These include the Macromedia Shockwave for Director control and the Adobe Acrobat control. The ActiveX technologies are enhancements to OLE, Microsoft's component software technology, which has become a well-established, industry standard since the introduction of OLE in 1994. ActiveX Controls are small, efficient modules that implement specific, specialized functions.

Activity Report

Provides a record of transmission time, date, size of the file, recipient's telephone number, transmission success or failure, the sender's name, and other pertinent information. This is a valuable management tool to get an overview of a company's fax traffic and costs.

A/D

Analog to digital conversion.

Adaptive Pulse Code Modulation

A way of encoding analog voice signals into digital signals by adaptively predicting future encodings by looking at the immediate past. The adaptive part of this method reduces the number of bits per second that another rival and more common method called PCM (Pulse Code Modulation) requires to encode voice. Adaptive PCM is not especially common because, even though it reduces the number of bits required to encode voice, the electronics to do it are expensive. See also PULSE CODE MODULATION.

ADC

Analog-to-Digital Converter.

Administrable Service Provider

A service provider which supports administrable services (e.g., SCR).

ADPCM

Adaptive Differential Pulse Code Modulation. A speech coding method which calculates the difference between two consecutive speech samples in standard PCM coded voice signals. This calculation is encoded using as adaptive filter and therefore, is transmitted at a lower rate than the standard 64 Kbps technique.

Typically, ADPCM allows an analog voice conversation to be carried within a 32k-kbit/s digital channel; 3 or 4 bits are used to describe each sample, which represents the difference between two adjacent samples. Sampling is done 8,000 times a second.

ADSI

Analog Display Services Interface. ADSI is a Bellcore standard defining a protocol on the flow of information between something (a switch, a server, a voice mail system, a service bureau) and a subscriber's telephone, PC, data terminal or other communicating device with a screen. The simple idea of ADSI is to add words to, and therefore a modicum of simplicity of use to a system that usually uses only touch tones.

Imagine a normal voice mail system. You call it. It answers with a voice menu. Push 1 to listen to your messages, 2 to erase them, 3 to store them, 4 to forward them, etc. It's confusing. You have to remember which is which. ADSI is designed to solve that. It's designed to send to your phone's screen the choices in words that you're hearing.

You then have the choice of responding to what you hear or what you see. Your response is the same -- a touch-tone button. ADSI's signaling is DTMF and standard Bell 202 modem signals from the service to your 202-modem equipped phone. From the phone to the service it's only touch-tone.

With ADSI, you don't hear the modem signaling because every time the service gets ready to send you information, it first sends a "mute" tone. ADSI works on every phone line in the world.

Advisory Tones

Signals such as dial tone, busy, ringing, fast-busy, call-waiting, camp-on-- and all the other tones which your telephone system uses to tell you that something is happening or about to happen.

AEB

Analog Expansion Bus. Dialogic's name for the analog electrical connection between its network interface modules and its analog resource modules. This bus is now "open." Technical specifications are available, thus enabling outsiders to make their own resource modules and/or network interface modules. The AEB interfaces DTI/124 and D/4x voice response component boards which fit in an AT-expansion slot of a PC. See also PEB, which is the more modern digital PCM expansion bus. See also SCSA, which is Dialogic's latest standard.

AGC

Automatic Gain Control. There are two electronic ways you can control the recording of something -- Manual or Automatic Gain Control (AGC). AGC is an electronic circuit in tape recorders, speakerphones, and other voice devices which is used to maintain volume. AGC is not always a brilliant idea, since AGC will attempt to produce a constant volume level.

This means it will try to equalize all sounds -- the volume of your voice, and, when you stop talking, the circuit static and/or general room noise which you undoubtedly do not want amplified. Sometimes it's better to have quiet, when you want quiet. Manual Gain Control is preferred in professional applications.

Alerting Signal

A ringing signal put on subscriber access lines to indicate there's an incoming call.

Algorithm

A set of ordered steps for solving a problem, such as a mathematical formula or instructions in a program. In the context of speech coding, it refers to the mathematical methods used to compress speech. Unique speech-coding algorithms are patentable. Specific implementations of an algorithm in the form of computer programs are also subject to copyright protection.

All Trunks Busy

When a user tries to make an outside call through a telephone system and receives a "fast" busy signal (twice as many signals as a normal busy in the same amount of time), he is usually experiencing the joy of All Trunks Busy. No trunks are available to handle that call.

The trunks are all being used at that time for other calls or are out of service. These days, many long distance companies are replacing a "fast" busy signal with a recording that might say something like, "I'm sorry. All circuits are busy. Please try your call later."

AMIS

See Audio Messaging Interchange Specification. AMIS is a standard for networking voice mail systems.

Amplitude

The distance between high or low points of a waveform or signal. Also referred to as the wave "height."

Amplitude Distortion

The difference between the output wave shape and the input wave shape.

Amplitude Equalizer

A corrective network that is designed to modify the amplitude characteristics of a circuit or system over a desired frequency range. Such devices may be fixed, manually adjustable, or automatic.

Amplitude Modulation

Also called AM, it's a method of adding information to an electronic signal in which the signal is varied by its height to impose information on it. "Modulation" is the term given to imposing information on an electrical signal. The information being carried causes the amplitude (height of the sine wave) to vary. In the case of LANs, the change in the signal is registered by the receiving device as a 1 or a 0. A combination of these conveys different information, such as words, numbers or punctuation marks.

AMX

Analog Matrix Switch. A Dialogic product. An 8 x 8 analog crosspoint switch on a board used for sharing resources among AEB-based products and for connecting to phones and external audio devices.

Analog

Comes from the word "analogous," which means "similar to." In telephone transmission, the signal being transmitted -- voice, video, or image -- is "analogous" to the original signal. In other words, if you speak into a microphone and see your voice on an oscilloscope and you take the same voice as it is transmitted on the phone line and ran that signal into the oscilloscope, the two signals would look essentially the same.

The only difference is that the electrically transmitted signal (the one over the phone line) is at a higher frequency. In correct English usage, "analog" is meaningless as a word by itself. But in telecommunications, analog means telephone transmission and/or switching which is not digital. See ANALOG TRANSMISSION.

Analog Bridge

A circuit which allows a normal two-person voice conversation to be extended to include a third person -- without degrading the quality of the call. Digital bridges work better.

Analog/Digital Converter

An A/D Converter. Pronounced: "A to D Converter." A device which converts an analog signal to a digital signal.

Analog Facsimile

Facsimile which can transmit and receive grey shadings -- not just black and white. "Analog" facsimile is usually transmitted digitally.

But it's called analog because of its ability to transmit what appear to be continuous shades of grey.

Analog Transmission

A way of sending signals -- voice, video, data -- in which the transmitted signal is analogous to the original signal. In other words, if you spoke into a microphone and saw your voice on an oscilloscope and you took the same voice as it was transmitted on the phone line and threw that signal onto the oscilloscope, the two signals would look essentially the same. The only difference between them is that the electrically transmitted signal would be at a higher frequency.

ANI

Automatic Number Identification. A phone call arrives at your home or office. At the front of the phone call is a series of digits which tell you the phonc number calling you. These digits may arrive in analog or digital form. They may arrive as touch-tone digits inside the phone call. They may arrive in a digital form on a separate circuit.

Whichever way they arrive, you will need some equipment to decipher the digits AND do "something" with them. That "something" might be throwing them in a database and bringing up your customer's record on a screen in front of your telephone agent as he answers the phone. "Good morning, Mr. Smith." Some large users say they could save as much as 20 seconds on the average IN-WATS call if they knew early on the phone number of the person calling them. They wouldn't need to ask their regular customers for their address, phone number, credit card number, etc.

Announcement Service

Allows a phone user to hear a recording when he dials a certain number. Could be an answering machine. Could be a voice response system. Could also be a solid-state digital announcer. Or a voice response system.

ANSI

American National Standards Institute. A standards- setting, non-government organization which develops and publishes standards for voluntary use in the U.S.

Answer Detect

The use of a digital signal processing technique to determine the presence of voice energy on a telephone line. It is used with call (answer) supervision, to identify an answered line. It is beginning to be used with computerized dialing equipment in business to consumer calling. This technique eliminates the need for a telephone representative to constantly monitor call set-up progress on each telephone line in the event a call is answered. See ANSWER SUPERVISION.

Answer Supervision

Follow this scenario: I call you long distance. My central office must know when you answer your phone, so my central office or long distance phone company can start billing me for the call. It works like this: when you, the called party, answer your phone, your central office sends a signal back to my central office (the originating CO). This tells my central office to start billing me for the call. This signal is called Answer Supervision.

Before the Divestiture of the Bell System in early 1984, most of the nation's long distance companies -- with the exception of AT&T -- did not receive Answer Supervision. They did not know precisely when the called party answered. So they started their billing cycle after some time -- 20 or 30 seconds after the caller completed dialing. These long distance companies presumed that after this time, some one will have answered and the call will be in progress (and they therefore start timing and billing for the call). Without Answer Supervision, their billing of calls is inaccurate. With the Divestiture of the Bell System, and the introduction of Equal Access, long distance companies now receive true Answer Supervision. See also SOFTWARE SUPERVISION.

Anti Aliasing

A computer imaging term. A blending effect that smoothes sharp contrasts between two regions of different colors. Properly done, this eliminates the jagged edges of text or colored objects. Used in voice processing, anti-aliasing usually refers to the process of removing spurious frequencies from waveforms produced by converting digital signals back to analog.

API

See Application Program Interface.

Application Class

A group of client applications that perform similar services, such as voice messaging or fax-back services.

Application Profile

A description of the kinds of resources and services required by a client application (or an application class). An application profile is defined once for an instance of an application; then system services such as the SCR will be able to fulfill the needs of the application without the application having to state its needs explicitly.

Application Programming Interface. "Hooks" into software. A set of standard software interrupts, calls, and data formats that application programs use to initiate contact with network services, mainframe communications programs, or other program-to-program communications.

For example, applications use APIs to call services that transport data across a network. A set of formalized software calls and routines that can be referenced by an application program to access underlying network services.

Standardization of APIs at various layers of a communications protocol stack provides a uniform way to write applications. NetBIOS is an early example of a network API. Also, applications use APIs to call services that transport data across a network. See also WINDOWS TELEPHONY.

Application

A software program that carries out some useful task. Database managers, spreadsheets, communications packages, graphics programs, and word processors are all applications.

Applications Generator

Basically, software that writes software. Applications generators are software tools that, in response to your input, write software code which a computer can understand. Applications generators have two major benefits:

First, they save time. An applications generator is perfect for demonstrating a quick, though not complete, programming application.

Second, they can often be used by non-programmers. Applications generators are often used in programming voice processors. Applications generators come in many flavors. They may be general purpose tools. Or they may provide support for specific application segments such as: connecting voice response units to mainframe databases; voice messaging system development; audiotex system development, etc.

One of applications generators' biggest advantages is that their ability to translate user specified screens and menus into programming code. In essence you produce the screen or menu using an interface as simple as a word processor. Then the applications generator translates that screen into programming code in a language, such as the widely-used "C." Once translated into C, a programmer proficient in C could go through the code and "improve" on it.

Applications Processor

A special purpose computer which attaches to a telephone system and allows the telephone system (and the people using it) to do different "applications" things, like voice mail, interactive voice response, etc. We think AT&T invented the term Applications Processor.

Appli/COM

See T.611.

Apps Gens

See APPLICATIONS GENERATORS.

ASCII

American Standard Code for Information Interchange. ASCII Character Set A character set consisting only of the characters included in the original 128-character ASCII standard.

Asynchronous Communication

A method of data communication in which the transmission of bits of data is not synchronized by a clock signal but is accomplished by sending the bits one after another, with a start bit and a stop bit to mark the beginning and end of the data unit. The two communicating devices must be set to the same speed-called the baud rate. Asynchronous communication normally is used for transmission speeds under 19,200 baud. Because of the lower communication speeds, normal telephone lines can be used for asynchronous communication.

Asynchronous Request

A request where the client does not wait for completion of the request, but does intend to accept results later. Contrast with synchronous request.

AT+V

A new ANSI standard for voice modems. It is a superset of the Hayes AT command set. AT+V combines pre-fixed Hayes AT commands with a new set of voice-related +V commands.

The specification is detailed in ANSI/TIA/EIA IS-101 "Facsimile Digital Interfaces – Voice Control Interim Standard for Asynchronous DCE." The TIA TR-29.2 subcommittee details the specification in their PN-3131. Rockwell's voice modem ship set does not comply with this standard, but uses another called AT#V, which is similar. In Windows 95, the variance between these command sets is ratified by the Win 95 system registry and vendor-supplied INF files.

AT WORK Protocol

Microsoft's integration protocol for linking peripheral office equipment, including telephony services, to Windows applications.

Audio

Sound you hear which may be converted to electrical signals for transmission. A human being who hasn't had his ears blown by listening to a Sony Walkman or a ghetto blaster can hear sounds from about 15 to 20,000 hertz.

Audio Frequencies

Those frequencies which the human ear can detect (usually in the range of 15 to 20,000 hertz). Only those from 300 to 3,000 hertz are transmitted through the phone. Which is why the phone doesn't sound "Hi-Fi."

Audio Menu

Options spoken by a voice processing system. The user can choose what he wants done by simply choosing a menu option -- hitting a touch-tone on his phone, or speaking a word or two. There are basically two ways of organizing computer or voice processing software -- menu-driven and non-menu driven. Menu-driven programs are easier for users to use. But they can only present as many options as can be reasonably spoken in a few seconds.

Audio menus are typically played to callers in automated attendant/voice messaging, voice response and transaction processing applications. See also MENU.

Audio Messaging Interface Specification (AMIS)

Issued in February 1990, AMIS is a series of standards aimed at addressing the problem of how voice messaging systems produced by different vendors can network or inter-network. Before AMIS, systems from different vendors could not exchange voice messages. AMIS deals only with the interaction between two systems for the purpose of exchange voice messages. It does not describe the user interface to a voice messaging system, specify how to implement AMIS in a particular systems or limit the features a vendor may implement. AMIS is really two specifications. One, called AMIS-Digital, is based on completely digital interaction between two voice messaging systems. All the control information and the voice message itself, is conveyed between systems in digital form. By contrast, the AMIS-Analog specification calls for the use of DTMF tones to convey control information and transmission of the message itself is in analog form. AMIS was discussed in detail in the October 1990 issue of Business Communications Review.

Audio Response Unit (ARU)

A device which translates computer output into spoken voice. Let's say you dialed up a computer and it said "If you want the weather in Chicago, push button "123," then it would give you the weather. But that weather would be "spoken" by an audio response unit.

Audiotex

A generic term for voice response equipment and services. Audiotex may be passive or interactive. In "passive," the classic applications fall into "Jokes, Scopes and Soaps." Dial a number, hear the joke of the day. Dial a number, hear your horoscope. Dial a number, hear what's happening on your latest TV soap. In its interactive form, Audiotex is basically another term for Interactive Voice Response. You call a phone number. A machine answers. It presents you several options, "Push 1 for information on Plays, Push 2 for information on movies, Push 3 for information on Museums." If you push 2, the machine may come back, "Push 1 for movies on the south side of town, Push 2 for movies on the north side of town, etc." Or it says, "To hear your present bank balance, touch-tone your bank account number in now." See also INTERACTIVE VOICE RESPONSE.

Auto attendant

"If you know the extension of your party, enter it now. If you would like to speak to an operator, press two." Auto attendants are being used more and more by businesses large and small to handle initial routing of incoming telephone calls.

Auto Fax Tone

Also called CNG, or Calling Tone. This tone is the sound produced by virtually all Group 3 fax machines when they dial another fax machine. CNG is a medium pitch tone (1100 Hz) that lasts 1/2 second and repeats every 3 1/2 seconds. A FAX machine will produce CNG for about 45 seconds after its dials. See also CNG.

Automatic Cover Letter

Allows the user to automatically attach a cover letter to the document being sent. This is especially convenient when sending material such as spreadsheets, for example.

Automatic Redial

Provides for the automatic redialing of a voice, data, or fax number in the event the receiving line is busy or an error occurred in sending the data. Some products allow the user to specify the redial attempts and correct specific errors to be used when redialing.

Automatic Routing

Allows incoming fax documents to be automatically routed to the addressed individual on a LAN or centralized system. Current technology provides several methods of accomplishing this using various techniques.

Automated Attendant

A specialized form of an Interactive Voice Response system. An IVR connected to a PBX. When a call comes in, this device answers it, and says something like: "Thanks for calling the ABC Company. If you know the extension number you'd like, push-button that extension now and you'll be transferred.

If you don't know it, push button "0" (zero) and the live operator will come on. Or wait a few seconds and the operator will come on, anyway."

Sometimes the automated attendant might give you other options -- like "dial 3" for a directory. Automated attendants are sometimes connected also to voice mail systems ("I'm not here. Leave a message for me."). Some people react well to automated attendants. Others don't.

A good rule: Before you spring an automated attendant on your people/customers/subscribers, etc., let them know. Train them a little. They'll then perceive the automated attendant device a lot more positively. Tip: When you reach an automated attendant and don't know the extension number, dial the person's last name on your touch-tone pad. Better automated attendants will recognize the name and translate it into the extension number.

Automatic Call Distributor (ACD)

A specialized phone system used for handling many incoming calls. Typically used by airlines, rent-a-car companies, hotels, etc. An ACD has four functions. First, it will recognize and answer an incoming call. Second, it will look in its database for instructions on what to do with that call. Third, based on that database's instructions, it will send the call to a recording that "somebody will be with you soon, please don't hang up!" Fourth, it will send the call to one operator of a group of operators -- as soon as that operator has completed their previous call, and/or the caller has heard the canned message.

The term Automatic Call Distributor comes from distributing the incoming calls in some logical pattern to a group of operators. That pattern might be Uniform (to distribute the work uniformly). It also might be Top-down. Distributing calls logically is the function most people associate with an ACD. But it's not the most important. Much more important is the management information which the ACD produces. This information typically is of two sorts -- 1: The arrival of incoming calls (when, how many, which lines, from where, etc.) and 2: How many callers were put on hold, told to wait and didn't. This is called information on ABANDONED CALLS.

This information is very important for staffing, for buying lines from the phone company and also for figuring what level of service to provide to the customer. And what different levels of service (how long for people to answer the phone) might cost.

Automatic Call Distributors are now combined with Interactive Voice Response Systems and ANI (automatic number identification) to give much effective and faster handling of incoming customer calls.

Automatic Call Sequencer

ACS. A device for handling incoming calls. Typically it performs three functions. 1. It answers an incoming call, gives the caller a message, and puts them on "Hold." 2. It signals the agent (the person who will answer the call) which call on which line to answer. Typically, the call which it signals to be answered is the call which has been on "hold" the longest. 3. It provides management information, such as how many abandoned calls there were, how long the longest person was kept on hold, how long the average "on hold" was, etc.

Automatic Callback

When a caller dials another internal extension and finds it busy, the caller dials some digits on his phone or presses a special "automatic callback" button. When the person he's calling hangs up, the phone system rings his number and the number of the original caller and the phone system automatically connects the two together. This feature saves a lot of time by automatically retrying the call until the extension is free.

Automatic Calling Unit

ACU. A device that places a telephone call on behalf of a computer.

Automatic Circuit Assurance

ACA is a PBX feature that helps you find bad trunks. The PBX keeps records of calls of very short and very long duration. If these calls exceed a certain parameter, the attendant is notified. The logic is that a lot of very short calls or one very long call may suggest that a trunk is hung, broken or out of order. The attendant can then physically dial into that trunk and check it.

Automatic Number Identification

ANI. Equipment at your central office which recognizes the telephone number of the person making the call so that information about the call can be sent to the call accounting (i.e. billing) system.

B

Backbone

The part of a communications network which carries network traffic between access devices.

Backplane

The high-speed communications line which individual components of a modern electronic system are connected to. For example, all the extensions of a PBX are connected to line cards (circuit boards), which slide into the PBX's cage. At the rear of the PBX cage, there is a connector. Each of these connectors is plugged into the PBX's Backplane. Also called a Backplane Bus. This backplane bus is typically very high speed, since it carries many conversations, much address information and considerable signaling. These days, the backplane bus is typically a time division multiplexed line -- somewhat like a train with many cars, each of which represent a time slice of another conversation. The backplane's capacity determines the overall capacity of the switch.

Baseboard

A Dialogic voice processing board without any daughterboards attached.

Battery

1. Term used to reference the DC power source of a telephone system. Often called "Talk Battery." 2. Storage battery used with central office switching systems and PBXs serving locations which cannot tolerate outages.

Batteries serve the following purposes: Act as a filter across the generator or power rectifier output to smooth out the current and reduce noise; provide a cushion against periodic overloads exceeding the generator/rectifier capacity; supply emergency power for a limited time in event of commercial power failure.

Baud

The number of changes in signal state (switching) per second in a signal sent by a modem. A baud may contain four or more bits. Sometimes confused with BPS, the bits per second transmitted on the channel.

Baud Rate

The transmission rate of a communications channel. Technically, baud rate refers to the maximum number of changes (switching) that can occur per second in the electrical state of a communications circuit. Under the RS-232C communications protocol, 300 baud is likely to equal 300 bps, but at higher baud rates the number of bits per second transmitted is actually higher than the baud rate because one change can represent more than one bit of data. For example, 1,200 bps is usually sent at 600 baud by sending two bits of information with each change in the electrical state of the circuit.

BISDN

Broadband ISDN (Integrated Services Digital Network) Also see ISDN.

Bit

Binary digit, the smallest amount of information in a binary system, a 0 or 1 condition.

BIT RATE

The number of bits that a communications channel can transmit in a fixed time interval, typically measured in kilobits (1000 bits) per second (kbps). Uncompressed, telephone-quality speech is 64 kbps. It is made up of 8,000 samples every second, with each sample being 8 bits. CD-quality voice requires approximately 706 Kbps (44,100 samples every second with each sample being composed of 16 bits).

Board

An SCSA definition. Any hardware module that controls its own physical interface to the SCbus or SCxbus. From a programming point of view, a board is an addressable system component that contains resources.

Bong

A tone that long distance carriers and value added carriers make in order to signal you that they now require additional action on your part -- usually dialing more digits.

BPS

Bits per second.

Broadcasting

This procedure allows the user to send a voice message, data file, or fax document to a group of people or companies. Groups can be temporarily or permanently stored in the telephone directories for repeated broadcasting.

Busy

In use. "Off-hook". There are slow busies and fast busies. Slow busies are when the phone at the other end is busy or off-hook. They happen 60 times a minute. Fast busies (120 times a minute) occur when the network is congested with too many calls. Your distant party may or may not be busy, but you'll never know because you never got that far.

Byte

A group of 8 bits, making up a single memory location. Most computers cannot address a bit, they can only address bytes.

C

The most common programming language in the voice processing industry. C operates under MS-DOS or UNIX, and other operating systems.

It is very powerful and is becoming a standard also for programming telecom switches, PBX and central office.

Cadence

In voice processing, cadence is used to refer to the pattern of tones and silence intervals generated by a given audio signal. Examples are busy and ringing tones.

A typical cadence pattern is the US ringing tone, which is one second of tone followed by three seconds of silence. Some other countries, such as the UK, use a double ring, which as two short tones within about a second, followed by a little over two seconds of silence.

Call

Two or more parties connected together for the purpose of communication. A party may be either a person or a client application, each respectively directing a group or telephone terminal equipment (phone, fax, etc.)

Call Center

A place where calls are answered and calls are made. A call center will typically have lots of people (also called agents), an automatic call distributor, a computer for order-entry and lookup on customers' orders. A Call Center could also have a predictive dialer for making lots of calls quickly. The term "call center" is broadening. It now includes help desks and service lines.

Call Control

The establishment and control of telephone links. This involves answering, originating, transferring, conferencing, and terminating telephone calls, and monitoring their status. The electronic signaling functions that control the setting up, monitoring and tearing down of telephone calls. The electronic signaling functions that control the setting up, monitoring, and tearing down of telephone calls. First-party call control is the view of directly controlling a telephone set; third-party call control takes the view of controlling the call through a switch (PBX) in a centralized fashion. Generally third-party call control also refers to the control of Other functions that relate to the switch at large, such as ACD queuing, etc.

Call Processing

Call processing is setting up, connecting, transferring, disconnecting, etc. a phone call. Call processing does not affect the content -- voice or otherwise -- of the conversation. That process is called "voice processing." And that is a broader concept, encompassing everything from compression, storage, editing to recognition.

Call Progress Tone

A tone sent from the telephone switch to tell the caller of the progress of the call. Examples of the common ones are dial tone, busy tone, ringback tone, error tone, re-order, etc. Some phone systems provide additional tones, such as confirmation, splash tone, or a reminder tone to indicate that a feature is in use, such as confirmation, hold reminder, hold, intercept tones.

Caller ID

Signals that identify the telephone number of the calling party; these signals are sent along the telephone line at the start of the call. Not all telephone network connections support caller ID signaling. In a more general sense, caller ID can be obtained by prompting the caller for an identity code.

Calling Party Identification

A new service being tested in some local areas which tells the person being called which number is calling them. They can then decide to answer or not answer the call. See ANI, which stands for Automatic Number Identification.

CALLPATH

IBM's announced telephone system link to IBM's computers. See CALLPATH SERVICES ARCHITECTURE.

CALLPATH CICS

Enabling software that connects your telephone systems with your IBM 370 or 390 (i.e. the mainframe version of CallPath/400, which works on the AS/400 platform).

CALLPATH Services Architecture

CSA. IBM's program for integrating voice and data technologies. It is both a strategy and set of open architecture commands and interfaces for integrating voice and database technologies.

The idea is that with CallPath a call will arrive at a computer terminal simultaneously with the database record of the caller. And such call and database record can be transferred simultaneously to an expert, a supervisor, etc. The first implementation is CallPath/400 which works with the IBM AS/400 minicomputer. CallPath CICS works with IBM mainframes. See OPEN APPLICATION INTERFACE and DIRECTTALK.

Canonical Address

A method for storing unique telephone numbers. Canonical addressing is used by TAPI for making telephone calls from a database of numbers. A canonical address describes all possible aspects of a telephone number. You can call a telephone number using canonical addressing independent of calling location or access method. A canonical address is stored in a database and preceded by an ASCII Hex (2B) to indicate its address type. It includes delimiters and strings for Country Code, Area Code, Subscriber Number, Subaddress and Name.

Carrier Frequency

The frequency of a carrier wave.

Carrier Wave

A wave having at least one characteristic that may be varied from a known reference value by modulation.

CAS

Communicating Applications Specification. A high-level API (application programming interface) developed by Intel and DCA that was introduced in 1988. CAS enables software developers to integrate fax capability and other communication functions into their applications.

CBF

Computer-Based Fax.

CBR
Case-based reasoning.

CCITT
Comite Consultatif Internationale de Telegraphique et Telephonique (Consultative Committee for International Telephone and Telegraph - CCITT). Now called the International Telecommunications Union (ITU), the world standards-setting organization. See aslo Telecommunication Standardization Sector.

CDE
Solaris Common Desktop Environment. Packaged with Solaris. Sun's windowing environment for Solaris. Provides transparent access to network resources and a common look and feel.

CED
Called Station Identification. 2,100-Hz tone when fax machine answers.

Central Office
(CO)Telephone company facility where subscribers' lines are joined to switching equipment for connecting other subscribers to each other, locally and long distance.

Centrex
Centrex is a business telephone service offered by a local telephone company from a local central office. Centrex is basically single line telephone service delivered to individual desks (the same as you get at your house) with features, i.e. "bells and whistles," added. Those "bells and whistles" include intercom, call forwarding, call transfer, toll restrict, least cost routing and call hold (on single line phones). Centrex is known by many names among operating phone companies, including Centron and Cenpac. Centrex comes in two variations -- CO and CU. CO means the Centrex service is provided by the Central Office. CU means the central office is on the customer's premises.

CFR
Confirmation to Receive frame.

Channel

1. An SCSA definition. A transmission path on the SCbus or SCxbus Data Bus that transmits data between two end points. 2. Typically what you rent from the telephone company. A voice-grade transmission facility with defined frequency response, gain and bandwidth. Also, a path of communication, either electrical or electromagnetic, between two or more points. Also called a circuit, facility, line, link or path.

Channel Bank

A multiplexer. A device which puts many slow speed voice or data conversations onto one high-speed link and controls the flow of those "conversations." Typically the device that sits between a digital circuit -- say a T-1 -- and a couple of dozen voice grade lines coming out of a PBX. One side of the channel bank will be connections for terminating two pairs of wires or a coaxial cable -- those bringing the T-1 carrier in.

On the other side are connections for terminating multiple tip and ring single line analog phone lines or several digital data streams. Sometimes you need channel banks. Sometimes, you don't. For example, if you're shipping a bundle of voice conversations from one digital PBX to another across town in a T-1 format -- and both PBXs recognize the signal -- then you will probably not need a channel bank.

You'll need a Channel Service Unit (CSU). If one, or both, of the PBXs is analog, then you will need a channel bank at the end of the transmission path whose PBX won't take a digital signal. See CHANNEL SERVICE UNIT and T-1.

Channel Service Unit

CSU. A device used to connect a digital phone line (T-1 or less) coming in from the phone company to either a multiplexer, channel bank or directly to another device producing a digital signal, e.g. a digital PBX, or data communications device. A CSU performs certain line-conditioning, and equalization functions, and responds to loopback commands sent from the central office. A channel service unit is also called a Data Service Unit. See CSU.

CIG

Calling Subscriber Identification. A frame that gives the caller's telephone number.

CGI

Common Gateway Interface

CLASS

1. Custom Local Area Signaling Services. It is based on the availability of channel interoffice signaling. Class consists of number-translation services, such as call-forwarding and caller identification, available within a local exchange of Local Access and Transport Area (LATA). CLASS is a service mark of Bellcore. Some of the phone services which Bellcore promotes for CLASS are Automatic Callback, Automatic Recall, Calling Number Delivery, Customer Originated Trace, Distinctive Ringing/Call Waiting, Selective Call Forwarding and Selective Call Rejection.
See also CALLING LINE IDENTIFICATION.

2. In an object-oriented programming environment, a class defines the data content of a specific type of object, the code that manipulates it, and the public and private programming interfaces to that code.

Class 1

The Class 1 interface is an extension of the EIA/TIA's (Electronics Industry Association and the Telecommunications Industry Association) specification for fax communication, known as Group III. Class I is a series of Hayes AT commands that can be used by software to control fax boards. In Class 1, both the T.30 (the data packet creation and decision making necessary for call setup) and ECM/BFT (error-correction mode/binary file transfer) are done by the computer. A specification being developed (fall of 1991) Class 2, will allow the modem to handle these functions in hardware. Industry analysts believe Class 2 will be the standard for the long haul, but approval is slow. Even so, some modem makers will shortly deliver data/fax modems.

Clear Channel

A channel which is used exclusively for data transmission, with no bandwidth required for administrative messages such as signaling or synchronization. All SCbus data channels are clear.

Client

The requesting communication element in a distributed computing system. Clients send requests to servers across a network and wait for indication from the server that the request is complete. Any object that uses the resources of another object.

Client Application

Any computer program making use of the processing resources of another program.

Client Operating System

Operating System running on the client.

Client-server

A software architecture which splits operation between two component programs and their respective communication elements: the client and the server. Client elements make requests of servers. Servers process the requests and often return processed data to the client.

CNG

Also called Auto Fax Tone, or Calling Tone. This tone is the sound produced by virtually all fax machines when they dial another fax machine. CNG is a medium tch tone (1100 Hz) that lasts 1/2 second and repeats every 3 1/2 seconds. A fax machine will produce CNG for about 45 seconds after it dials. The CNG tone is useful for owners of fax/phone/modem switches. Such switches answer an incoming call. If they hear a CNG tone, they will transfer the call to a fax machine. If they don't, they'll transfer the call to a phone, answering machine or perhaps a modem. Depends on how they're set up.

CO

Central Office. In North America, a CO is that location which houses a switch to serve local telephone subscribers. Sometimes the words "central office" are confused with the switch itself. In Europe and abroad, the words "central office" are not known. The more common words are "public exchange." But those words tend to refer more to the switch itself, rather than the site, as in North America. See also CENTRAL OFFICE or PUBLIC EXCHANGE.

CO Lines

These are the lines connecting your office to your local telephone company's Central Office which in turn connects you to the nationwide telephone system.

COM

Component Object Model. COM is Microsoft's cornerstone of the ActiveX platform. COM is a language independent component architecture (not a programming language). It is meant to be a general purpose, object-oriented means to encapsulate commonly used functions and services. The COM architecture provides a platform independent and distributed platform for multi-threaded applications.

COM also encompasses everything previously known as OLE Automation (Object Linking and Embedding). OLE Automation was originally designed to allow higher level programming languages access to COM objects. An object is a set of functions collected into interfaces. Each object has data associated with it. The source of the data itself is called the data object. With COM, the transfer of the data itself is separated from the protocol for doing so.

Compression

Reducing the representation of the information, without losing the information itself. Reducing the bandwidth or bits necessary to encode information. Compression saves transmission time and storage requirements on storage devices such as hard disks, tape devices, and floppy disks.

Compression Algorithm

The arithmetic formulae which convert a signal into a smaller bandwidth or fewer bits.

Compression Ratio

Measurement of compressed data. For example, a file compressed to 1/4th its original size can be expressed as 4:1. In telecommunications, compression ratio also refers to the amount of bandwidth-reduction achieved. For example, 4:1 compression of a 64 kbps channel is 16 kbps.

CO Simulator

A desktop device which pretends to act like a mini-central office. The smallest version will consist of two lines and two RJ-11 jacks. Plug a phone into both jacks. Pick up one phone. You hear dial tone. Dial or touch-tone two or three digits. Bingo, the second phone rings.

You pick up the second phone. You can have a conversation with yourself or with a machine -- like a voice processing system. Most central office simulators can simulate normal on-hook, off-hook, dialing, answering, speaking, etc. Some now can simulate caller ID features -- including number of person calling.

Compiler

A computer program used to convert symbols meaningful to a human into codes meaningful to a computer. For example, taking instructions written in a "higher" level language such as C, BASIC, COBOL or ALGOL and converting them into machine code which can be read and acted upon by a computer.

Compression Algorithm

The arithmetic formulae which convert a signal into smaller bandwidth or few bits.

Computer-Aided Dialing

Another term for Predictive Dialing. See PREDICTIVE DIALING.

Computer-Telephone Integration

CTI. The merging of the computer and telephone, which will transform the personal computer from being an information processing device to also being a powerful platform for communications. Linking telecommunications to the ever-increasing processing capabilities and rich user interface of the computer will enable new forms of communications and richer access to existing types of communication, including voice, asynchronous data, fax, remote access to LANs, Internet access, on-line services and more. "The Communicating PC" will redefine how we share ideas and information, and provide a portal to other people, computers and network services, anywhere in the world. See also Computer Telephony.

Computer Telephony

Computer telephony represents the cooperative merger between telephony and data communications. It is a form of telecommunication that concentrates on the movement of both encoded and voice-related information from one point to another for the purpose of automating transactions between machines and humans, or between machines.

Interactive voice response, voice mail, and every type of host computer-to-switch integration scheme are examples of computer telephony. Because computer telephony requires both computers and telephones to work cooperatively, data processing is employed. Computer Telephony is a $4 billion, 30%-a year growth industry, with many areas growing at over 100% a year. Computer Telephony has two basic goals: to please customers and to enhance corporate productivity.

In order to please customers, developers of this technology must orchestrate a broad range of technologies, each of which is changing and growing more powerful every minute. The top disciplines being attacked by computer telephony are:

1. Messaging. This includes voice, fax and electronic mail, fax routers, paging and unified messaging (also called integrated or mixed media messaging). We use it to control our fax blasters, fax servers and Internet Web-vectored phones.

2. Real-time Connectivity. This means handling both inbound and outbound call calls. Your automated attendant and "predictive" or "preview" dialing system are the most used in this area. Today, "follow me" numbers, video, audio and text-based conferencing, "PBX in a PC," and collaborative computing are gaining acceptance. Real-time connectivity is at work when you use one number calling and when calls are routed with the help of LAN-based call director programs like Phonetastic or CallWare.

3. Transaction Processing and Information Access via the Phone. We live in a world of self-navigation over the phone.

Every day, we encounter Interactive Voice Response, Audiotex, and Other means to gain customer access to enterprise data. This is the classic "giving data a voice," model. You encounter it when you order documents over the phone with a fax on demand system or when you shop on the World Wide Web. Transactions as simple as hearing the time or weather over the phone are included.

4. Adding Intelligence to Phone Calls. When a "screen pop" of customer records collides at an agent's desk coincident with inbound or outbound phone calls - intelligence is added. This includes mirrored Web page "pops," smart agents, skills-based call routing and virtual (geographically distributed) call centers. Companies like GeoTel, Lansys, Teloquent and Edify really shine in this area. These folks are pioneering the concept of computer telephony groupware, smart help desks and "AIN" (intelligent network-based) computer telephony services.

5. Core Technologies. There are dozens of them. We use voice recognition, text-to-speech, applications generators (of all varieties -- GUI to forms-based to script-based), VoiceView, DSVD, computer-based fax routing /binary file transfer, USB (Universal Serial Bus), GeoPort, video and audio compression, call progress, dial pulse recognition, Caller ID and ANI, digital network interfaces (T-1, E-1, ISDN BRI and PRI, SS7, frame relay and ATM), voice modems, client-server telephony, logical modem interfaces, multi-PC telephony synchronization and coordination software, the Internet,

the Web and the "Intranet." We owe most of the recent advances in these technologies to DSPs (Digital Signal Processors).

With DSPs, we can manipulate, crunch, analyze, decode, reformat and in general do just about anything we want with LAN,WAN, or POTS phone line transmissions. DSP algorithms are on the lowest level of the computer telephony food chain - but the most fundamental building block we have.

Conference Bridge

A telecommunications facility or service which permits callers from several diverse locations to be connected together for a conference call.

The conference bridge contains electronics for amplifying and balancing the conference call so everyone can hear each other and speak to each other. The conference call's progress is monitored through the bridge in order to produce a high quality voice conference and to maintain decent quality as people enter or leave the conference.

Configuration Manager

A service which manages configuration information and controls system startup.

Connection

A TDM data path between two Resources or two Groups. It connects the inputs and outputs of the two Resources, and may be unidirectional (simplex) if either of the Resources has only an input or an output; Otherwise it is bi-directional (dual simplex). It usually has a bandwidth that is a multiple of a DS0 (64kbit) channel. Inter-group connections are made between the Primary Resource of each Resource Group.

Cover Page

The cover page is the first page of a fax message. It generally includes a header, typically the sender company's logo; the recipient's name and fax telephone number; the sender's fax and voice telephone numbers; the system's date and time; a message; and a footer.

CPE

Customer Provided Equipment, or Customer Premise Equipment. Originally it referred to equipment on the customer's premises which had been bought from a vendor who was not the local phone company.

Now it simply refers to telephone equipment -- key systems, PBXs, answering machines, etc. -- which reside on the customer's premises. "Premises" might be anything from an office to a factory to home.

CPU

The Central Processing Unit. The computing part of a computer. The "brain" of the computer. It manipulates data and processes instructions coming from software or a human operator. See CENTRAL PROCESSING UNIT.

CRP

Command repeat.

CSI (CSID)

Called Subscriber Identification. An identifier whose coding format contains a number, usually the telephone number from the remote terminal used in fax.

CSTA

Computer Supported Telecommunications Applications. This is a set of API calls agreed upon by the ECMA.

CSU

Channel Service Unit. Also called a Data Service Unit. A device to terminate a digital channel on a customer's premises. It performs certain line-conditioning, and equalization functions, and responds to loopback commands sent from the central office.

A CSU sits between the digital line coming in from the central office and devices such as channel banks or data communications devices. See CHANNEL SERVICE UNIT.

CTI

Computer Telephone Integration. Basically a polite term for connecting a computer to a telephone switch and having the computer issue the switch commands to move calls around. The most classic application for CTI is in call centers.

Picture this: A call comes in. That call carries some form of caller ID -- either ANI or Class Caller ID. The switch "hears" the calling number, strips it off, sends it to the computer. The computer does a lookup, sends back the switch instructions on what to do with the call. The switch follows orders. It might send the call to a specialized agent or maybe just to the agent the caller dealt with last time.

Customer Premises Equipment

CPE. Terminal equipment, supplied by either the telephone common carrier or by a competitive supplier, which is connected to the nationwide telephone network and resides on the customer's premises.

CVSD

Continuously Variable Slope Delta modulation. A method for coding analog voice signals into digital signals.

D/A

Digital to analog conversion.

DAA

Data Access Arrangement. A device required to hook up CPE (Customer Provided Equipment), usually modems and other data equipment, to the telephone network.

Database

A data file and its associated index files. (Occasionally the term is used to refer to multiple data files, each with its own index files.)

Data File

The file used to store the data portion of a database. Data files are usually given the .DBF extension when using any of the xBASE family of database managers.

Data Communications

This is a form of telecommunications that concentrates on the movement of encoded information from one point to another. This information is readable by machines, such as modems, computers, and fax machines as opposed to telephony, which is "readable" by the human ear.

Data Stream

A continuous flow of call processing data.

dB

Decibel. A unit of measure of signal strength, usually the relation between a transmitted signal and a standard signal source. Therefore, 6 dB of loss would mean that there is 6 dB difference between what arrives down a communications circuit, and what was transmitted by a standard signal generator.

DBMS

Database Management System. An inclusive term for a database format and its management and development software tools.

DCN

Disconnect frame. Indicates the fax call is done. The sender transmits it before hanging up; it does not wait for a response.

DCS

Digital Command Signal. Signal sent when the caller is transmitting, which tells the answerer how to receive the fax. Modem, speed, image width, image encoding, and page length are all included in this frame.

DDS

Digital Dataphone Service

Device Class

A group of related physical devices or device drivers through which applications send and receive the information or data that makes up a call.

Every device class has a *device class name* that uniquely identifies the class and provides information about the programming interface and commands that can be used to open and communicate with the devices in the class. TAPI associates devices from one or more device classes to each line or phone device.

To access one of these devices, you must retrieve the device ID for the device by using the lineGetID or phoneGetID function. You supply the device class name and the function returns the specific port name, device name, device handle, or device identifier that you need to open and access the device.

Device Driver

A piece of software that expands MS-DOS's ability to work with peripherals. The routines that make peripherals (a mouse, a RAM disk, a print spooler) work. They may be part of another program (many applications include device drivers for printers) or separate programs.

DOS comes with two device drivers: ANSI.SYS and VDISK.SYS. These are things DOS is not used to. The first is to expand the capabilities of the keyboard. The second is used to carve a RAM disk out of the 640K of RAM that MS-DOS can address. In short, device drivers can be used to expand the abilities of DOS. They are loaded from your CONFIG.SYS file when you boot up the computer. Some network operating systems run as device drivers.

They do this to get loaded before anything else so they can get the memory space they want. By being loaded as a device driver they are able to tell DOS how to operate certain "network" things, like virtual drives, shared printers, spoolers, print queues and so on. See DRIVER and VOICE DRIVER.

Dialable Address

A method for dialing unique telephone numbers based on their stored (canonical) digit strings. A dialable address contains the dialing and routing information to place calls from a specific location on a specific telephone line. TAPI supplies the means to translate a canonical address to a dialable address.

Dial-It 900 Service

A special one-way mass calling service that allows prospects, customers and others to reach you from anywhere across the country. In contrast to 800-service, the caller pays the 900 charge. The caller generally pays one charge for the first minute, with a lesser charge for each additional minute. DIAL-IT 900 Service is a great way to involve your customers and prospects in a promotion! Premium Billing lets you select a rate above standard DIAL-IT 900 rates. The long distance carriers (through their deals with local phone companies) handle the billing. You, the information provider, may split the revenues with the long distance provider. International DIAL-IT 900 service is currently available from a growing number of countries.

Dial Pulsing

A means of signaling consisting of regular momentary interruptions of a direct or alternating current at the sending end in which the number of interruptions corresponds to the value of the digit or character. In short, the old style of rotary dialing. Dial the number "five" and you'll hear five "clicks." See DIAL SPEED and DTMF.

Dial Speed

The number of pulses that a rotary dial can send in a given period of time, typically 10 pulses per second. A Hayes modem with a communications package, like Crosstalk, can send 20 pulses a second.

Dial Tone

The sound you hear when you pick up a telephone. Dial tone is a signal (350 + 440 Hz) from your local telephone company to you that it is alive and ready to receive the number you dial. If you have a PBX, dial tone will typically be provided by the PBX. Dial tone comes from the switch to which your phone is connected to.

Dictation Access and Control

A telephone system feature which allows a user to dial a dictation machine and use that machine (giving it instructions by push button) as if that machine were in his office.

Typically the material on that dictation machine is taken off by one or several typists out of a centralized word processing pool and word processed into letters, reports, legal briefs, etc.

Telephone suppliers usually don't supply the dictation equipment. Newer telephone dictation machinery is, in reality, a specialized application of voice processing equipment. See VOICE PROCESSING.

DID

Direct Inward Dialing. You can dial inside a company directly without going through the attendant. This feature used to be an exclusive feature of Centrex but it can now be provided by virtually all modern PBXs and some modern hybrids.

DID Trunks

DID trunks are employed to reduce the number of channels between the PBX and the telephone company central office. DID trunks are one-way trunks. A PBX perceives the DID trunk as one of its single-line telephones and can interpret four-digit dialing.

Digital Facsimile

A form of fax in which densities of the original are sampled and quantized as a digital signal for processing, transmission, or storage.

Digital Recording

A system of recording in which musical information is converted into a series of pulses that are translated into a binary code intelligible to computer circuits, and stored on magnetic tape of magnetic discs. Also called PCM - Pulse Code Modulation.

Digital Signal Processor (DSP)

A specialized digital microprocessor that performs calculations on digitized signals that were originally analog and then sends the results on. The big advantage of DSP lies in the programmability of digital microprocessors. DSPs can be used for compression of voice signals to as few as 4,800 bps. DSPs are an integral part of all voice processing systems and facsimile machines.

Digital-To-Analog Conversion

A circuit that accepts digital signals and converts them into analog signals. A modem typically has such a circuit. It also has other circuits, such as those doing with signaling. See MODEM.

Digital Transmission

The use of a binary code to represent information. Analog signals, like voice or data, are encoded digitally by sampling the signal many times a second and assigning a number to each sampling. Unlike an analog signal which picks up noise along the way, a digital signal can be reproduced precisely.

Digital Trunk

Generic name for a telephone connection that uses digital rather than analog transmission. Common examples are T-1 in North America and E-1 in Europe.

Digital Voice Coding

Technology by which linear audio (voice) samples are collected and then compressed using an encoding algorithm. Typically used to store voice data for future decoding.

Digitize

Converting an analog or continuous signal into a series of ones and zeros, i.e. into a digital format.

Digitized Voice

Analog voice signals represented in digital form. There are many ways of digitizing voice. See Pulse Code Modulation for the most common.

Direct Inward Dialing

DID. The ability for a caller outside a company to call an internal extension without having to pass through an operator or attendant. In large PBX systems, the dialed digits are passed down the line from the CO (central office). The PBX then completes the call. Direct Inward Dialing is often proposed as Centrex's major feature.

But automated attendants (a specialized form of interactive voice response systems) also provide a similar service.

DirectTalk

A family of IBM voice processing products introduced in the summer of 1991. According to IBM, its IBM CallPath DirectTalk product line lets businesses automate routine operations and also provide callers with easy access to many kinds of information over the telephone -- at any hour of the day and with greater accuracy. Businesses can raise the level of service they provide and do it with fewer people and with greater efficiency.

DirectX

Audio components of Microsoft DirectSound and Direct3Dsound are called DirectX. These include tools for interactive-media and game programmers. DirectX addressees sound-accelerator hardware to improve performance and minimize CPU usage. DirectX wave audio playback services are designed to support interactive-media applications for Win 95 and Win NT. DirectSound and Direct3DSound allow you simultaneously to play multiple wave files and move sound sources within a simulated 3-D space.

DIS

Digital identification signal.

Disconnect Supervision

The change in electrical state from off-hook to on-hook. This indicates that the transmission connection is no longer needed.

DMX

Digital Matrix Switch. Dialogic board that provides digital switching among four PEB spans -- for a maximum of 96 channels.

DNIS

Dialed Number Identification Service. DNIS is a feature of 800 lines. Let's say you subscribe to several 800 numbers. You use one line for testing your advertisements on TV stations in Phoenix; another line for testing your advertisements on TV stations in Chicago; and yet another for Milwaukee.

Now you get an automatic call distributor and you terminate all the lines in one group on your ACD. You do that because it's cheaper to man and run one group of incoming lines.

One queue is more efficient than several small ones, etc. You have all your people answering all the calls. You now need to know which calls are coming from where. So your long distance carrier sends you the call's DNIS -- the numbers the person dialed to reach you.

Those DNIS digits might come to you in many ways, depending on the technical arrangement you have with your long distance company. In-band or out-of-band. ISDN or data channel, etc. Make sure you understand the difference between DNIS and ANI. DNIS tells you the number your caller called. ANI is the number your caller called from.

Domain

Internet name of a computer on the 'Net. Flatiron Publishing's domain name is FlatironPublishing.com.

DOS

Disk Operating System. As in MS-DOS, which stands for MicroSoft Disk Operating System. DOS is the software that organizes how a computer reads, writes and reacts with its disks -- floppy or hard -- and talks to its various input/output devices, including keyboards, screens, serial and parallel ports, printers, modems, etc.

DPI

Dots per inch. A measure of output device resolution. The number of dots a printer can place in a horizontal inch.

Driver

Also called a Device Driver. A piece of software which lets applications software tell a computer's various parts what to do. Those parts might be everything from Dialogic boards to laser printers.

Drop And Insert

That process wherein a part of the information carried in a transmission system is demodulated (dropped) at an intermediate point and different information is entered (inserted) for subsequent transmission.

Dry T-1

T-1 with an unpowered interface.

DS-0

Digital Service, level 0. It is 64,000 bps, the worldwide standard speed for digitizing one voice conversation. There are 24 DS- channels in a DS-1.

DS-1

Digital Service, level 1. It is 1.544 megabits per second in North America, 2.048 Mbps elsewhere. Why there's no consistency is one of those wonderful questions. The 1.544 standard is an old Bell System standard. The 2.048 standard is a ITU standard. Standard for 1.544 per second is 24 voice conversations. Standard for 2.048 is 30 conversations.

DSP

Display System Protocol or Digital Signal Processor. A Digital Signal Processor is a specialized computer chip designed to perform speedy and complex operations on digitized waveforms. Useful in processing sound and video.

DTC

1. Digital Trunk Controller. 2. Digital transmit command.

DTMF

Dual Tone Multi Frequency. A fancy term for describing push button or touch-tone dialing (touch-tone is a former trademark of AT&T).

In DTMF, when you touch a button on a push button dial, it makes a tone, which is actually the combination of two tones, one high frequency and one low frequency. Thus the name Dual Tone Multi Frequency. In U.S. telephony, there are actually two types of DTMF signaling -- one that is used on normal business or home push button/touchtone phones, and one that is used for signaling within the telephone network itself.

When you go into a central office, look for the testboard. There you'll see what looks like a standard touch-tone pad.

Next to the pad there'll be a small toggle switch that allows you to choose the sounds the touch-tone pad will make -- either normal touch-tone dialing or the network version. The eight possible tones that comprise the DTMF signaling system were specially selected to easily pass through the telephone network without attenuation and with minimum interaction with each other. Since these tones fall within the frequency range of the human voice, additional considerations were added to prevent the human voice from inadvertently imitating or "falsing" DTMF signaling digits. One way this was done to break the tones into two groups, a high frequency group and a low frequency group. A valid DTMF tone has only one tone in each group. Here is a table of the DTMF digits with their respective frequencies. One Hertz (abbreviated Hz.) is one cycle per second of frequency.

	Low frequency	High frequency
1	697 Hz.	1209 Hz.
2	697	1336
3	697	1477
4	770	1209
5	770	1336
6	770	1477
7	852	1209
8	852	1336
9	852	1477
0	941	1336
*	941	1209
#	941	1477

There are four other digits defined in the DTMF system and usable for specialized applications that cannot be generated by standard telephones. They are:

Digit	Low frequency	High frequency
A	697 Hz	1633 Hz.
B	770	1633
C	852	1633
D	941	1633

DTMF Cut-Through

The capability of a voice response system to receive DTMF tones while the voice synthesizer is delivering information, i.e. during speech playback. This capability of DTMF cut-through saves the user waiting until the machine has played the whole message (which typically is a menu with options).

The user can simply touch-tone his response anytime during the message -- when he first hears his selection number, when the message first starts, etc. When the voice processor hears the touch-toned selection (i.e. the DTMF cut-through), it stops speaking and jumps to the chosen selection. For example, the machine starts to say, "If you know the person you're calling, touch-tone his extension in now.." But before you hear the "If you know" you push button in 230, which you know is Joe's extension. Bingo, the message stops and Joe's extension starts ringing.

Dumb Switch

A slang word for a telecommunications switch that contains only basic switching software and relies on instructions sent it by an outside computer. Those instructions are typically fed the "dumb" switch through a cable from the computer to one or more RS-232 serial ports which the dumb switch sports. The switch makes no demands on what type of computer it talks to, but simply insists that it be able to feed the computer questions and promptly receive responses in a form that it (the switch) can understand. Plain ASCII is OK. For example, the dumb switch might signal the computer, "A call is coming in on port 23, what do I do now?" The computer might reply:

"Answer it and transfer it to extension 23." Or it might say "answer it and put it on hold," or "answer it, put it on hold and play it recording number three." In essence, a dumb switch is anything but. It is in reality an empty cage containing whatever network interface cards the user has chosen. Each of these network interface cards is designed to "talk" to one type of telephone line. That line might be a T-1 line. It might be a normal tip and ring loop start line. It might be a tie trunk with E&M signaling. The card may handle one or many lines, but always of the same type. The card knows how to answer a call or pulse out a call on that particular type of line. It has all the telephony smarts.

What it lacks is the intelligence of what to do with the calls. That is provided by the outside computer. Well, almost. Most "dumb" switches do contain rudimentary intelligence -- a small computer and some memory.

That computer is usually programmed to handle "default" calls -- and to handle calls should the link to the outside computer fail, or the outside computer itself fail. Dumb switches come in flavors all the way from residing in their own cabinet to being printed circuit cards which reside in one or more of the personal computer's slots.

Dumb switches are programmed to do "specialized" telecom applications, for example emergency 911, added value 800 services, cellular switching, automatic call distributors, predictive dialers, etc. They can, of course, be programmed to be "normal" PBXs. The question increasingly being asked is "If I want to program a specialized telecom application should I use a dumb switch or should I use an open PBX?" And the answer is "It depends." Depends on what you want to do. Depends on what software is available, etc. See also OAI.

800 Service
See 800 SERVICE, spelled as EIGHT HUNDRED SERVICE.

E & M Leads

The pair of wires carrying signals between trunk equipment and a separate signaling equipment unit. The "M" lead transmits a ground or battery conditions to the signaling equipment. The "E" lead receives open or ground signals from the signaling equipment. These leads are also known as Ear and Mouth Leads. The Ear lead typically means to receive and the Mouth lead typically means to transmit.

Changes of voltage on these leads convey such information as seizure of circuit, recognition of seizure, release of circuit, dialed digits, etc. In the old days it was the PBX operators who originated trunk calls by asking the long distance carrier for free trunks using their mouth or M lead.

If the carrier had a free trunk, the PBX heard about it through its ear or E lead. See also E & M SIGNALING.

E & M Signaling

In telephony, an arrangement that uses separate leads, called respectively the "E" lead and "M" lead, for signaling and supervisory purposes. The near end signals the far end by applying -48 volts dc (vdc) to the "M" lead, which results in a ground being applied to the far end's "E" lead. When -48 vdc is applied to the far end "M" lead, the near-end "E" lead is grounded. The "E" originally stood for "ear," i.e., when the near-end "E" lead was grounded, the far end was calling and "wanted your ear."

The "M" originally stood for "mouth," because when the near-end wanted to call (i.e., speak to) the far end, -48 vdc was applied to that lead. When a PBX wishes to connect to another PBX directly or to a remote PBX or extension telephone over a leased voice grade line, a channel on T-1, the PBX uses a special line interface which is quite different from that which it uses to interface to the phones it's attached directly to (i.e. with in-building wires).

The basic reason for the difference between a normal extension interface and the long distance interface is that the signaling requirements differ -- even if the voice signal parameters such as level and two-wire, 4-wire remain the same.

When dealing with tie lines or trunks it is costly, inefficient and too slow for a PBX to do what an extension telephone would do, i.e. go off hook, wait for dial tone, dial, wait for ringing to stop, etc. The E&M tie trunk interface device is the closest thing there is to a standard that exists in the PBX, T-1 multiplexer, voice digitizer telco world. But even then it comes in at least five different flavors. See E & M LEADS.

Eight Hundred Service

800-Service. A generic and common (and not trademarked) term for AT&T's, MCI's, US Sprint's and the Bell operating companies' IN-WATS service. All these IN-WATS services have "800" as their "area code."

Dialing an 800-number is free to the person making the call. The call is billed to the person or company being called.

The telephone company suppliers of 800 services use various ways to configure and bill their 800-services. One way: you can buy an 800 line which will ring on your normal phone line. You'll only pay per call, but you won't receive any incoming call if you're making an outgoing one. (You can even terminate an 800 number on your cellular phone.) For other 800 services you might pay a flat monthly rate plus "so-much" (i.e. timed usage) per call. That timed usage may include some calculation for the distance the incoming call traveled. 800-Service is now available for calls from Canada and some countries in Europe.

More and more long distance companies are introducing 800-service. 800 Service works like this: You're somewhere in North America. You dial 1-800 and seven digits. Your local central office sees the "1" and recognizes the call as long distance. It ships that call to a bigger central office (or perhaps processes the call itself). At that central office it's processed, a machine will recognize the 800 "area code" and examine the next three digits. Those three digits will tell which long distance carrier to ship the call to. Each long distance company has been assigned specific 800 three digit "exchanges." For example, MCI has the exchange 999. AT&T has the exchange 542. If you want a phone number beginning with 800-999, then you must subscribe to MCI 800 service. If you want a phone number beginning with 800-542, you must subscribe to AT&T 800 service.

Once the 800-call is "passed off" to whichever carrier it belongs to, that carrier sends the call to a switch attached to a huge "translation" database. The call arrives at the switch. The database says the call 800-NNN-XXXX is really 212-555-1234, and sends the 800 line to that number.

As a real-life example, Flatiron Publishing, publishers of this book, has an 800 number, namely 800-LIBRARY (or 800-542-7279). When you call that number, the following number -- 212-206-6870 -- in New York City rings. Dialing 800-542-7279 is effectively the same as dialing 212-206-6870, except that if you dial 212-206-6870 you'll pay. If you dial 800-542-7279, I'll pay. Because 800 long distance service is essentially a database lookup and translation telephone service, there are endless "800 services" you can create.

E-1

Another name given to the CEPT digital telephony format devised by the ITU that carries data at the rate of 2.048 Mbps (DS-1 level). CEPT format consists of 30 voice channels, one signaling channel, and one framing (synchronization) channel. Since robbed-bit signaling is not used (as it is for T-1 in North America) all 8 bits per channel are used to code the waveshape sample. E-1 is the European version of North American T-1, though T-1 is 1.544 Mbps. See T-1.

Echo

A wave reflected or otherwise returned with sufficient magnitude and delay to be perceived. This is an effect sometimes experienced with long distance calls. It is usually associated with relatively long round-trip delay in the four-wire portion of the circuit so that the unwanted sound is perceived as being separate in time from the wanted one. The phenomenon of echo primarily affects the talker and not the listener. Ghosts on a television picture are a type of echo.

Echo Canceller

The echo suppressor used on circuits with long transmission time. It attenuates the direction of transmission which is not active so that any echoes that are fed into a circuit do not return to the starting point and cause confusion.

Echo cancellation stops a received signal from being transmitted back to its origin by constructing a signal closely approximating the echo component and subtracting this from the locally transmitted signal. This avoids double-talking problems. ECM Error Correction Mode. Encapsulated data within HDLC frames providing the receiver with an opportunity to check for, and request retransmission of garbled data.

ECMA

European Computer Manufacturers Association. An independent standards group of computer and PBX manufacturers. Their standards are recognized internationally.

EDI

Electronic Data Interchange. A series of standards providing automated computer-to-computer exchange of business documents (structured business data, editable documents, or electronic transactions such as invoices, purchase order, etc.) between different companies and computers over telephone lines.

E-Mail

Electronic mail. A popular application on both LANs and WANs which provides communication among users. There is a variety of systems which vary considerably in their level of sophistication. E-mail services can include simple message handling as well as complex file sharing.

E-Mail Fax Gateway

With an E-mail fax gateway, all E-mail users can send and receive faxes from within the company's E-mail package. Because the fax messaging is integrated directly into the mail system, fax users have virtually no new software to learn and can take advantage of E-mail features.

This includes workgroup distribution lists, attaching graphics files to text messages, support for different user operating systems on the network, and instant notification of received messages. With an E-mail fax gateway, it is possible to send the same message to both mail users and fax recipients.

Electronic Mail Integration

Integrating computer fax and E-mail technology allows the user to send and receive fax documents using a company's E-mail program. The most popular method uses MHS for Novell networks.

Electronic Voice Mail

A system which stores messages spoken by a user usually over a telephone, which can be retrieved by the intended recipient when that person next calls into the system. Also called Voice Mail, it operates just like a touch-tone controlled answering machine.

Email-to-fax

Conversion of electronic mail (ASCII text) into fax image format, suitable for sending to a fax machine.

Emulate

To imitate a computer or computer system by a combination of hardware and software that allows programs written for one computer or terminal to run on another. The most common data terminal is a DEC VT-100. Our communications program, Crosstalk, allows us to "emulate" a DEC-VT100 on our IBM PCs and PC clones.

ENOS

Enterprise Network Operating Systems. Part of Sun's Networking Solutions. This provides the foundation for Sun's networking environment. It uses NFS (de facto standard for global file sharing), and TCP/IP protocol. TCP/IP is Sun's vehicle of choice for transferring data between database clients and server. Sun uses WebNFS here. It's is the de facto standard TCP/IP protocol for remote file access.

EOM

End of Message frame. A frame from the sender indicating that the message is done, and that Phase B can be repeated. See also EOP.

EOP

End of Procedure frame. A frame indicating that the sender wants to end the call.

ESP

Enhanced Service Provider. A vendor who adds value to telephone lines using his own provided software and hardware. Also called an IP, or Information Provider. An example of an ESP is a public voice mail box provider or a database provider, say one giving the latest airline fares. An ESP is an American term, unknown in Europe, where they're most called VANs, or Value Added Networks. See also INFORMATION PROVIDER.

Ethernet

A LAN used for connecting computers, printers, workstations, terminals, etc., within the same building. Ethernet operates over twisted pair wire and over coaxial cable at speeds up to 10 Mbps.

Event

An unsolicited, asynchronous signal from a device or device driver that reports on a change in the status of the device. Events are generally attention-getting messages, allowing a process to know when a task is complete or when an external event occurs. A message that reports on a change in the status of an object.

Expansion Slots

In a computer there are card slots for adding accessories such as voice processing interface cards, internal modems, extra drivers, hard disks, monitor adapters, hard disk drivers, etc.

Most modern PBXs are actually cabinets with nothing but expansion slots. And into these slots we fit trunk cards, line cards, console cards, etc. Some phone systems have "universal" slots, meaning you can put any card in any slot. Some phone systems have dedicated expansion slots, meaning that they expect only a certain card in that slot.

Extensibility

This means that it's easy to add new technologies without re-inventing the wheel.

Facsimile

Facsimile, or fax, equipment allows information (written, typed, or graphic) to be transmitted through the switched telephone system and printed at the other end. The sending fax scans the material to be sent, digitizing it into binary bits and sending those through a modem to the receiving fax, which essentially reverses the process and outputs it through a printer.

Fast Busy

A busy signal which sounds at twice the normal rate (120 interruptions/minute vs. 60/minute). A "fast busy" signal indicates all trunks are busy.

Fax

Abbreviation for facsimile.

Fax-Back

You go to your fax machine. You dial a voice response unit. It says "Punch in 23 if you want to receive the latest specs on our new XYZ machine. Now when you are ready, hit the 'start' button on your fax machine and hang up." A few seconds later, your fax machine disgorges paper containing the latest specs on the XYZ machine. Welcome to "fax-back."

FaxBIOS

API used for in-application faxing. Developed by WordPerfect Corp. and Everex Systems, Inc.

Fax Board

A specialized synchronous modem designed to transmit and receive facsimile documents. Many also allow for binary synchronous file transfer, and V.22 bis communication.

Fax Mailbox

Companies can send facsimiles of documents to be stored for later retrieval to a fax mailbox -- a cousin to a voice mailbox. Travelers can check their fax mailboxes and have the faxes sent to convenient locations, like a hotel front desk.

Fax Server

In a LAN, a PC or a self-contained unit that has fax circuitry accessible to all the network's workstations. The server receives requests for fax services and manages them so that they are answered in an orderly, sequential manner. It is used to send and receive faxes by any network user, sharing the common resource of one or more fax boards.

Depending on application, a fax server may have a specialized interactive voice response system that routes faxes to a fax machine the user designates by touch-tone numbers. The receiving unit may be the user's or one designated by him.

FCC Registration Number

A number assigned to specific telephone equipment registered with the FCC, as set forth in FCC docket 19528, part 68. The presence of this number affixed to a device indicates that the FCC has approved it as being a compatible device for direct connection to telephone line facilities.

Field

A component of a record. Fields may contain either numeric, string, or date values. Fields are numbered starting with 0 as the first field.

Firewall

Hardware and\or software which limits the exposure of a computer or group of computers to an attack from an external location. Routers and other internetworking devices use their access control capabilities to build firewalls that can, for example, keep fault from propagating throughout the entire internet.

Firmware

Programs kept in semi-permanent storage, such as various types of read-only memory. These programs can be altered, but with difficulty.

Firmware is used in conjunction with hardware and software. It also shares the characteristics of both. Firmware is usually stored on PROMS (Programmable Read Only Memory) or EPROMs (Electrical PROMS).

Firmware contains software which is so constantly called upon by a computer or phone system that it is "burned" into a chip, thereby becoming Firmware. The computer program is written into the PROM electrically at higher than usual voltage, causing the bits to "retain" the pattern as it is "burned in". Firmware is non-volatile. It will not be "forgotten" when the power is shut off. Hand-held calculators contain firmware with instructions for doing their various mathematical operations.

First Party Call Control

See call control.

Flash

Quickly depressing and releasing the plunger in or the actual handset-cradle to create a signal to a PBX or Centrex that special instructions will follow such as transferring the call to another extension.

Flash Hook

A brief on-hook period. A common use of the flash-hook is in the residential call waiting feature. A new call comes in while a conversation is in progress.

The central office, however, doesn't give the caller a busy signal. The caller hears ringing and the called party, who's having a conversation, hears a special "beep" tone. If he or she chooses, the called party can quickly push the hook-switch on their telephone, sending what the industry calls a "flash hook." The central office puts the original caller on hold and the new dial tone, allowing a three-way conference or call transfer.

Flowchart

A graphic or diagram which shows how a complex operation, e.g. programming, takes place. The flowchart breaks that operation down into its smallest, and easiest-to-understand events.

FOD

Fax On Demand. Dial up a number. hear an voice response unit say, "Would you like a timetable?" You say "Yes." It says put in your fax number. You punch in your fax number and bingo, your fax machine starts receiving a fax of the timetable.

Formant

A point of excitation, or high energy, in a speech waveform caused by resonance in the human vocal tract. Formants are responsible for the unique timbre of each individual's voice.

Frame

A group of data bits in a specific format, with a flag at each end to indicate the beginning and end of the frame. The defined format enables network equipment to recognize the meaning and purpose of specific bits.

Frame Relay

Frame relay switching is a form of fast packet switching, but uses smaller packets and requires less error checking than traditional forms of packet switching. Like traditional X.25 packet networks, frame relay networks use bandwidth only when there is traffic to send.

Frequency

The number of complete oscillations per second of an electromagnetic wave.

Frequency Domain

Waveforms, such as speech signals, are typically viewed in the time domain, i.e. as power levels or voltages varying over time.

The 19th century French mathematician Fourier demonstrated an algorithm called "Fast Fourier Transform," or FFT, which can express any complex waveform over a fixed interval as the sum of a series of sine waves of different energy levels. Analyzing signals in the frequency domain has proven an extremely powerful technique with diverse applications, including filtering, recognition, and speech modeling.

Frequency Response

The variation (dB) in relative strength between frequencies in a given frequency band, usually the voice frequency band of an analog telephone line.

ft-SPARC

Suns' new line of SPARC-based, fault tolerant computers. Resulting from Sun's acquisition earlier this year of Integrated Micro Products (IMP). Fully redundant design has full fault tolerance in processor modules, I/O subsystems and mirrored disks. Has hot-pluggable modules and redundant power systems.

FTT

Failure-to-train signal.

FTP

File Transfer Protocol. Software that can transfer files from one computer to another across the 'Net.

Full Duplex

A communications protocol in which the communications channel can send and receive signals at the same time.

FSK

Frequency Shift Keying. A modulation technique for translating 1s and 0s into something that can be carried over telephone lines, like sounds. A 1 will be assigned a certain frequency of tone, and a 0 another tone. The transmission of the bits keys the sounds to shift from one frequency to the other.

G

G.711

ITU (International Telecommunications Union) standard for audio codecs. Provides encoding at 64 kilobits per second (mu-law and A-law).

G.723

ITU standard for audio codecs. This standard is for compressed digital audio over POTS lines -- Plain Old Telephone Lines. It is the voice part of H.324.

GIF

Graphic Interchange Format. A popular graphics file format used on the Web. It's sort of compressed. JPEG is another.

Gigaplane

Center plane bus used in Sun's Ultra Enterprise Server line. Uses separate paths of address, data, and control lines. Communicates with several subsystems concurrently. With a 167MHz UltraSPARC CPU, the Gigaplane can do rates of 2.5Gbytes per second.

GlareGlare occurs when both ends of a telephone line or trunk are seized at the same time for different purposes or by different users. Most embarrassing.

Gopher

A menued interface to Internet files.

Ground Start

A way of signaling on subscriber trunks in which one side of the two wire trunk (typically the "Ring" conductor of the Tip and Ring) is momentarily grounded to get dial tone. There are two types of switched trunks one can typically lease from a local phone company -- ground start and loop start.

PBXs work best on ground start trunks, though many will work -- albeit intermittently -- on both types. Normal single line phones and key systems typically work on loop start lines. You must be careful to order the correct type of trunk from your local phone company and correctly install your telephone system at your end -- so that they both match. A ground start trunk initiates an out-going trunk seizure by applying a maximum local resistance of 550 ohms to the tip conductor. See LOOP START.

Group

An associated set of one or more Resource Objects. Groups encapsulate the functionality of the Resource Objects that are associated with them. Resource Objects within a Group have defined connectivity. The Group provides three services to the application: implicit management of connectivity between group members; representation of a single entity to the applications (group ID); and reservation of all physical resources (CPU, memory, time slots) required to provide the application with exclusive use of configured resources.

Group1

Analog fax equipment, according to Recommendation T.2 of the ITU.

It sends an ISO A4 or (U.S. letter size) 8.5 x 11- inch page in six minutes over a voice-grade telephone line using frequency modulation with 1,300 Hz corresponding to white and 2,100 Hz to black of the original. Because North American six-minute equipment uses a different modulation scheme, it is not compatible with Group 1 equipment.

Group2

Analog facsimile equipment, according to Recommendation T.3 of the ITU. It sends an A4 or 8.5 x 11-inch page in three minutes over a voice-grade telephone line using 2,100-Hz AM-PM- VSB.

Group3

A digital high-speed fax standard for reliable transmission over ordinary telephone lines. Based on ITU Recommendation T.4.

Group4

ITU fax standard primarily designed to work with ISDN. Written before a real machine was made, it is generally considered difficult to implement. Although it offers a transmission rate of about five seconds per page, it requires an ISDN or leased line. Currently, there is no way to use it over ordinary voice grade telephone lines, and equipment from different manufacturers cannot interface.

GSM

GSM stands for Groupe Speciale Mobile, now known as Global System for Mobile Communications, is the standard digital cellular phone service you will find in Europe and Japan. GSM actually is a set of ETSI standards specifying the infrastructure for a digital cellular service. To ensure interoperability between countries, these standards address much of the network wireless infrastructure, including the radio interface (900 MHz), switching, signaling, and intelligent network. An 1,800 MHz version, DCS1800, has been defined to facilitate implementation in some countries, particularly the UK. Since GSM is limited to technical standards, an association of GSM operators called the Memorandum of Understanding (MoU) ensures service interoperability, allowing subscribers to roam across networks.

GSM has gained widespread acceptance in several parts of the world, most notably Europe, with deployment in 52 countries by mid year '94. GSM subscriber data is carried on a Subscriber Identity Module (SIM) or "smartcard" which is inserted into the phone to get it going. As a result, the subscriber potentially has the option of either SIM card mobility or terminal mobility across multiple networks.

One of GSM's major problem is that people wearing hearing aids can't use GSM cell phones, because GSM phones produce unpleasant high-pitched squealing the closer they get to the cell phone.

GSM technical Characteristics: Receiver frequency: 935.2 -- 959.8 MHz. Transmitter frequency: 890.2 -- 914.8 MHz. Access method: mixed TDMA & FDMA with optional frequency hopping. Security: Optional radio interface encryption.

Carrier frequency division: 200 KHz. Users per carrier frequency: 8. Speech bit rate (transfer rate): full rate (13 kbps) or half rate. Total bit rate: 21 Kbps. Bandwidth per channel: 25 KHz

H.221

ITU standard for ISDN conferencing. Defines frame structure for 64 to 1,920 kilobits per second channels in audiovisual teleservices. Related ITU standards are H.223 (multiplexing), H.245 (controlling), and H.261 for video codecs.

H.261

ITU standard for video encoding. It was the watershed standard (1990) that made it possible for video codecs from different makers to successfully communicate with each other. H.263 is another ITU standard for video encoding. It makes improvements on H.261.

H.320

The most common family of ITU-T videoconferencing standards for transmitting audio and video over circuit-switched digital networks (primarily ISDN).

H.323

ITU standard for taking the H.320 video standard and running it over LANs and IP networks (e.g. the Internet). It defines the bridge between LAN-based terminal, equipment and services to systems connected to circuit-switched networks.

H.324

ITU standard for taking the H.320 standard to transmit audio and video over analog lines. Uses 28.8 kilobits per second (V.34) modems. Uses H.263 video compression algorithm and G.723 voice compression algorithm. Uses H.223 and H.245 multiplexing and control protocols.

Half Duplex

A communications protocol in which the communications channel can handle only one signal at a time. The two stations alternate their transmissions.

Handle

A string of characters (alphabetic, numeric, or any combination including the underscore), userdefined in the database configuration file, used by Pro/Found to reference database and index files.

Handoff

The change of ownership of a Group (and therefore, typically, a call) from one session to another. For example, if a call center application discovers that a caller wishes to access a technical support Audiotex database, it hands off the call to an application servicing that database.

Handshaking

An exchange of signals between the fax transmitter and the fax receiver to verify that transmission can proceed, determine which specifications will be used, and to verify reception of the documents sent.

HDLC

High-Level Data-Link Control Standard. It always contains a frame called the Digital Identification Signal (DIS), which describes the standard ITU features of the machine. It can also contain two other frames: a Non-Standard Facilities (NSF) frame, which tells the caller about vendor-specific features, and, usually, a Called Subscriber Identification (CSI) frame, which contains the answerer's telephone number.

High Level Languages

Essentially any of the computer languages whose code is not unique to the hardware or architecture of a particular computer. High level languages are more like human language than the machine language which computers talk. High level languages translate human instructions into the machine language computers can understand, but which humans don't have to (in order to tell the computer what to do).

Computer languages such as C, BASIC, FORTRAN, COBOL and Pascal are high level languages. They are a number of levels (at a High Level) away from the actual bit manipulation (machine language, also called "bit twiddling" by the Hackers).

HIVR

Host Interactive Voice Response. Tying a voice response unit into a mainframe computer which has lots of data. Applications which can be produced include bank-by-phone, reservations-by-phone, etc.

Hook Flash

See FLASH

Hookswitch

Also called SWITCHHOOK. It's typically the place on your telephone instrument where you lay your handset. When you lift the handset, you are said to be going "off hook." When you place your handset back, you are said to be "on hook." When you lift the handset (i.e. go off hook) you are, in effect, signaling your central office that you'd like to make a call, or that you have answered the incoming ringing call.

Host Computer

A computer attached to a network providing primarily services such as computation, data base access or special programs of special programming languages. The central computer in a time-sharing operation.

Host Interactive Voice Response

Voice Response system communicating with a host computer See HIVR.

Host Processor

Same as HOST COMPUTER.

Hot Key

Refers to TSR utilities in DOS or filters in a Windows environment that allow users to fax without leaving their present application. The ability to send a fax from within an application is one of the most important features of computer fax technology.

HTML

HyperText Markup Language. The formatting language of the World Wide Web. HTML is how you actually "write" or "program" Web pages. It's a primitive language, really. Everything is in ASCII. You put codes in to define fonts, layout, embed graphics and hypertext links.

HTTP

Hypertext Transfer Protocol is the transport protocol in transmitting hypertext documents around the Internet.

Huffman Encoding

A popular lossless data compression algorithm that replaces frequently occurring data strings with shorter codes. Some implementations include tables that predetermine what codes will be generated from a particular string. Other versions of the algorithm build the code table from the data stream during processing. Huffman encoding is often used in image compression. (Q.v., Modified Huffman Code.)

Hunt

Refers to the progress of a call reaching a group of lines. The call will try the first line of the group. If that line is busy, it will try the second line, then it will hunt to the third, etc. See also HUNT GROUP.

Hunt Group

A series of telephone lines organized in such a way that if the first line is busy the next line is hunted and so on until a free line is found. Often this arrangement is used on a group of incoming lines. Hunt groups may start with one trunk and hunt downwards. They may start randomly and hunt in clockwise circles. They may start randomly and hunt in counter-clockwise circles. Inter-Tel uses the terms "Linear, Distributed and Terminal" to refer to different types of hunt groups. In data communications, a hunt group is a set of links which provides a common resource and which is assigned a single hunt group designation. A user requesting that designation may then be connected to any member of the hunt group. Hunt group members may also receive calls by station address.

Hyperchannel

A data path on the SCbus or SCxbus Data Bus made up of more than one time slot. By bundling time slots into a hyperchannel, data paths with a bandwidth greater than 64 Kbps can be created.

Hypertext

A method of writing and displaying text that allows the text to be linked to something else. Click on it. Go elsewhere. Hypertext can contain photos, audio and video.

I

Idle

A state of the SCbus or SCxbus Message Bus where no information is being transmitted and the bus line is pulled high.

IGMP

Internet Group Management Protocol. IGMP is used by IP hosts to report their host group memberships to any immediately-neighboring multicast routers. IGMP is an asymmetric protocol and is specified here from the point of view of a host, rather than a multicast router. (IGMP may also be used, symmetrically or asymmetrically, between multicast routers. GMP is a integral part of IP.

In-Band Signaling

Signaling made up of tones which pass within the voice frequency band, and are carried along the same circuit as the talk path that is being established by the signals. Virtually all signaling -- request for service, dialing, disconnect, etc. -- in the U.S. today is in-band signaling.

Most of that signaling is MF -- multi-frequency dialing. The more modern form of signaling is out-of-band.

Indexed Database

A database indexed on a key field. Indexing allows for rapid retrieval of records through an index field.

Index Field

The field to be used when indexing a database.

Index File

An (optional) file used for indexing the data in a database. Index files are usually given extensions which identify them as index files. For example, when using dBASE III+, the index files are given the NDX extension.

Index Handle

See Handle.

Index Order Retrieval

Implies a sequence in which records are read from the database in an order based on the contents of each record's index field.

Information Center Mailboxes

A voice bulletin board on a voice mail system. Here's their explanation: Multiple callers can access, directly or indirectly, recorded announcements containing information that would otherwise have been given live by employees. Callers are frequently "outside" users of the system. One type of "listen only" mailbox simply plays the messages to the callers. This technology, sometimes known as audiotex, makes it possible to create a verbal database so callers can select which information they want to hear.

Another type of Information Center Mailbox prompts callers to reply to announcements. Callers wanting further information can be given the opportunity to leave their names and phone numbers after listening to a product description. They can also be transferred to a designated employee who can immediately take an order. If desired, a password can be required before confidential or controlled access information can be heard.

Information Provider

A business or person providing information to the public for money. The information is typically selected by the caller through touch tones, delivered using voice processing equipment, and transmitted over tariffed phone lines, e.g., 900, 976, 970. Typically, billing for information providers' services is done by a local or long distance phone company.

Sometimes the revenues for the service are split by the information provider and the phone company. Sometimes the phone company simply bills a per minute or flat charge. A typical "information provider" is American Express, which provides a service -- 1-900-WEATHER. By dialing that number you can touch-tone in city names and find out temperatures, weather forecasts, etc. Calling 1-900-WEATHER costs several dollars a minute.

Integrated (or *unified)* messaging

Also called unified messaging, is the concept of unifying all incoming messages onto a single computer screen. A telephony application that offered unified messaging would present a user with a single list showing all new electronic mail, telephone messages, and faxes. Clicking on a telephone message could result in either the telephone ringing with the message, or with the computer's sound card relaying the message. Unified messaging gives users the luxury of scanning their messages, and deciding which to deal with first.

Interactive Voice Response

IVR. Picture a standard personal computer.

You enter information through the keyboard. You see the results of your work on your screen. With interactive voice response, the keyboard, (or data entry device) becomes the touch-tone pad of a telephone (either local, but most likely remote) and the screen (the output device) becomes a voice synthesizer which electronically converts the computer's output (what you'd normally see on the screen) into spoken words. You call a phone number. A machine answers. It presents you with several options, "Push 1 for information on Plays, Push 2 for information on movies, Push 3 for information on Museums."

If you push 2, the machine may come back, "Push 1 for movies on the south side of town, Push 2 for movies on the north side of town, etc." That's a very simple interactive voice response (also called Voice Processing) application.

There are literally thousands of others. A bank might say "Thanks for calling the XYZ Bank, if you'd like your balance, punch in your account number now." A cable TV company might say "Thanks for calling ABC TV, if you'd like to see tonight's feature movie, push 1. $5.00 will be billed to your account." There are applications for Interactive Voice Response in every industry, in every company. Hint: Think customer!

What can we do to make our customers' lives easier? Can we help them find our nearest store? Can we tell them where to get the stuff we sell them repaired? Would they like to know where the order is they placed a week ago?

Intercept Service

A service of the local phone in which a phone call is redirected by an operator or a recording to another phone number or a message. Intercept could also be sold to companies who have moved and would like to retain a recording and/or a phone advertisement for many months after they've moved.

Interconnect Companies

Companies which sell, install and maintain telephone systems for end users, typically businesses.

Inter-LATA

Telecommunications services that originate in one and terminate in another Local Access and Transport Area (LATA). Under provisions of Divestiture, the Bell operating companies cannot provide Inter-LATA service, but can provide Intra-LATA service. Some LATAs are very large. So some "local" phone companies provide the equivalent of long distance service. And some of these phone companies have different pricing packages. Some of these packages are cheap, but not highly-publicized. See also LATA.

Internet

Internet is a computer network which joins many government and university and private computers together over phone lines (mostly T-1s and T-3s). In 1995 the Government Accounting Office (GAO) said that Internet linked 59,000 networks, 2.2 million computers and 15 million users in 92 countries. Internet traces its origins to a network set up in 1969 by the Defense Department. In 1991 it was running off $20 million a year in federal subsidies and managed by the National Science Foundation. An IBM/MCI venture known as Advanced Network and Services manages a network called NSFnet, which connects hundreds of research centers and universities. NSFnet also manages links to dozens of other countries.

All these networks are collectively known as Internet. NSFnet was founded by the National Science Foundation, a Federal Government agency and is composed of leased telephone lines that link special computers called routers, which transmit packages of data to three million users in 33 countries. See various INTERNET definitions following.

Internet Address

A unique, 32-bit identifier for a specific TCP/IP host on a network. Also called an Internet Protocol or IP address. IP addresses are normally printed in dotted decimal form, such as 128.127.50.224.

Internet Gateway

Internet gateways are devices which typically sit on a local area network and handle all the translations between IPX traffic on your LAN

(IPX is the NetWare protocol) and the TCP/IP traffic on the Internet. TCP/IP is the protocol used on the Internet.

Internet Group Name

In Microsoft networking, a name registered by the domain controller that contains a list of the specific addresses of computers that have registered the name. The name has a 16th character ending in 0x1C.

Internet Mib Subtree

A tree-shaped data structure in which network devices on a local area network and their attributes can be identified within the confines of a network management scheme. The name of an object or attribute is derived from its location on this tree. For example, an object in MIB-I might be named 1.2.1.1.1.0. the first 1 indicates the object is on the Internet. The 2 denotes that it falls within the Management category.

The second 1 shows the object is part of the first fully defined MIB, known as MIB-I. The third 1 indicates which of the eight object groups is being referenced. And the fourth 1 is a textual description of the network component. The O indicates there is only one object instance. An object instance links a particular object to a specific node on the network. The numbering system is infinitely extendible to accommodate additions to this base identification scheme. This common naming structure permits equipment from a variety of vendors to be managed by a single management station that uses SNMP. The four main categories of the tree are Directory, Management, Experimental and Private/Enterprises.

Internet Number

The dotted-quad address used to specify a certain system. The Internet number for cs.widener.edu is 147.31.130. A resolver is used to translate between hostnames and Internet addresses.

Internet Packet Exchange IPX

Novell NetWare's native LAN communications protocol, used to move data between server and/or workstation programs running on different network nodes.

Internet Protocol

IP. Part of the TCP/IP family of protocols describing software that tracks the Internet address of nodes, routes outgoing message, and recognizes incoming messages. Used in gateways to connect networks at OSI network Level 3 and above.

Internet Protocol Suite

A suite of network protocols which have been adopted as the main de facto protocols for LANs.

Internet Relay Chat IRC

Sort of like CB radio, but run on the Internet, and far more confusing than CB radio.

Internet Server

An Internet server is a device which users on the Internet access to get services. Such services might be electronic mail, news, a Web page, etc. A company will have one or more Internet servers attached to the Internet when it wants to deliver services to people on the Internet. Such Internet servers could be called e-mail servers, FTP servers, News servers and World Wide Web servers. Internet servers most commonly run on Unix. But Microsoft Windows NT is increasingly gaining popularity.

Intranet

An Internet-like network within an organization. Intranets use web server and browser technology and other Internet emerging standards to share data on Internet-like networks within an organization (the "intraprise") rather than just for external connection to the Internet.

IP Multicasting

The transmission of an IP datagram to a "host group", a set of zero or more hosts identified by a single IP destination address. A multicast datagram is delivered to all members of its destination host group with the same "best-efforts" reliability as regular unicast IP datagrams, i.e., the datagram is not guaranteed to arrive intact at all members of the destination group or in the same order relative to other datagrams.

IPX

Internet Packet eXchange. A lowlevel network communications protocol uscd by Novell networks. In most networks, IPX is supplemented and extended by SPX.

ISDN

Integrated Services Digital Network. A collection of standards which define interfaces for, and operation of, digital switching equipment, developed by carriers, equipment manufacturers, and international standards organizations. It is intended to form the basis for the next generation telephone network and is currently being implemented by carriers throughout the world. Instead of one analog telephone line, there would be two 64 kbps bearer lines and one 16 Kbps data line.

Each bearer line could carry voice, video, data, images or combinations of these. As the name implies, it would be a point-to-point digital system. BISDN stands for Broadband Integrated Services Digital Network. It is a packet switching technique which uses packets of fixed length, resulting in lower processing overhead and higher speeds. Also known as cell switching and ATM.

ISO

The International Standards Organization in Paris, devoted to developing standards for international and national data communications. The U.S. representative to the ISO is ANSI.

ISP

Internet Service Provider. This company rents lines into the Internet and typically sells dial-up service to you and me. Sometimes, an ISP will let a company put up its Web page on ISP equipment.

ITU

International Telecommunications Union, the world standards-setting organization. Formerly the Comite Consultatif Internationale de Telegraphique et Telephonique (Consultative Committee for International Telephone and Telegraph - CCITT). See Telecommunication Standardization Sector.

IVR

Interactive Voice Response.

IXC

InterExchange Channel, or InterExchange Carrier -- as contrasted to the LEC -- the Local Exchange Carrier, a new word for a local phone company. InterExchange Carriers used to be called "Other Common Carriers," except that didn't include AT&T. Now, AT&T, MCI, Sprint and all the Other Common Carriers are called InterExchange Carriers.

J

Java

Sun's general-use programming language introduced in 1995. It's object-oriented and portable. Allows for development of "applets" to be distributed on Intranets and the Internet. Sun Microelectronics is developing Java chips, so Java applets will run on embedded processors. Nortel was first to announce it will put these chips in its next-generation phones.

Java Telephony API

(JTAPI) is a set of modularly-designed, application programming interfaces for Java-based computer-telephony applications. JTAPI is designed to serve a broad audience, from call control centers to Web-page designers, by offering telephony interface extension packages that can be assembled in various combinations with a core package in configurations that match the complexity of the application. The JTAPI is easily extended to accommodate custom applications, while ensuring applications written to the JTAPI are portable across platforms with no modification. Applications written to the JTAPI are independent of platform or phone system. JTAPI interfaces will soon be available for other computer-telephony integration applications, such as SunXTL, Microsoft and Intel TAPI, Novell and Lucent TSAPI and IBM Call Path.

JScript

Microsoft's open implementation of JavaScript. JScript is a high-performance scripting language designed to create active online content on the World Wide Web.

JScript allows developers to link and automate a variety of objects in Web pages, including ActiveX Controls and "applets" created using the Java language from Sun Microsystems Inc. Both Visual Basic Script and JScript are integrated with Microsoft Internet Explorer 3.0 using the ActiveX Scripting interface. ActiveX Scripting is a published standard that any application or scripting language can use. Because JScript uses the ActiveX Scripting interface, developers can script Java Applets using the JavaScript language. The ActiveX Scripting interface allows developers to orchestrate the interaction among a variety of software components on the Internet, including Java Applets and ActiveX Controls written in other languages, using any scripting language that supports this interface.

JTAPI

See Java Telephony API

JTAPI Address Object

Part of the JTAPI Core call model. The Address object represents a telephone number. It is an abstraction for the logical endpoint of a phone call. This is distinct from a physical endpoint. In fact, one address may correspond to several physical devices.

JTAPI Call Model

The JTAPI Core call model is defined in the Core API package. A call model describes a set of software objects that correspond to physical and conceptual entities in the telephony world. These objects fit together in a specified way to represent a telephone call. The Core API objects are: Provider Object, Call Object, Connection Object, TerminalConnection Object, Terminal Object and Address Object

In the physical view, each Core object represents a tangible property or telephony equipment. From a logical view, the call model represents an abstraction of telephony software entities or the functional properties of the objects. In describing these objects, it is difficult to separate the objects' physical representation from their logical properties, therefore, the description of these objects changes perspective frequently.

JTAPI Call Object

Part of the JTAPI Core call model. The Call object represents a telephone call, the information flowing between the service provider and the call participants. A telephone call comprises a Call object and zero or more connections. In a two-party call scenario, a telephone call has one Call object and two connections. A conference call is three or more connections associated with one Call object.

JTAPI Connection Object

Part of the JTAPI Core call model. A Connection object models the communication link between a Call object and an Address object. Relationships between Call and Address like connected, disconnected, and alerting are modeled by the Connection object as states. The Connection object also serves as a container for zero or more TerminalConnection objects. Connection objects model the logical aspects of a call connection.

JTAPI Core Package

All JTAPI implementations make use of the Core package. Many application developers will only need basic telephony, in which case they will only need to use the Core API package. The Core API package provides basic telephony: placing calls, answering calls, and dropping calls. It defines the basic call model that the extension packages follow in design.

JTAPI Provider Object

Part of the JTAPI Core call model. The Provider object is an abstraction of telephony service provider software. The provider might manage a PBX connected to a server, a telephony/fax card in a desktop machine, or a computer networking technology, such as IP. A Provider hides the service-specific aspects of the telephony subsystem and enables Java applications and applets to interact with the telephony subsystem in a device-independent manner.

JTAPI Standard Extension Packages

The JTAPI specification defines standard extension packages. The core telephony package, Call Control, Call Center, Private Data and Terminal Set Management extension packages are at version 1.0.

The specifications for Media, and Capabilities extension packages are at version 0.3. Mobile, and Synchronous are still under consideration. There are currently eight standard extension packages: Call Control, Call Center, Private Data, Terminal Set Management, Capabilities, Media Services, Mobile Phones and Synchronous.

JTAPI TerminalConnection Object

Part of the JTAPI Core call model. TerminalConnection objects model the relationship between a call and physical entities, represented by the Terminal object. The TerminalConnection object signals a Terminal when there is an incoming call and monitors the Terminal's activity during the process of a call. This object also closely communicates with the Connection object to receive information on a Call's change in state and to send information on the Terminal's state change. It models the physical aspects of a call connection.

JTAPI Terminal Object

Part of the JTAPI Core call model. The Terminal object represents a physical device like a telephone and its associated properties. Each Terminal object may have one or more Address Objects (telephone numbers) associated with it, as in the case of some office phones capable of managing multiple call appearances. Additionally, Terminal objects that have more than one phone line may share a telephone number and a single Address object with another Terminal in an adjacent office. However, each Terminal has a unique TerminalConnection associated with a Call even though it may share an Address.

KBPS

Kilobits per second. It is a bandwidth equal to 1000 bits.

L

LAN

Local Area Network. A short distance network (typically within a building or campus) used to link together computers and peripheral devices under some form of standard control. See RING and ETHERNET.

LATA

Local Access and Transport Area; one of 161 local telephone service areas in the United States.

As a result of the Bell divestiture that now distinguishes local from long-distance service, switched calls with both endpoints within the LATA (intraLATA) are generally the sole responsibility of the local telephone company, while calls that cross outside the LATA (interLATA) are passed on to an interexchange carrier.

LEC

Local Exchange Company. The local phone companies, which can be either a Bell Operating Company (BOC) or an independent (e.g. GTE) which provides local transmission services. Prior to divestiture, the LECs were called telephone companies or telcos.

Line

See Channel.

Line Powered

Telephone equipment that is powered solely by the CO talk battery supplied in a standard phone line.

Local Loop

The physical wires that run from the subscriber's telephone set, or PBX or key telephone system, to the telephone company central office.

Increasingly, the local loop now goes from the main distribution frame in the basement to the phone company. And the subscriber is responsible for getting his/her wires from the box in the basement to his phone system.

Loop

1. Typically a complete electrical circuit. 2. The loop is also the pair of wires that winds its way from the central office to the telephone set or system at the customer's office, home or factory, i.e. "premises" in telephones. 3. In computer software. A loop repeats a series of instructions many times until some prestated event has happened or until some test has been passed.

Loop Current Detection

When a voice, modem, or fax resource seizes the line (i.e., completes the connection between tip-and-ring terminals of the telephone cable), current flows from the positive battery supply in the telephone central office, through the twisted pair in the loop, through the board, and back to the central office negative terminal where it is detected, showing that this telephone line is off hook. The fax board also detects the loop current and can detect problems such as disconnects, shutting down the connection or a busy signal, making it wait and redial.

Loop Start

You "start" (seize) a phone line or trunk by giving it a supervisory signal. That signal is typically taking your phone off hook. There are two ways you can do that -- ground start or loop start. With loop start, you seize a line by bridging through a resistance the tip and ring (both wires) of your telephone line.

Loop Timing

A way of synchronizing a circuit that works by taking a synchronizing clock signal from incoming digital pulses.

Lossless Compression Coding

Coding designed not to lose any data when compressing or restoring an image.

Low Level Language

A programming language that uses symbols -- one step away from the machine language of a computer. Low level computer languages such as Assembler and C which actually manipulate the bits in computer registers. Higher level languages such as Basic and Fortran will take care of the piddling details of doing specific functions when you give it a broad command like "PRINT". In a lower level language, you must provide all the details of instruction necessary in the code (program) to perform the operation. It is possible to do this by calling standard routines, but still takes up the programmers' time in deciding which routines, and keeping the registers straight as he designs the program.

MAE

Metropolitan Area Ethernet. This is a basically a fiber-optic data ring around the city. It is used to inexpensively connect companies and offices to this citywide network.

Mailbox

Messages belonging to a single owner in a voice mail system. Today, these will be recorded voice messages, but increasingly mailboxes will include E-mail and fax documents.

MAPI

The Microsoft Windows Messaging Application Programming Interface, a set of API functions that enable messaging. Part of the Windows Open Services Architecture (WOSA). In any messaging application, there are common functions such as sending, receiving, saving, and reading mail. Each function is controlled within the application by a command or function call. These commands are the Application Programming Interfaces, or APIs. They tell the underlying messaging system what to do; that is, they trigger an action. MAPI has a dual API approach(client and messaging-service). Windows APIs provide the link for Windows-based applications.

Windows service provider interfaces provide the link for service providers. Together they enable Windows-based applications to access a wide range of services across multiple environments. The second level of MAPI is called the MAPI Messaging Subsystem. It is built into the Windows operating system. The MAPI Messaging Subsystem and a MAPI dynamic-link library (DLL) are responsible for handling multiple transport-service providers.

Drivers for each transport exist in the form of a Windows DLL to provide the interface between the MAPI messaging subsystem message spooler and the underlying back-end messaging system(s) or services. The message spooler is similar to a print spooler. It assists with the routing of messages instead of print jobs. The spooler, MAPI.DLL, and the transport drivers work together to handle the sending and receiving of messages.

MBONE

Multicast Backbone. Circa 1992 IETF (Internet Engineering Task Force) effort. Came out of earlier ARPA DARTnet experiments. Supports multicast audio and video across the Internet. Provides one-to-many and many-to-many network delivery services for apps like videoconferencing and audio. Supports simultaneous communication between several hosts.

MCA

Micro Channel Architecture.

MCF

Message Confirmation Frame. Confirmation by the receiver that it is ready to receive the next fax page, starting Phase C again.

Media Processing

The processing of transactions during a telephone call; these transactions may include fax operations, speech recognition and synthesis, Touch Tone recognition, voice and fax store-and-forward messaging, and the conversion of messages from one format to another (such as from text to voice, or from fax to text).

Menu
Options displayed on a computer terminal screen or spoken by a voice processing system. The user can choose what he wants done by simply choosing a menu option -- either typing it on the computer keyboard, hitting a touch tone on his phone, or speaking a word or two. There are basically two ways of organizing computer or voice processing software -- menu-driven and non-menu driven.

Menu-driven programs are easier for users to use. But they can only present as many options as can be reasonably crammed on a screen or spoken in a few seconds. Non-menu driven screens can allow more any alternatives. But they're much more complex and frightening.

It's the difference between receiving a bland "A" or "C" prompt on the screen -- as in MS-DOS and receiving a menu of "Press A if you want Word Processing," "Press B if you want Spread Sheet," etc. It's very easy to write menus in MS-DOS using BATch files. See also AUDIO MENUS.

Message
The transport container for SCSA requests, replies and events. Assumes a set of conventions for directing the delivery of the message to the proper entity, either a client or service provider. See also SCSA Message Protocol.

Method
The specific implementation of an operation for a class; code that can be executed in response to a request.

Micro Channel
A proprietary 32-bit bus developed by IBM for its PS/2 family of computers' internal expansion cards. Also offered by Tandy and other vendors.

Microcode
Programmed instructions that typically are unalterable. Usually synonymous with firmware and programmable read-only-memory (PROM).

microSPARC I and II

Sun SPARC chips capable of handling 50 and 110Mhz and above, respectively. microSPARC processors are at the heart of Sun's SPARCstation 1+/IPC and 2+/IPC workstations. They are also used in many third-party motherboards and workstations.

MIDI

Musical Instrument Digital Interface. The standard protocol for the interchange of musical information between musical instruments, synthesizers, and computers.

Mini/Mainframe Fax Servers

Many mini/mainframe fax servers function in the same way as LAN fax servers: users can send and receive faxes from their terminals, saving both time and money while they increase the quality of their fax transmissions.

By being on the same computer system as an organization's data, mini/mainframe servers can provide an EDI-like component for large, vertical applications such as accounting or purchase order systems, broadening the electronic reach of these programs to any recipient with a fax machine.

Modem

Acronym for MOdulator/DEmodulator. Equipment that converts digital signals to analog signals and vice-versa. Modems are used to send data signals (digital) over the telephone network, which is usually analog. A modem modulates binary signals into tones that can be carried over the telephone network. At the other end, the demodulator part of the modem converts the tones back to binary code.

Modified Huffman Code (MH)

A one-dimensional data compression technique that compresses data in an horizontal direction only and does not allow transmission of redundant data. Huffman encoding is a lossless data compression algorithm that replaces frequently occurring data strings with shorter codes. Often used in image compression.

Modified READ (MR)

Relative Element Address Differentiation code. A two-dimensional compression technique for fax machines that handles the data compression of the vertical line and that concentrates on space between the lines and within given characters.

Modified Modified READ (MMR)

A two-dimensional coding scheme for Group 4 fax, but now finding use with Group 3 machines.

Modulation

The process of varying some characteristic of the electrical carrier wave as the information to be transmitted on that carrier wave varies. Three types of modulation are commonly used for communications, Amplitude Modulation, Frequency Modulation, and Phase Modulation. And there are variations on these themes called Phase Shift Keying (PSK) and Quadrature Amplitude Modulation (QAM).

MPI

Media Platform Interface libraries. Part of the SunXTL Teleservices architecture. MPIs provide a layer of abstraction between details of the system services, applications and providers. The system services include a message passing "server," a STREAMS multiplexor driver, a provider configuration database and a database administration tool.

MPS

Multi-Page Signal. A frame sent if the sender has more pages to transmit.

MSI

Modular Station Interface. A Dialogic board that interfaces analog phones to SCbus and PEB-based products.

MTBF

Mean Time Between Failure. The length of time a user may reasonably expect a devicc or system to work before an incapacitating fault occurs.

MTTR

Mean Time to Repair. The average time required to return a failed device or system to service.

Multimedia WAVE API

A Microsoft Win 32 API used to exchange audio data after TAPI sets up a hone call. The API talks to the Unimodem V TSPI to set-up a media stream for playing a WAVE file. It works alongside TAPI and the Win 32 COM API. The Wave API does not handle call control, but rather audio data media streams. It synchronizes WAVE output in parallel with TAPI calls with Unimodem for devices with separate serial and wave output ports.

Multiplexer

Electronic equipment which allows two or more signals to pass over one communications circuit. That "circuit' may be a phone line, a microwave circuit, a through-the air TV signal. That circuit may be analog or digital. There are many multiplexing techniques to accommodate both.

MVIP

Multi-Vendor Integration Protocol. A standard for digital switching technology that permits the design of sophisticated applications such as voice, fax, speech recognition, and others. The technology was developed by Mitel, and the standard was proposed by Natural Microsystems.

NAP

Network Applications Platform (Unisys, Blue Bell, PA, term). A public telephone network service with the ability to send and receive voice messages and facsimile documents.

NetBIOS

The Network Basic Input Output System. An IBM-developed standard for network communication. Most non IBM networks require a NetBIOS compatibility module or utility to use NetBIOS communication and protocols. (See also SPX.)

Network

1. Networks are common in our lives. Think about trains and phones. A networks ties things together. Computer networks connect all types of computers and computer-related things together -- terminals, printers, modems, door entry sensors, temperature monitors, etc. The networks we're most familiar with are long distance ones, like phones and trains.

But there are also Local Area Networks (LANs), which exist within a limited geographic area -- like the few hundred feet of a small office, or they can span an entire building, or even touch a "campus," such as a university, or industrial park. There are also Metropolitan Area Networks (MANs). See LAN.

Network Board

A board device designed to act as an interface between a computer-based signal processing system and a telephone network.

Network Fax Servers

A network fax server allows users to send faxes from their network workstations. Many fax servers allow the faxing of documents from the applications that created them, leaving users free to continue with their work. Incoming faxes can be directed back to your workstation as well.

Different fax server products offer different ways of handling incoming faxes, but most use one of the following: the fax can be received by an operator who forwards it to the proper workstation, a fax extension can be assigned to particular workstations for touch-tone routing, or there can be a direct fax telephone line accessed through a Direct Inward Dial (DID) line.

Network Interface

The point of inter-connection between Telephone Company communications facilities and terminal equipment, protective apparatus or wiring at a subscriber's premises. The network interface or demarcation point is located on the subscriber's side of the Telephone Company's protector.

Network Interface Module

Electronic circuitry connecting a workstation (typically a personal computer) to the telephone network. Network interface modules come in as many flavors as there are ways of connecting to the telephone network -- from simple loop start phone lines to complex primary rate interfaces (PRI) on ISDN. Usually, the network interface module slides into one of the expansion slots inside a personal computer. The card transmits and receives messages from the resource modules connected to the telephone network.

NNX/NXX

A three-digit code to identify the central office in which N is any digit 2 to 9 and X is any digit. The first three digits of a North American telephone number. Originally only NNX codes were used. Now subscriber dials 1+ in making a direct distance dialed toll call and the code may be NXX. Area codes are also being changed to NXX codes.

Node

1. A point of connection into a network. In LANs, it is a device on the network. In packet switched networks, it is one of the many packet switches that form the network's backbone. 2. An independent SCSA unit in a distributed processing SCSA network (MNA - Multinode Architecture) consisting of one or more resource and/or network boards, and one or more SCxbus adapter boards. Communication between nodes take place via the SCxbus. From a device programming point of view, a node is simply an addressable system unit which contains boards connected by an SCbus.

Noise

Unwanted electrical signals introduced into telephone lines by circuit components or natural disturbances that tend to degrade the line's performance.

Novell Telephony Services (TSAPI)

Novell's telephony server product, providing a single standard client-server call control interface (including an API for programmers) to many of the diverse PBX systems on the market today.

NSC
Non-Standard Facilities Command. A response to the called fax DIS response.

NSF
Non-Standard Facilities.

NSS
Non-Standard Facilities Setup command, a response to an NSF frame.

NVP
Network Voice Protocol. Circa 1973 ARPANET protocol. Old voice protocol. Used to support real-time voice over the ARPANET. Both LPC and CVSD encoding schemes were successfully implemented by Culler-Harrison, Inc., the Information Sciences Institute, Lincoln Laboratory and Stanford Research Institute.

OAI
Open Application Interface. Basically an opening in a telephone system that lets you link a computer to that phone system and lets the computer command the phone system to answer, delay, switch, hold etc. calls. The term is also called PHI -- as in PBX-Host-Interface. The term OAI was first used by PBX makers, NEC and InteCom. And now the term has become somewhat generic, like all good things. Essentially every manufacturer of phone systems is evolving towards open application interfaces of their own.

OA&M
Operations, Administration and Maintenance. This term refers to PBX management.

Octet
The ITU standard term for byte.

Off-Hook

When the handset is lifted from its cradle it's Off-Hook. Lifting the hookswitch alerts the central office that the user wants the phone to do something like dial a call. A dial tone is a sign saying "Give me an order." The term "off-hook" originated when the early handsets were actually suspended from a metal hook on the phone.

When the handset is removed from its hook or its cradle (in modern phones), it completes the electrical loop, thus signaling the central office that it wishes dial tone. Some leased line channels work by lifting the handset, signaling the central office at the other end which rings the phone at the other end. Some phones have autodialers in them. Lifting the phone signals the phone to dial that one number. An example is a phone without a dial at an airport, which automatically dials the local taxi company. All this by simply lifting the handset at one end -- going "off-hook."

Off-Hook Signal

In telephone switching, a signal indicating seizure, request for service, or a busy condition.

ONC+

Part of Sun's ENOS Networking Solutions. ONC+ makes the network seamless. Automates and centralizes PC network administration.

One-Dimensional Coding

A data compression scheme that considers each scan line as being unique, without referencing it to a previous scan line. One-dimensional coding operates horizontally only.

On-Hook

When the phone handset is resting in its cradle. The phone is not connected to any particular line. Only the bell is active, i.e. it will ring if a call comes in. See ON-HOOK DIALING and OFF-HOOK.

On-Hook Dialing

Allows a caller to dial a call without lifting his handset. After dialing, the caller can listen to the progress of the call over the phone's built-in speaker. When you hear the called person answer, you can pick up the handset and speak or you can talk hands-free in the direction of your phone, if it's a speakerphone.

Critical: Many phones have speakers for hands-free listening.
Not all phones with speakers are speakerphones -- i.e. have microphones, which allow you to speak, also.

Operator Agent

A Win 95 applet included with Unimodem V. Operator Agent answers telephone calls and routes them either on a an application/device specific basis (i.e., fax calls are routed to the fax app and voice calls are routed to a voice app), or it will route calls based on interaction with the caller (i.e., touch-tone input from the caller). A third method for routing calls, distinctive ringing, is handled by Unimodem itself.

Operator Services

Any of a variety of telephone services which need the assistance of an operator. For example, such services typically include collect calls, third-party billed calls, person-to-person calls.

Out-Of-Band Signaling

Signaling that is separated from the channel carrying the information -- the voice, data, video, etc. Typically the separation is accomplished by a filter. The signaling includes dialing and other supervisory signals. Out-band-band hardware signaling takes place on its own data path, without the necessity of "bit robbing" or other methods of mixing signals into a data stream.

P

Packet

A bundle of data, usually in binary form, organized in a specific way for transmission.

Packet Switching

Sending data in packets through a network to a remote location. The data are subdivided into individual packets of data, each with a unique identification and individual destination address. This way each packet can take a different route and may arrive in a different order than it was shipped. The packet ID allows the reassembling of data in the proper sequence. This is an efficient way to move digital data. Although it has been used with fax messages, is not yet useful for voice.

PABX (PBX)

Private Automatic Branch Exchange. Originally, PBX was the word for a switch inside a private business (as against one serving the public). PBX means a Private Branch Exchange. Such a "PBX" was typically a manual device, requiring operator assistance to complete a call.

Then the PBX went "modern" (i.e. automatic) and no operator was needed any longer to complete outgoing calls. You could dial "9." Thus it became a "PABX." Now all PABXs are modern. And a PABX is now commonly referred to as a "PBX." Some manufacturers have tried to make their PBX appear different by calling it something else. Siemens/Rolm calls theirs the "CBX" (Computerized Branch Exchange). Some others call theirs the "EPABX" (the Electronic Private Automatic Branch Exchange. Then there are the special ones, like SRX's SRX, NEC's IMS (Information Management System), etc.

PAGE

A chunk of information, such as a document or a file, on the Web. You write them in HTML code. They can contain graphics, audio and video.

PAM

Pulse Amplitude Modulation. The process of representing a continuous analog signal (a voice conversation) with a series of discrete analog samples. This concept is based on the information theory which suggests that the signal can be accurately recreated from a sufficient sample. Why bother? Sampling allows several signals to then be combined on a channel that otherwise would only carry one telephone conversation.) PAM was used as part of a method of switching phones calls in several PBXs. It is not a truly "digital" switching system. PAM is the basis of PCM, pulse code modulation. See PCM.

Part 68 Requirements (Registration)

Specifications established by the FCC as the minimum acceptable protection which communications equipment must provide the telephone network. Meeting these requirements does not certify that equipment performs any task. Equipment which is sold for connection to the public network in the US must conform to Part 68.

PBX (PABX)

Private Branch eXchange. A private (i.e. you, as against the phone company owns it), branch (meaning it is a small phone company central office), exchange (a central office was originally called a public exchange, or simply an exchange). In other words, a PBX is a small version of the phone company's larger central switching office.

A PBX is also called a Private Automatic Branch Exchange, though that has now become an obsolete term. In the very old days, you called the operator to make an external call. Then later someone made a phone system that you simply dialed nine (or another digit -- in Europe it's often zero), got a second dial tone and dialed some more digits to dial out, locally or long distance. So, the early name of Private Branch Exchange (which needed an operator) became Private AUTOMATIC Branch Exchange (which didn't need an operator). Now, all PBXs are automatic. And now they're all called PBXs, except overseas where they still have PBXs that are not automatic.

PCM

Pulse Code Modulation. PCM is a method of taking an analog voice signal and encoding it into a digital bit stream. First, the amplitude of the voice conversation is sampled. This is called PAM, Pulse Amplitude Modulation. This PAM sample is then coded into a binary (digital) number. This digital number consists of zeros and ones. It can then be switched and transmitted digitally. There are three basic advantages to PCM voice. They are the three basic advantages of digital switching and transmission.

First, it is less expensive to switch and transmit a digital signal. Second, by making an analog voice signal into a digital signal, you can interleave it with other digital signals -- such as those from computers or facsimile machines. Third, a voice signal which is switched and transmitted end-to-end in a digital format will usually come through "cleaner," i.e. have less noise, than one transmitted and switched in analog. The reason is simple: An electrical signal loses strength over a distance. It must then be amplified. In analog transmission, everything is amplified, including the noise and static the signal has collected along the way.

In digital transmission, the signal is "regenerated," i.e. put back together again, by comparing the incoming signal to a logical question: Is it a one or a zero? Then, the signal is regenerated, then it is "amplified and sent along its way. PCM refers to a technique of digitization. It does not refer to a universally accepted standard of digitizing voice. The most common PCM method is to sample a voice conversation at 8000 times a seconds. The theory is that if the sampling is at least twice the highest frequency on the channel, then the result sounds OK. Thus, the highest frequency on a voice phone line is 4,000 Hertz. So one must sample it at 8,000 times a second. Many PCM digital voice conversations are typically put on one communications channel. In North America, the most typical channel is called the T-1 (also spelled T1). It places 24 voice conversations on two pairs of copper wires (one for receiving and one for transmitting). It contains 8000 frames each of 8 bits of 24 voice channels plus one framing (synchronizing bit) bit which equals 1,544,000 bits per second. i.e. 8000 x (8 x 24 + 1) equals 1.544 megabits.

Europe uses a different scheme for multiplexing voice conversations. It is based not on 24 voice channels, but on 32 channels. This scheme keeps two of the 32 channels for control and actually transmits 30 voice conversations at a data rate of 2,048,000 bits per second. The European system is calculated as 8 bits x 32 channels x 8000 frames per second. European PCM multiplexing is not compatible with North American multiplexing.

The two systems cannot be directly connected. Some PBXs in the U.S. conform to the U.S. standard only. Some (very few) conform to both. Both the European and North American T-1 "standards" have now been accepted as ISDN "standards." In addition to PCM, there are many other ways of digitally encoding voice. PCM remains the most common. See E-I, T-1 and VOICE COMPRESSION.

PCM-30

Short name of international 2.048 Mbps T-1 (also known as E1) service derived from the fact that 30 channels are available for 64 Kbps digitized voice each using pulse code modulation (PCM).

PEB

PCM Expansion Bus. Dialogic's name for the digital electrical connection between its network interface modules and its resource modules. This bus is now "open." Technical specifications are available, thus enabling outsiders to create their own resource modules and/or network interface modules. See also AEB, which is Dialogic's Analog Expansion Bus.

Pel

Picture Element. A pel contains only black and white information, no grey shading.

Personal call manager

Brings the management of telephone calls onto the computer screen where users can initiate a phone call with the click of the mouse, chose to respond to calls via caller ID, selectively forward calls to a conference room or other site, and in other ways personalize their response to telecommunications.

Personal Identification Number

PIN number. 1. An AT&T term meaning the last four digits of your AT&T, MCI Bell operating company Credit Card -- the card you use for making long distance numbers. 2. Some banks and financial institutions issue credit cards for machine, teller-less banking. These machines, called Automated Teller Machines, ask you for a password consisting of several numbers or characters. These are not on your credit card. These numbers or characters, called PIN numbers, are designed to make sure the right person is using your card. It's not a good idea to use your birthday as you PIN number.

Phase A

In a fax machine's call process, the call establishment, when the transmitting and receiving units connect over the telephone line, recognizing one another as fax machines. This is the start of the handshaking procedure.

Phase B

In a fax machine's call process, the pre-message procedure, where the answering machine identifies itself, describing its capabilities in a burst of digital information packed in frames conforming to the HDLC standard.

Phase C

In a fax machine's call process, the fax transmission portion of the operation. This step consists of two parts-C1 and C2-which take place simultaneously. Phase C1 deals with synchronization, line monitoring, and problem detection. Phase C2 includes data transmission.

Phase D

In a fax machine's call process, this phase begins once a page has been transmitted. Both the sender and receiver revert to using HDLC packets as during Phase B. If the sender has further pages to transmit, it sends an MPS frame, and the receiver replies with an MCF and Phase C recommences for the following page.

Phase E

In a fax machine's call process, the call release portion. The side that transmitted last sends a DCN frame and hangs up without awaiting a response.

Phonemail

A ROLM term for Voice Mail. Rolm's PhoneMail is a voice messaging system that provides (1.) telephone answering (with the user's own greeting), (2.) the capability to store and forward voice messages, and (3.) the capability to turn on a message waiting light or message on the recipient's phone. PhoneMail can be used positively -- to speed the flow of information. It can also be used negatively -- to allow the user to "hide behind" the system and avoid the outside world and anyone in the outside world who might actually want to buy something. See also VOICE MAIL.

PIM

Personal information managers are like computerized appointment books, and are used on many personal computers, to help a person keep track of appointments, telephone numbers, and other contact information. A telephony enabled PIM could provide point and click telephone dialing, automated conference calling, or automated access to databases and other applications.

PIN

Procedure interrupt negative.

PIN Number

Personal Identification Number. A group of characters entered as a secret code to gain access to a computer system, such as the one that completes long distance calls. See PERSONAL IDENTIFICATION NUMBER.

Pixel

Picture Element. The smallest area of an original sampled and represented by an electrical signal. A pixel has more than two levels of greyscale information.

Phone Book

This is a special file for fax telephone numbers and other transmission information for people with whom the user regularly communicates. It has Catcgory and Class fields to organize contacts into groups and subgroups.

With the Phone Book it is not necessary to type numbers and names each time a fax is sent. It can be used to send to a single number or to more than one fax number for broadcasting purposes.

PIP

Procedure interrupt positive.

Playback

Retrieval, decoding and transmission of encoded data.

Player

A resource object that plays TVM data. The audio data can come from a voice or audio encoded file, or from text that has passed through a text-to-speech service. The output of a player can be analog audio, TDD, ADSI, etc.

Polling

Refers to some form of data or fax network arrangement whereby a central computer or fax machine/board very quickly asks each remote location in turn whether they want to send some information. The purpose is to give each user or remote data terminal an opportunity to transmit and receive information on a circuit or using facilities which are being shared. Polling is typically used in a multipoint or multidrop line. It is done mostly to save money on telephone lines.

POP

Post Office Protocol. A system on the Internet you can use to get your e-mail. You use a mail server to get the mail and then download it to your computer.

Pop-Up

See Hot Key.

Port

An entrance to or exit from a network, the physical or electrical interface through which one gains access.

The interface between a process or a program and a communications or transmission facility. A point in the computer or telephone system where data may be accessed. Peripherals are connected to ports.

Portability

This means that application software can be dragged across different computing platforms and operating systems. With the JTAPI spec, developers can write an application that will run on a PC, Mac, or UNIX host.

PostScript

A page description language for PCs, which is standard for many graphics and desktop publishing applications. It allows the creation of elaborate documents. Other applications such as spreadsheets, word processors, and databases rarely require this sophistication.

POTS

Plain Old Telephone Service. The basic service supplying standard single line telephones, telephone lines and access to the public switched network. Nothing fancy. No added features. Just receive and place calls. Nothing like Call Waiting or Call Forwarding. They are not POTS services. Pronounced POTS, like in pots and pans.

POINT-TO-POINT PROTOCOL

PPP. An implementation of TCP/IP which is intended for transmission using telephone lines. PPP is a new standard which supersedes SLIP.

Predictive Dialer

See PREDICTIVE DIALING.

Predictive Dialing

An automated method of making many outbound calls without people and then passing answered calls to a person as the calls are answered. Here's the story: Imagine a bunch of operators having to call a bunch of people. Those calls may be for collections. They may be for employee call-ups. They may be for alumnae fund raising. When it's done manual, here's how it works:

Before each call operators spend time reviewing paper records or computer terminal screens, selecting the person to be called, finding the phone number, dialing the numbers, listening to rings, listening to phone company intercepts, busy signals and answering machines.

Operators also spend time updating the records after each call. Predictive dialing automates this process, with the computer choosing the person to be called and dialing the number and only passing it to an operator when a real live human being answers. There are enormous productivity gains made by screening out answering machines, busy signals, network busy signals, non-completed calls, operator intercepts etc. The result is productivity increases of 300% to 600%.

According to generally accepted industry lore, a well-run manual dialing center can get its people talking on the phone for 25 minutes an hour. With a predictive dialer you can get them on the phone making sales, collecting money, etc. for 55 minutes an hour. It's a major productivity gain. True predictive dialing should not be confused with automated dialing.

True predictive dialing has complex mathematical algorithms that consider, in real time, the number of available telephone lines, the number of available operators, the probability of getting no answer, a busy signal, a disconnected number, operator intercept or an answering machine, the time between calls required for maximum operator efficiency, the length of an average conversation and the average length of time the operators need to enter the relevant data. Some predictive dialing systems constantly adjust the dialing rate by monitoring changes in all these factors. Some people don't like the term "predictive dialing," since they think it's getting "a bad rap" in Washington, DC by being associated with junk phone calls. As a result some e people would prefer to call it Computer Aided Dialing.

Preview Dialing

Preview dialing is a term used to describe an automatic dialer.

Preview dialing is also called "screen dialing" or "cursor dialing." Typically the prospect's account information and/or phone number appears on the screen BEFORE the call is made. Thus the agent can "preview" the number, the screen, the customer. If the agent wants to make the call, the agent hits a key, such as "Enter" and the computer dials the number. In some preview dialing equipment, the agent must hit a key if he/she DOESN'T want the number dialed. Contrast preview dialing with Predictive Dialing where the computer makes all the dialing decisions and presents the calls to the agent only after they are connected. Predictive dialing is a lot faster than preview dialing. See PREDICTIVE DIALING.

PRI
See PRIMARY RATE INTERFACE.

PRI-EOM
Procedure interrupt-end of message.

PRI-EOP
Procedure interrupt-end of procedures.

PRI-MPS
Procedure interrupt-multipage signal.

Primary Rate Interface
PRI. The ISDN equivalent of a T-1 circuit. The Primary Rate Interface (that which is delivered to the customer's premises) provides 23B+D (in North America) or 30B+D (in Europe) running at 1.544 megabits per second and 2.048 megabits per second, respectively. There is another ISDN interface. It's called the Basic Rate Interface. It delivers 2B+D over either one or two pairs. In ISDN, the "B" stands for Bearer, which is 64,000 bits per second, which can carry PCM-digitized voice or data.

Primary Resource
The main resource around which a Group is constructed. Typically, the primary resource will be an interface to the telephone network, but it may also be a switch port.

Printed Circuit Board

PCB. Flat material (fiberglass/epoxy) on which electronic components mount. A PCB also provides electrical pathways, called traces, that connect components. Printed circuit boards are what PBXs and computers are made of these days.

Be careful when you're replacing PCBs. They're usually very sensitive to static electricity. Handle them only when you're attached to a static electricity strap that is properly grounded. Lay them down only on a surface you're sure is static electricity free. And don't touch the components on PCBs whatever you do.

Printer Emulation

Enables the mimicking of a printer-generated document. This way, the outgoing fax will look as if it has come from the printer attached to a computer. This can include full formatting, as well as letterhead, signature, and different graphic images.

Private Branch Exchange

PBX. Term used now interchangeably with PABX. See PABX.

Processing Power

The number of computations that a computer, microprocessor, or digital signal processor can complete in a fixed time interval. May be measured in MIPS (millions of instructions per second) or MFlops. Typical low-end DSP chips provide up to 10 MFlops; high-end chips 30 or more.

Program Logic

The particular sequence of instructions in a program.

Programming Language

A language used by a programmer to develop instructions for the computer. It is translated into machine language by language software called assemblers, compilers and interpreters. Each programming language has its own grammar and syntax.

Prompts

1. Recorded instructions delivered by voice processing units. Prompts may include MENUS or other information that is played each time you get into the system. 2. Messages from the computer instructing the user on how to use the system. See MENU and AUDIO MENU.

Prosody

Intonation. In text to speech, prosody refers to how natural it sounds -- the ups and downs of the sentence.

Protocol

A specified set of rules, procedures, or conventions, relating to format and timing of data transmission between two devices. A standard procedure that two data devices must accept and use to be able to understand each other.

PSK

Phase Shift Keying. A method of modulating the phase of a signal to carry information.

PSTN

Public Switched Telephone Network. An abbreviation used by the ITU.

PTT

Post Telephone & Telegraph Administration. The PTTs, usually controlled by their governments, provide telephone and telecommunications services in most countries.

Pulse Amplitude Modulation

PAM. A technique for placing binary information on a carrier to transmit that information. PAM is a technique for analog multiplexing.

The amplitude of the information being modulated controls the amplitude of the modulated pulses. Samples of each input voltage are placed between voltage samples from other channels. The cycle is repeated fast enough so the sampling rate of any one channel is more than twice the highest frequency transmitted. See also PAM and PCM.

Pulse Code Modulation

PCM. The most common and most important method in North America in which a telephone system samples a voice signal and converts that sample into an equivalent digital code. PCM is a digital modulation method that encodes a Pulse Amplitude Modulated (PAM) signal into a PCM signal. See PCM and T-1.

Pulse Density

In T-1, since "O"s are represented by no pulse and "1"s by alternating pulses, pulse density refers to the number of no pulse ("O") periods allowed before a pulse ("1") must occur. Typically, no more than 15 no pulse periods ("O"s) are allowed before a pulse ("1") must occur.

Pulse Dialing

One or two types of dialing that uses rotary pulses to generate the telephone number.

Pulse Dispersion

The spreading out of pulses as they travel along an optical fiber.

Pulse Duration Modulation

PDM. That form of modulation in which the duration of the pulse is varied in accordance with some characteristic of the modulating signal.

Pulse Link Repeater

A signaling set that interconnects the E and M leads of two circuits. In E & M signaling, a device that interfaces the signal paths of concatenated trunk circuits. Such a device responds to a ground on the "E" lead of one trunk by applying -48Vdc to the "M" lead of the connecting trunk, and vice versa. This function is a built-in, switch-selectable option in some commercially available carrier channel units.

Pulse Overshoot

In T-1, the amount of signal voltage that can remain at the trailing end of a pulse. It can be no more than 10-30% of the pulse amplitude. Also called afterkick.

Pulse-Position Modulation
PPM. That form of modulation in which the positions in time of the pulses are varied in accordance with some characteristic of the modulating signals, without modifying the pulse width.

Pulse Repetition Frequency
PRF. In radar, the number of pulses that occur each second. Not to be confused with transmission frequency which is determined by the rate at which cycles are repeated within the transmitted pulse.

Pulse Stuffing
When timing signals on digital circuits get out of whack, some method of allowing mismatches must be provided. In time division multiplexing, this is called pulse stuffing. One stream of data has bits added to it so its final rate is the same as the master clock.

Pulse Train
The resulting electronic impulses that transmit encoded information.

Pulse Width
In T-1, refers to the width (at half amplitude) of the bipolar pulse (typically 324 + or -45 nsec).

PVC
Permanent Virtual Circuit

QAM
Quadrature Amplitude Modulation. A modulation technique that uses variations in signal amplitude, allowing data-encoded symbols to be represented as any of 16 or 32 different states. Some QAM modems allow dial-up data rates of up to 9,600 bps.

Quantization

Breaking down into discrete values for the purposes of sampling or transmission. In speech-compression algorithms, quantization is the art and science of encoding a set of model parameters about a speech signal with as few bits of information as possible.

Take an analog voice signal, sample it and put numbers on those samples. That's called quantization. Here's an explanation from Understanding Telephone Electronics: "A circuit called a quantizer takes in the analog signal and produces an equivalent number.

Threshold levels are established and numbers are assigned to the analog samples as their amplitudes fall within the bands formed by the threshold numbers. (For example, everything within the range 1000.05 and 1000.06 Hz is given the number 14568.) The assigned number in most costs is an approximation rather than a true value because the true value would require many more bits (i.e. more threshold limits) in the binary code...

This approximation causes an error which is the difference between the approximate number and the true sample. This quantization error adds noise to the signal, called quantization noise, which is heard in the telephone as hissing.

Quantization noise can be reduced by making the threshold bands narrower. However, providing more intervals requires more bits in the binary code. Therefore, more bandwidth is needed. There is a tradeoff between small quantizing intervals (higher bandwidth, lower noise), and fewer intervals (lower bandwidth, higher noise).

Quantize

The process of encoding a PAM signal (Pulse Amplitude Signal) into a PCM signal (Pulse Code Modulation). See QUANTIZING and QUANTIZATION.

Quantizing
The second stage of pulse code modulation (PCM). The waveform samples obtained from each communication channel are measured to obtain a discrete value of amplitude. These quantized values are converted to a binary code and transmitted to a distant location to reconstruct the original waveform. See QUANTIZATION.

Quantization Noise
Signal errors which result from the process of digitizing (and therefore ascribing finite quantities to) a continuously variable signal. See QUANTIZATION.

Queuing
The act of stacking or holding calls to be handled by a specific person, trunk, or trunk group.

R

RAM
Random Access Memory. The primary memory in a computer. Memory that can be overwritten with new information. The random access part of the name comes from the fact that the next bit of information in RAM can be located-no matter where it is-in an equal amount of time, making access to it considerably faster than to information in other storage media, such as a hard disk.

Raster Scanning
The method of scanning in which the scanning spot moves along a network of parallel lines, either from side to side or top to bottom.

Real Time
A voice telephone conversation is conducted in Real Time. That is, there is no perceived delay in the transmission of the voice message or in the response to it. This concept often applies to interaction between a computer and a terminal. See also REAL TIME CAPACITY.

In data processing or data communications, real time means the data is processed the moment it enters a computer, as opposed to BATCH processing where the information enters the system, is stored and is operated on a later time.

Record

An entry in a database. Records contain fields. Records may be retrieved in either recordorder or keyorder. The sampling, encoding and storage of data.

Reference Line

The first scanning line in memory. The location of each black pixel of this line is kept in memory for the next scanned line. Depending on the compression technique used, more or fewer scan lines are necessary.

Registration Number (FCC Part 68)

Approval number given to telephone equipment to certify that a particular device passes the tests defined in Part 68 of the FCC Rules. These tests certify the phone won't cause any harm to the public network. They do not attest to the commercial value of the product, nor whether it will (or won't) sell.

Relational Database

A database composed of multiple table files, linked or related by the contents of a single "key" field duplicated in each.

Remote Diagnostics

You own a phone system. You have a service company. There's some problem with it. Instead of sending a technician out, your service company dials your PBX from a data terminal or PC and "asks" your PBX in computerese what's wrong with it. If it isn't too broken, it will come back and give you some indication. This is called remote diagnostics.

Remotely-hosted

The Client and the Server are different (i.e., the application is on a different physical box than the service provider).

Remote Programming

Dial your phone system with your friendly personal computer, modem and a communications software package and you can change the telephone system's programming remotely. This feature is great for companies with telephone systems in many locations. They can all be run from one central point. This feature is also great. You may want some changes made on your system. It's obviously a lot cheaper for your vendor to make those changes from his office than having to visit yours. It's also a lot faster. See REMOTE DIAGNOSTICS.

Reply

An event which is a service providers response to a synchronous or asynchronous request.

Resource

The abstraction of a standardized vendor-independent interface of a physical device used for call processing as seen by the Application. All Resources have common methods across all implementations. Examples of Resources are voice store and forward, fax send and receive, text to speech conversion, voice recognition, etc. Resources are assumed to have at a minimum one input or output of circuit switched TDM on the internal switch fabric of the system.

Resources are shared among multiple applications. Once a Resource has been allocated to an application it is locked from the use of any Other application until freed. It is assumed that applications specify at some level their resource requirements to the server prior to accessing them.

This may either be through explicitly attaching them to a Group, or having the server implicitly allocate them based on usage.

Resource Class

A set of methods (in object-oriented terms, a class) for controlling resource instances (or a resource). May be abbreviated as just "class."

Resource Module

A Dialogic term referring to devices which perform specific voice processing functions, such as voice compression, voice recognition, facsimile transmission and reception and conversion of computer text to spoken words over the phone. Resource modules are typically connected to the telephone network through devices known as "network interface modules."

Resource Object

An instance of a Resource Class.

Resource Unit

An entity that communicates directly with a resource device. From the client application's point of view, it is the Resource Unit that provides services such as voice store-and-forward, fax send and receive, etc.

Ring

1. As in Tip and Ring. One of the two wires (the two are Tip and Ring) needed to set up a telephone connection.

2. Also a reference to the ringing of the telephone set. 3. The design of a Local Area Network (LAN) in which the wiring loops from one workstation to another, forming a circle (thus, the term "ring"). In a ring LAN, data is sent from workstation to workstation around the loop in the same direction. Each workstation (which is usually a PC) acts as a repeater by re-sending messages to the next PC in the ring. The more PC's, the slower the LAN.

Network control is distributed in a ring network. Since the message passes through each PC, loss of one PC may disable the entire network. However, most ring LANs have a way of recovering very fast should one PC die or be turned off. If it dies, you can remove it physically from the network. If it's off, the network senses that and the token ignores that machine. In some token LANs, the LAN will close around a dead workstation and join the two workstations on either side together. If you lose the PC doing the control functions, another PC will jump in and take over. This is how the IBM Token- Passing Ring works.

Ringing Voltage

In addition to talk battery, a Central Office provides ringing signaling. Ring Voltage is generally 70 to 90 volts at 17 Hz to 20 Hz.

RJ-11

RJ-11 is a six conductor modular jack that is typically wired for four conductors (i.e. four wires). Occasionally it is wired for only two conductors -- especially if you're only wiring up for tip and ring. The RJ-11 jack (also called plug) is the most common telephone jack in the world.

The RJ-11 is typically used for connecting telephone instruments, modems and fax machines to a female RJ-22 jack on the wall or in the floor. That jack in turn is connected to twisted wire coming in from "the network" -- which might be a PBX or the local telephone company central office. RJ-22 wirings typically flat. None of its conductors (i.e. wires) are twisted. You cannot use flat cable for high-speed data communications, like local area networks. See also RJ-22 and RJ-45.

RJ-14

A jack that looks and is exactly like the standard RJ-11 that you see on every single line telephone. Whereas the RJ-11 defines one line -- with the two center, red and green, conductors being tip and ring, the RJ-14 defines two phone lines. One of the lines is the "normal" RJ-11 line -- the red and green center conductors. The second line

RJ-45

The RJ-45 is the 8-pin connector used for data transmission over standard telephone wire. That wire could be flat or twisted. And it's very important that you know what you're working with. You can easily use flat wire for serial data communications up to 19.2 Kbps. If you wish to connect to a 10BaseT local area network, which you also do with a RJ-45, you must use twisted wire. You can typically tell the difference by looking at the cable. If it's flat grey satin (like a typical phone wire, only bigger) than it's probably untwisted. If it's circular, then it's probably twisted and therefore good for LANs.

RJ-45 connectors come into two varieties -- keyed and non-keyed. Keyed means that the male RJ-45 plug has a small, square bump on its end and the female RJ-45 plug is shaped to accommodate the plug. A keyed RJ-45 plug will not fit into a female, non-keyed (i.e. normal) RJ-45. See RJ-11 and RJ-22.

Robbed-Bit Signaling

This explanation from Gary Maier of Dianatel: ISDN is the key to future sophisticated telephone network services with its dynamic, highly configurable T-1 connection (also called PRI connection). Since T-1 is a common method of carrying 24 telephone circuits, many wonder about the uses for ISDN, especially when they learn ISDN signaling requires an entire voice channel, reducing today's T-1 from 24 voice channels to 23.

But the popular signaling mechanism of "robbed bit" signaling in T-1 has serious limitations. Robbed bit signaling typically uses bits known as the A and B bits. These bits are sent by each side of a T-1 termination and are buried in the voice data of each voice channel in the T-1 circuit. Hence the term "robbed bit" as the bits are stolen from the voice data.

Since the bits are stolen so infrequently, the voice quality is not compromised by much. But the available signaling combinations are limited to ringing, hang up, wink, and pulse digit dialing. In fact, the limitations are obvious when one recognizes DNIS and ANI information are sent as DTMF tones. This introduces a problem: time.

Each DTMF tone requires at least 100 milliseconds to send, which in a DNIS and ANI situation with 20 DTMFs will take at least two full seconds. There is also a margin for error in transmission or detection, resulting in DNIS or ANI failures. With the explosion of telephone related services, the telephone companies are turning to ISDN PRI to provide the more complicated and exact signaling required for new services. ISDN employs a more robust method of signaling. ISDN uses a T-1 circuit as 23 voice channels and one signaling channel.

The term 23B plus D refers to 23 bearer (voice) channels and 1 Data (signaling) channel. The data channel carries the signaling information at a rate of 64 kilobits per second. This speed is many times greater than some of the most powerful modems available. Because of this high speed, telephone calls can be placed more quickly, and because of the protocol used, DNIS or ANI transmission failures are impossible.

Additionally, since no bits are "robbed" from the voice channels, the voice quality is better than that of Robbed Bit signaling on today's T-1 circuits. Also, computer modems and high speed faxes can use the voice channel for sending digital data instead of the traditional analog bit "noise." Therefore, ISDN PRI offers the end user countless new service capabilities. One channel could be used for faxing, another for modem data, several for video, another for a LAN and the remainder for voice. Suddenly, the average T-1 circuit becomes a pipeline for all communications! Increasingly long distance carriers are using ISDN PRI to provide inbound 800 calls with ANI and DNIS and re-routing skills. Dianatel makes some of the most sophisticated ISDN interface equipment around.

Rotary Dial

The circular telephone dial. As it returns to its normal position (after being turned) it opens and closes the electrical loop sent by the central office.

Thus it generates pulses for each digit dialed. You can hear the "clicks". The number "seven," for example consists of seven "opens and closes," or seven clicks. You can dial on a rotary phone without using the rotary dial. Simply depress the switch hook quickly, allowing pauses in between to signify that you're about to send a new digit. It's a good party trick.

Routing

The process of selecting the correct path for a message.

RS-232

A set of standards specifying various electrical and mechanical characteristics for interfaces between computers, terminals, and modems. It applies to synchronous and asynchronous binary data transmission.

RSVP

ReSerVation Protocol. Circa 1993 resource reservation protocol, as contrasted with simple set-up and tear down controls. RSVP is based on receiver-oriented simplex control. It uses "reservation styles" to control the aggregation of reservations in real-time multipoint-to-multipoint apps. RSVP uses a soft-state approach in maintaining receiver-initiated resource reservations. Allows for decoupling the routing and reservation elements of transactions.

RTC

See Runtime Control.

RTN

Retrain negative.

RTCP

Real Time Conferencing Protocol. Supports real-time conferencing for large groups on the Internet. It has source identification and support for audio and video bridges/gateways. Supports multicast-to-unicast translators. 2. Retrain positive.

RTP

Real Time Protocol. Supports transport of real-time data like interactive voice and video over packet switched networks. A thin protocol providing support for content identification, timing reconstruction, loss detection and security. The ARPA DARTnet transcontinental IP network experiments lead to RTPs popularity. Now championed by the Audio/Video Transport (AVT) Working Group. AVT is part of the IETF (Internet Engineering Task Force). RTP does not do resource reservation or quality of service control. It relies on resource reservation protocols like RSVP.

Runtime Control

The mechanism by which one Resource Object can influence the behavior of another. Typically used for things such as terminating conditions and speed/volume control. Runtime Control is often abbreviated RTC.

S

Sampling Rate

The number of times per second that an analog signal is measured and converted to a binary number -- the purpose being to convert the analog signal to a digital analog. The most common digital signal -- PCM -- samples 8,000 times a minute.

SAPI

The Windows Speech API, a WOSA element, provides services that enable the use of speech recognition and/or text-to-speech in applications. This API uses the OLE Component Object Model (COM) architecture under Win32, Win 95 and Win NT 3.51. The component objects can be accessed through C/C++ or Visual Basic's OLE Automation. The speech architecture is divided into two levels, a high level that is designed for ease and speed of implementation, and a low level that allows applications complete control of the technology. The Speech API lets you add simple speech recognition and text-to-speech to an application using C++ and the architecture involved.

SBus

Sun's resource sharing and expansion bus interface for the SPARC architecture. SBus expansion cards can communicate with each other through this interface. SBus competes with VME, PCI and EISA/ISA as industry standard I/O buses for computing platforms. Currently, there are five computer telephony vendors selling SBus compliant DSP and network interface cards for SBus. You can develop cards with an both SBus computer interface and a mezzanine for MVIP and SCSA busses.

Scalability

This is "Forklift Upgrade" shark repellent. It means you can expand the application code to handle more lines without having to trash your existing investment in the hardware. The SCSA application model can accommodate system expansion without requiring changes to application code.

Applications are insulated from the underlying hardware implementation. Developers can add or subtract resources in a system without affecting the application.

Scanner

A device used to input graphic images in digital form. A fax machine's scanner determines the brightness of a document's pixels for transmission.

Scheduled Transmission

A feature allowing the user to schedule a fax transmission at a specific date or time in the future. Key benefits are convenience and cost savings. Scheduling jobs at a period of low telephone rates can have immediate considerable savings.

SCR

See SCSA Call Router

SCSA

Signal Computing System Architecture. A new standard for the telecom/voice processing/computing industry announced by Dialogic in early Spring 1993. SCSA incorporates virtually every other standard in PC-based switching -- including the most popular ones, Mitel's ST-Bus, MVIP, Siemens PCM Highway, AEB and PEB.

SCSA Call Router

A system service of SCSA which provides the basic necessities of inbound and outbound call processing and call sharing to client applications, without those applications needing to be aware of the underlying telephony interface operations.

SCSA Message Protocol

The open communications protocol by which entities communicate with one another in an SCSA system. The SCSA Message Protocol (SMP) is independent of the transport layers it is built upon, computer hardware, operating system, network topology (or lack thereof), and technology vendor.

All SCSA-compliant AIAs will translate the functions called by client applications (via the API) into SMP messages; these are transmitted to service providers regardless of their location. Therefore, applications written to the API will be portable from one call-processing environment to another.

SCSA Server

A collection of service providers (objects) which in the aggregate implement the minimum set of services required for SCSA system conformance. The assumption is that these services are at a minimum provided to remote-hosted client applications via common transports such as LANs, but may also be provided to client applications which are hosted on the SCSA server itself (see Self-hosted). Note that this is a logical image which may be implemented through multiple nodes (machines).

SDK

Software Development Kit

SDLC

Synchronous Data-Link Control.

Secondary Resource

Any resource that is attached to a Group after the Group has been created around a primary resource.

Self-hosted

The Client and the Server are on the same computer platform. A server with one or more self-hosted applications may be a standalone unit which is not connected to any Other system.

Serial Wave Driver

Microsoft's standard driver for modems using a serial port for voice communication (as opposed to a hardware audio port on a sound card). It supports both AT+V and AT#V standards. This provides a standard means for modem makers to be compatible with Win 95. Unimodem V communicates with the Wave Driver through DLL called vmocdctl.dll. The driver supports several versions of Adaptive Differential Pulse code Modulation (ADPCM) at 4.8, 7.2 and 8 KHz and 8KHz PCM.

Server

A communications element providing a service such as shared access to a file system, a printer or an E-mail system to LAN users. Usually a combination of hardware and software. A process in a distributed computing system that provides a service in response to requests from clients.

Service Control Points

SCPs. The local versions of the national SMS/800 number database. SCPs contain the intelligence to screen the full ten digits of an 800 number and route calls to the appropriate, customer-designated long distance carrier.

Service Provider

An addressable entity providing application and administrative support to the client environment by responding to client requests and maintaining the operational integrity of the server.

Shell Accounts

Text-based-only interfaces controlled by the host servers which normally don't allow for the use of graphic Web browsers.

Sidetone

A part of the design of a telephone handset which allows you hear your own voice while talking on the telephone. The idea of doing this is to give you a little feedback that the telephone you're speaking on is working. Too much sidetone becomes an echo and is bad. Too little sidetone makes the channel unerring.

Signal Errors

Signal errors indicate the return status of a Do/Change function. Despite the name, signal errors do not always indicate an error condition.

If a signal error is encountered and the k command line option specifies a Level to go to, the application will continue from the specified Level.

Signaling

Pertains to the transmission of electrical signals to and from the user's premises and the telephone company central office.

Examples of central office signals to the user's premises are ringing (audible alerting) signals, dial tone, speech signals, etc.

Signals from the user's telephone include off-hook (request for service), dialing (network control signaling), speech to the distant party, on-hook (disconnect signal), etc.

Signal-To-Noise Ratio

The ratio of the usable signal being transmitted to the noise or undesired signal. Usually expressed in decibels. This ratio is a measure of the quality of a transmission.

Signal Processing Component

An atomic bundle of signal processing functionality which can be allocated to a single group. It can be capable of supporting the functionality of one or more resources or a simple coder.

Signal Processing Element

That part of a Signal Processing Component which is associated with a single Resource.

Signal Processing Platform

This is a software component that supports a specific hardware package. This is typically an executable program and may control one or more instances of the vendor-specific hardware package. Each SPP may support several different types of signal processing functionality clustered into SPCs.

SIR (ASR)

Speaker Independent Recognition. See Speaker Independent Recognition.

SIT Tones

1. Standard Information Tones. These are tones sent out by a central office to a pay phone to indicate that the dialed call has been answered by the distant phone, etc. 2. Special Information Tones. These are tones for identifying network provided announcements.

Here's Bellcore's explanation: Automated detection devices cannot distinguish recorded voice from live voice answer unless a machine-detectable signal is included with the recorded announcement.

The ITU, which specifies signals that may be applied to international circuits, has defined Special Information Tones for identifying network provided announcement. The SIT used to precede machine-generated announcements also alerts the calling customer that a machine-generated announcement follows. Since SIT consists of a sequence of three precisely defined tones, SIT can be machine-detected, and therefore machine-generated announcements preceded by a SIT can be classified.

Many SIT encodings have been defined including: Vacant Code (VC), Intercept (IC), Reorder (RO) and No Circuit (NC). With the exception of some small Stored Program Control Systems (SPCSs) and some customer negotiated announcements, Bell operating companies in North America now precede appropriate announcements with encoded SITs to detect and classify announcements.

SLIP
Serial Line IP; a protocol that allows a computer to use the Internet protocols (and become a full-fledged Internet member) with a standard telephone line and a high-speed modem. SLIP is being superseded by PPP, but still is commonly used. A software technique for connecting a computer to the 'Net. With SLIP you can dial up an ISP that offers SLIP to place yourself directly on the 'Net. You are no longer just a terminal. You can telnet and FTP to other computers and pull their files directly back to your computer, not to the ISP's computer. Other 'Net surfers with SLIP can do the same to you.

SMP
Symmetric Multiprocessing. Computer systems designed to use many processors simultaneously. It is a means to achieve modular scaling of a platform, load sharing and redundancy.

SMS/800

The national database Service Management System that will retain all 800 records. This database provides long distance carriers a single interface for 800 number reservations and record maintenance. Developed by Bellcore, the database has been in use by various Regional Bell Operating Companies (RBOCs) since 1988. The FCC has mandated that a neutral third party administer the database.

SMTP

Simple Mail Transfer Protocol. Internet mail is routed using this software protocol.

Software Supervision

"Answer Supervision" is knowing when the person at the other end answers the phone. The main reason for wanting to know this is so that a phone company can start billing the call. There are two ways of doing answer supervision. You can get it from the nation's phone system, i.e. the distant office signals back across the country when the called person picks up the phone. Or you can fake it with "software supervision."

Essentially this means there's electronics which "listens" to the call. If it "hears" voice or something like voice, it assumes the conversation has started and it's time to start billing the call. Software supervision is not accurate. But when you haven't got access to real answer supervision (for whatever reason) it's better than the previous alternative, which was "timeout."

In timeout answer supervision, the carrier simply assumed the call had begun after a certain number of seconds -- like 30 -- had elapsed with the calling person hanging up. This meant, for example, if you called Grandma and she wasn't there, but you left it ringing, 'cause you knew she took time to answer the phone, then you'd be charged for the call -- even though she didn't answer phone! See also ANSWER SUPERVISION.

SoftWindows

Sun software for running PC applications on SPARC platforms.

Solaris

Sun's Unix-based 32-bit operating environment for network application servers, Internet servers, and high-end desktop systems. Solaris has multithreading, symmetric multiprocessing, integrated TCP/IP-based networking and centralized network administration tools. Solaris has an open systems architecture. It supports Sun SPARC, Intel and PowerPC processor-based platforms.

Solaris supports IPX/SPX protocol stack to connect to Novell networks. Security and system administration uses NFS with compliance to the POSIX 1003.6 standard. Solaris is scalable from desktop computers to Internet servers; from single processor computers to 64-processor systems.

Source Code

A computer program which can be translated by another program into a program that will run on a particular computer. The program which finally runs on that computer is known as the object code.

Span

1. Refers to that portion of a high speed digital system than connects a C.O. (Central Office) to C.O. or terminal office to terminal office. 2. Also called a T-Span Line. A repeatered outside plant four-wire, two twisted-pair transmission line.

Span Line

A T-1 link.

SPARC

Sun's open, RISC-based (Reduced Instruction Set Computer) architecture for microprocessors. SPARC is the basis for Sun's own computer platforms and it's licensed to third parties.

SPARCengine

Sun's standard microprocessors supporting SBus expansion modules and I/O. 50Mhz and above speeds. Up to 512 MB memory on-board.

SPARC V9

Sun's latest microprocessor architecture. It handles real time MPEG-2 decompression and other on-chip multimedia.

Speaker Independent (Voice) Recognition (ASR)

SIR. Technologies for the automated conversion of speech to accurate and meaningful textual information (typically ASCII). SIR is typically used to accept input from callers to voice processors where the callers are using rotary dial phones instead of touch-tone -- Dual Tone Multi-Frequency (DTMF) -- phones. SIR can substitute for the numbers on the DTMF keypad and can add the benefit of a few basic voice commands, e.g., Yes, No, Help, etc.

Because computer processing demands are formidable with speaker independent recognition, accurate speaker independent products are created with limited vocabularies. In contrast, trainable or speaker dependent recognizers can feature larger vocabularies at lower prices. SIR has been slowly gaining acceptance in telephone applications.

SIR is increasingly used in automated operator assistance applications. SIR will see increased use as system builders respond to pressures to provide voice processing functions to the enormous rotary phone installed base domestically and abroad.

Spectral Energies

The energy intensity of a waveform at specific frequencies, as analyzed by an FFT. See "Frequency Domain."

Speech Concatenation

A term used in voice processing for economical digitized speech playback that uses independently recorded files of phrases or file segments linked together under application program control to produce a customized response in natural sounding language. For example, order status, bank balances, bus schedules or lottery results, etc. Concatenation is done for speed and economy. It lends itself to limited and structured vocabularies that are best stored in RAM (Random Access Memory) or speedily accessible from disk.

Concatenation does not replace Text-To-Speech (TTS) as a method of getting the voice processor to deliver its responses. Concatenation, however, can be an excellent complement to TTS when a voice application demands broad, real time vocabulary production. See TEXT-TO-SPEECH.

Speech Synthesis

See TEXT-TO-SPEECH.

SPI

See Service Provider Interface.

Spooler

A program that controls spooling. Spooling, a term mostly associated with printers, stands for Simultaneous Peripheral Operations On Line. Spooling, temporarily stores programs or program outputs on magnetic tape, RAM, or disks for output or processing. On a LAN, a printer is controlled by a spooler. The spooler places each print request on the LAN in the print queue and prints it when it reaches the top of the queue.

SPX

Sequenced Packet eXchange. A connectionoriented session protocol for network data exchange that uses the IPX (Internetwork Packet Exchange) communication protocol. (See also NetBIOS.)

Standards

Agreed principles of protocol. Standards are set by committees working under various trade and international organizations. RS standards, such as RS-232-C are set by the EIA, the Electronics Industries Association. ANSI standards for data communications are from the X committee.

Standards from ANSI would look like X3.4-1967, which is the standard for the ASCII code. The ITU does not put out standards, but rather publishes recommendations, owing to the international personalities involved.

V series recommendations refer to data transmission over the telephone network, while X series recommendations, such as X.25, refer to data transmission over public data networks.

State
The instantaneous properties of an object that characterize that object's current condition.

State Machine Programming
To control multiple telephone lines in a single voice processing program, a new program structure is required. Dialogic calls this technique state machine programming. Computer Science called state machines "Deterministic Finite State Automata."

Station
A dumb word for a telephone. Also called an instrument, or a telephone instrument. An extension station is one connected "behind" a PBX or key system. In other words, the PBX or key system is between the station and the telephone central office. We tried to remove the word "station" from this dictionary, but failed. We suspect the word comes from the very old days when the telephone industry was regulated by the Interstate Commerce Commission, (the ICC) which also regulated the railroad industry.

Station Adapters
Cables and interface assemblies for connecting Dialogic network interface and switching products to telephones or analog telephone lines.

Station Set
Another word for a common desk telephone.

Store And Forward
In communications systems when a message is transmitted to some intermediate relay point and stored temporarily. Later the message is sent the rest of the way. Not very convenient for voice conversations, but useful for telex type, and other one way transmissions of messages. Telephone answering machines, as well as voice mailboxes are considered forms of Store and Forward message switching.

SunPC
Hardware/software solutions from Sun for MSDOS on SPARC platforms.

SunXTL API

Part of the SunXTL Teleservices architecture. An object-oriented interface accessed using the C++ language. The API includes XtlProvider, XtlCall, XtlCallState, and XtlMonitor objects. These base objects define the command and callback methods of SunXTL teleservices. There are three methods types on these objects: synchronous requests, asynchronous commands, and asynchronous events.

SunXTL Provider

Part of the SunXTL Teleservices architecture. Device-specific software entity. Enables a telecommunications device to communicate with and be controlled by the SunXTL platform and its applications.

SunXTL Provider Configuration Database

Part of the SunXTL Teleservices architecture. Means of registering third-party "providers" in an SunXTL-based system. The database shows the existence of each provider, how to invoke it (used by the SunXTL system internally) and describes the specific capabilities of that provider.

SunXTL Provider Interface

Part of the SunXTL Teleservices architecture. The means for third-party developers and Independent Hardware Vendors (IHVs) to integrate provider, technology or call-type-specific features into a SunXTL environment. Developers use the SunXTL Provider Library (MPI) to do this.

SunXTL Provider Library

Part of the SunXTL Teleservices architecture. Isolates the provider code from the intricacies of the XTL system services. The library supplies interfaces to the provider database, the server messaging system for commands and callbacks, and the means to make device data streams accessible to applications.

SunXTL Server

Part of the SunXTL Teleservices architecture. Provides multi-client and multi-device support. The server is the central point of contact for all teleservices services.

Resource management and security are provided by the server. Communicates with the SunXTL provider to place and receive telephone calls. An application may access the data associated with a call by acquiring a data stream from the API.

SunXTL System Services

Part of the SunXTL Teleservices architecture. Acts as the intermediary between the application view of a call object and the provider's implementation of the call. The SunXTL server, along with the application and provider libraries, handles the interprocess message passing, object identification and creation, call ownership, security, and asynchronous event notification. The SunXTL subsystem also manages a database of available providers and helps manage data stream routing and access.

SunXTL Teleservices

Sun's "Teleservices" architecture for Solaris. SunXTL is the foundation library for applications using or controlling telecom data streams. It is used for computer telephony application development. The SunXTL subsystem and API includes call control functions. It establishes a call or connection, and data stream access methods to control the flow of data over that connection.

SuperSPARC II

Sun SPARC 80 MHz processors. Used in workstations and in multiple configurations for ft-SPARC fault tolerant systems.

Supervised Transfer

A call transfer made by an automatic device such as voice response unit which attempts to determine the result of the transfer -- answered, busy, ring no answer -- by analyzing call progress tones on the time.

Supervision

Supervision of a phone call is detecting when a called party has picked up his phone and when that party has hung up. Supervision is used primarily for billing purposes.

Not all long distance carriers have supervision capability. It depends on how "equal accessed" they have chosen to be. See ANSWER SUPERVISION and SOFTWARE SUPERVISION.

SVC

Switched Virtual Connections

Switch

A mechanical, electrical or electronic device which opens or closes circuits, completes or breaks an electrical path, or selects paths or circuits.

Switch Domain

A single instance of a particular technology-specific connection type.

Switched Network

A network providing switched communications service; that is, the network is shared among many users, any of whom may establish communication between desired points when required.

Switchhook

A switchhook was originally an electrical "switch" connected to the "hook" on which the handset (or receiver) was placed when the telephone was not in use.

Switch Hook Flash

A signaling technique whereby the signal is originated by momentarily depressing the switch hook.

Switch Port

A resource that allows a Group to communicate with anOther Group. All Groups implicitly possess a Switch Port as a secondary resource, but in order to use it, the application must explicitly connect the Switch Ports of two Groups.

Synchronous Request

A request where the client blocks until the completion of the request. Contrast with asynchronous request.

Synthesized Voice

Human speech approximated by a computer device that concatenates basic speech parts (or phonemes) together.

System Service Provider

An entity that provides system-wide services, such as session management and security, and the allocation and tracking of resources and groups.

T

T-1

Also spelled T1. A digital transmission link with a capacity of 1.544 Mbps (1,544,000 bits per second). T1 uses two pairs of normal twisted wires, the same as you'd find in your house. T1 normally can handle 24 voice conversations, each one digitized at 64 Kbps. But, with more advanced digital voice encoding techniques, it can handle more voice channels. T1 is a standard for digital transmission in North America. It is usually provided by the phone company and used for connecting networks across remote distances.

Bridges and routers are used to connect LANs over T1 networks. There are faster services available, such as T2 and T4, but these are not used much. T1 links can often be connected directly to new PBXs and many new forms of short-haul transmission, such as short-haul microwave systems. It is not compatible with T1 outside the United States and Canada. In Europe T-1 is called E-1 or E1. Outside of the United States and Canada, the "T-1" line bit rate is usually 2,048,000 bits per second. Japan, France and West Germany impose slight variations that make their formats unique.

Only one element remains constant -- the DS-0. The 64 kilobit per channel is universal. Most often it represents a PCM voice signal sampled at 8,000 times per second. However, the form of PCM encoding differs between T-1 (mu-law) and E-1 (A-law companding).

According to Bill Flanagan's book, the differences are not so great that a multiplexer cannot convert between them. Conversion of E-1 to T-1 involves both the compression law and the signaling format. At the higher rate of 2,048,000, 32 time slots are defined at the CEPT interface, but two are used for signaling and other housekeeping chores.

Typically 30 channels are left for user information -- voice, video, data, etc. CEPT is the Conference of European Postal and Telecommunications administrations. Standards-setting body whose membership includes European Post, Telephone, and Telegraphy Authorities (PTTs). For a full explanation of T1 see Bill Flanagan's book The Guide to T-1 Networking.

T.120

The most important transmission protocol standard for document conferencing over transmission media ranging from analog phone lines to the Internet. It's Network independent (Internet, PSTN, LAN, ISDN). T.120 does conference control, application and data sharing. It's for documents and files what H.320 is to video. Creating products based on the both standards is what Intel (Internet Phone) and Microsoft (NetMeeting) are doing.

T.122

ITU standard for multipoint communication service for audiographic conferencing.

T.123

ITU standard for protocol stacks in audiographic teleconferencing.

T.124

ITU standard for generic T.120 conference control.

T.125

ITU standard for the T.120 Multipoint Communication Service Protocol Specification.

T.126

ITU standard for T.120 Multipoint Still Image and Annotation Protocol.

T.127
ITU standard for T.120-based Multipoint Binary File Transfer.

T.128
ITU standard for Audio Visual Control for Multipoint Multimedia Systems.

T2
A digital transmission link with a capacity of 6.312 Mbps. T2 can handle at least 96 voice messages.

T3
A digital transmission link with a capacity of 44.736 Mbps. T3 can handle 672 voice messages simultaneously. T3 runs on fiber optic cable.

T.4
ITU recommendation for Group 3 machines, using T.30 and various V series standards. It also describes compression methods (Modified Huffman and Modified READ).

T.6
ITU recommendation for Group 4 machines. It defines the facsimile coding schemes and their associated coding control functions for black and white images.

T.30
ITU recommendation. This handshake protocol describes the overall procedure for establishing and managing communication between two fax machines. It covers five phases of operation: call setup, pre-message procedure (selecting the communication mode), message transmission (including phasing and synchronization), post-message procedure (EOM and confirmation), and call release.

T.35
ITU recommendation proposing a procedure for the allocation of ITU members' country or area codes for non-standard facilities in telematic services.

T.611

Also known as Appli/COM. A messaging standard proposed by France and Germany, defining a programmable communication interface (PCI) for Group 3 fax, Group 4 fax, teletext, and telex service.

Table

See data file.

Table-Driven

Describing a logical computer process, widespread in the operation of communications devices and networks, where a user-entered variable (e.g. a password) is matched against an array of predefined values (a table of acceptable passwords) and checked for access, authorization, etc. A frequently used process in least cost routing, in network routing, in access security, and in modem operation.

Tag

Some database formats use individual DOS files for each index. Others, such as dBASE IV and FoxPro, combine multiple indexes into a single file. Each such index within the index file is known as a tag. Two parameters are required to access a tag within an index file: the name and extension of the index file, and the name of the desired tag within it.

Talk Battery

The DC voltage supplied by the central office to the subscriber's loop to operate the carbon transmitter in the handset.

Talk Path

The tip and ring conductors of a telephone circuit.

Talk-Off

Talk-off is one hazard of in-band signaling. Talk-off occurs when your voice has enough 2600 Hz energy to activate the 2600 Hz tone- detecting circuits in the central office. The 2600 Hz tone is used for in-band signaling.

TAO

See TELEPHONY APPLICATION OBJECT FRAMEWORK

TAPI

See Windows Telephony API

TCF

Training Check Frame. Last step in a series of signals called a training sequence, designed to let the receiver adjust to line conditions.

TCP

Transmission Control Protocol. ARPAnet-developed transport layer protocol. Corresponds to OSI layers 4 and 5, transport and session. TCP is a transport layer, connection-oriented, end-to-end protocol. It provides reliable, sequenced, and unduplicated delivery of bytes to a remote or local user. TCP provides reliable byte stream communication between pairs of processes in hosts attached to interconnected networks. It is the portion of the TCP/IP protocol suite that governs the exchange of sequential data. See also TCP/IP.

TCP/IP

Transmission Control Protocol/Internet Protocol. A set of protocols developed by the DOD to link dissimilar computers across many kinds of networks. Including unreliable ones and connected to dissimilar LANs. Developed in the 1970s by the U.S. Department of Defense's Advanced Research Projects Agency (DARPA) as a military standard protocol, its assurance of multi vendor connectivity has made it popular among commercial users. Consequently, TCP/IP now is supported by many manufacturers of minicomputers, personal computers, mainframes, technical workstations and data communications equipment. It is also the protocol commonly used over Ethernet (as well as X.25) networks and the Internet. It has been implemented on everything from PC LANs to minis and mainframes. Although committed to an eventual migration to an OSI architecture, TCP/IP currently divides networking functionality into only four layers:

A Network Interface Layer that corresponds to the OSI Physical and Data Link Layers. This layer manages the exchange of data between a device and the network to which it is attached and routes data between devices on the same network.

An Internet Layer which corresponds to the OSI network layer. The Internet Protocol (IP) subset of the TCP/IP suite runs at this layer. IP provides the addressing needed to allow routers to forward packets across a multiple LAN inter network.

In IEEE terms, it provides connectionless datagram service, which means it attempts to deliver every packet, but has no provision for retransmitting lost or damaged packets. IP leaves such error correction, if required, to higher level protocols, such as TCP.

IP addresses are 32 bits in length and have two parts: the Network Identifier (Net ID) and the Host Identifier (Host ID). Assigned by a central authority, the Net ID specifies the address, unique across the Internet, for each network or related group of networks. Assigned by the local network administrator, the Host ID specifies a particular host, station or node within a given network and need only be unique within that network.

A Transport Layer, which corresponds to the OSI Transport Layer. The Transmission Control Protocol (TCP) subset runs at this layer. TCP provides end-to-end connectivity between data source and destination with detection of, and recovery from, lost, duplicated, or corrupted packets – thus offering the error control lacking in lower level IP routing.

In TCP, message blocks from applications are divided into smaller segments, each with a sequence number that indicates the order of the segment within the block. The destination device examines the message segments and, when a complete sequence of segments is received, sends an acknowledgement (ACK) to the source, containing the number of the next byte expected at the destination.

An Application Layer, which corresponds to the session, presentation and application layers of the OSI model. This layer manages the function required by the user programs and includes protocols for remote log-in (Telnet), file transfer (FTP), and electronic mail (SMTP).

TDM

Time-division multiplex. This is the means whereby a single electric circuit may be used by multiple devices: use of the circuit is divided into "time slots" which are allocated to the contending devices in a rotating fashion. Each device gets exclusive use of the circuit for the duration of its time slot.

Telco

The local telephone company. Often a term of endearment.

Telecommunications

Telecommunications, or telecom for short, is the transmission or reception of signals, images, sounds, or intelligence of any nature by wire, radio, light pulse, or electromagnetic system. That just about covers virtually any means of transmission, regardless of source, channel, or media type. Telephony and data communications are subsets of telecommunications that are specific to computer-type transmissions and telephone-type transmissions, respectively.

Telefax

European term for fax.

Telecommunication Standardization Sector (TSS)

One of four permanent parts of the International Telecommunications Union (ITU), based in Geneva, Switzerland. It issues recommendations for standards applying to modems, packet switched interfaces, V.24 connectors, etc. Although it has no power of enforcement, the standards it recommends are generally accepted and adopted by industry. Until March 1, 1993, the TSS was known as the ITU (see above).

Telecopier

European term for fax.

Telemarketing

Marketing and sales conducted via the telephone. There are two sides to telemarketing -- incoming and outgoing.

Incoming telemarketing is largely run through 800 toll-free IN-WATS numbers and local FX (foreign exchange) lines. Outgoing telemarketing is organized over OUT-WATS lines. Everything from computers to clothes is now sold over the phone.

An expanding range of telecom gadgetry is being developed to automate telemarketing -- including automated outbound dialers, voice processing technology, and automatic call distributors.

The tone recognition, voice detection and transaction Audiotex and transaction processing capabilities of voice processing gear can be used to enhance all telemarketing applications.

Telephone Manager

Apple's telephony API for the Macintosh environment.

Telephony

Taken from Greek root words meaning "far sound," telephony is the discipline of converting or transmitting voice or other signals over a distance, and then re-converting them to an audible sound at the far end.

Telephony Application Object (TAO) Framework

TAO is the software part of the Signal Computing System Architecture (SCSA). The TAO Framework is made up of four components:

1) A set of standard SCSA Application Programming Interfaces (SCSA APIs) that defines the way client and server applications access the Message Protocol to communicate with server resources and system services. 2) The Message Protocol that defines a standard way for passing messages among applications, system resources, and system services. 3) A set of core System Services that includes resource management, runtime control, switching, event and error handling, configuration management, session management, and security. 4) A set of Service Provider Interfaces (SPIs) that defines the way System Services and Resources (the Service Providers) access the Message Protocol.

Other key features of the SCSA Telephony Application Object Framework are: It is an open software architecture; It is designed specifically for multiple application client server environments, where multiple distributed applications use the same server resources; It has sophisticated application management capabilities; It facilitates both simple and complex application development; Its core system services and technology-specific resource APIs are interoperable; It is object-oriented and operating system and hardware independent.

Telephony Database Transactions

The use of telephony terminal equipment (phone and/or fax) as input/output devices to operate on a computer database.

Telephony Interface Control

A Telephony Interface Control resource is any resource that interfaces with the telephone network (public or private). This is usually claimed as the primary member of a group.

Telephony Server

A server software program that allows client applications to take advantage of call control hardware remotely. See Novell Telephony Server for an example.

Telephony Service Provider.

TSP. This is written by vendors of telephony equipment for TAPI-compliant applications. TSPs are a service layer beneath the TAPI APIs. The way TAPI and Unimodem work together illustrates this. Unimodem includes both a TSP (unimodem.tsp) and a low-level virtual device driver (unimodem.vxd). Modem functions represented by the TSP are implemented in the unimodem.tsp DLL. TAPI APIs are implemented in a DLL called tapi.dll. The TSP portion of Unimodem translates TSPI calls from TAPI and feeds them into the Unimodem Virtual Device Driver (unimodem.vxd) as AT commands. In this fashion, TAPI program developers are abstracted from the low-level AT, AT+V and AT#V command set.

Telephony Service Provider Interface

TSPI. This is an interface implemented by a service provider to make its functions available to apps through TAPI. See Telephony Service Provider.

Telephony Workgroup

A group of people with a common telecomputing requirement: for example, a technical support group might need an application that automatically links incoming calls to client records on a database.

TELNET

A program that lets you log into other computers on the Internet.

Terminal Emulation

A microcomputer or personal computer mimics, pretends to be (i.e. emulates) a data terminal. It does this with special printed circuit boards inserted into its motherboard and/or special software.

Text-To-Speech (Speech Synthesis)

TTS. Technologies for converting textual (ASCII) information into synthetic speech output. This technology is used in voice processing applications that require the production of broad, unrelated and unpredictable vocabularies, e.g., products in a catalog, names and addresses, etc.

This technology is appropriately used when system design constraints prevent the more efficient use of speech concatenation alone. See SPEECH CONCATENATION.

Third party Call Control

See Call Control.

TIFF

Tagged Image File Format. TIFF provides a way of storing and exchanging digital image data. Current TIFF specifications support three main types of image data: black-and-white data, halftones or dithered data, greyscale data and, more recently, color.

Timbre

The quality of tone distinctive to a particular voice

Time-Varying Media

Time-varying media, such as audio data (as opposed to space-varying media, such as image data).

Tip & Ring

An old fashioned way of saying "plus" and "minus," or ground and positive in electrical circuits. Tip and Ring are telephony terms. They derive their names from the operator's cordboard plug. The tip wire was connected to the tip of the plug, and the ring wire was connected to the slip ring around the jack. A third conductor on some jacks was called the sleeve. That's it. Nothing more sinister. Nothing more interesting.

Tone

An audio signal consisting of one or more superimposed amplitude modulated frequencies with a distinct cadence and duration.

Tone Dial

A push button telephone dial that makes a different sound (in fact, a combination of two tones) for each number pushed. The correct name for tone dial is "Dual Tone MultiFrequency" (DTMF). This is because each button generates two tones, one from a "high" group of frequencies -- 1209, 1136, 1477 and 1633 Hz -- and one from a "low" group of frequencies -- 697, 770, 852 and 841 Hz.

The frequencies and the keyboard, or tone dial, layout have been internationally standardized, but the tolerances on individual frequencies vary between countries.

This makes it more difficult to take a touch tone phone overseas than a rotary phone. You can "dial" a number faster on a tone dial than on a rotary dial, but you make more mistakes on a tone dial and have to redial more often. Some people actually find rotary dials to be, on average, faster for them. The design of all tone dials is stupid. Deliberately so.

They were deliberately designed to be the exact opposite (i.e. upside down) of the standard calculator pad, now incorporated into virtually all computer keyboards. The reason for the dumb phone design was to slow the user's dialing down to the speed Bell central offices of early touch tone vintage could take.

Today, central offices can accept tone dialing at high speed. But sadly, no one in North America makes a phone with a sensible, calculator pad or computer keyboard dial. On some telephone/computer workstations you can dial using the calculator pad on the keyboard. This is a breakthrough. It a lot faster to use this pad.

The keys are larger, more sensibly laid out and can actually be touch-typed (like touch-typing on a keyboard.) Nobody, but nobody can "touch-type" a conventional telephone tone pad. A tone dial on a telephone can provide access to various special interactive voice response services and features -- from ordering your groceries over the phone to inquiring into the prices of your (hopefully) rising stocks.

Tone Set

A collection of tones which are customarily used as a set for the purposes of call setup and teardown (e.g., DTMF, R1 MF, R2 MF). In the case of DTMF, the tone set can also be used by the Client Application during the conversation portion of a call.

Touch Tone

A term coined by AT&T for Tone Dialing.

TSP

See Telephony Service Provider.

TTS

Text-To-Speech. A term used in voice processing. See TEXT-TO-SPEECH.

Trellis Coding

A method of forward error correction used in certain high-speed modems where each signal element is assigned a coded binary value representing that element's phase and amplitude. It allows the receiving modem to determine, based on the value of the preceding signal, whether or not a given signal element is received in error.

Trellis Code Modulation (TCM)

A version of quadrature amplitude modulation (QAM) that enables relatively high bit rate signals to be used on ordinary voice-grade circuits. When used as a modem modulation technique in which algorithms are used to predict the best fit between the incoming signal and a large set of possible combinations of amplitude and phase changes. TCM provides for transmission speeds of 14.4 Mbps and above on single voice-grade telephone lines.

Trunk

A communication line between two switching systems. The term switching systems typically includes equipment in a telephone company's central office and PBXs. A tie trunk connects PBXs. Central office trunks connect a PBX to the switching system at the central office.

TSI

Transmitting Subscriber Information. A frame that may be sent by the caller, with the caller's telephone number (may be used to screen calls).

T-Span

A telephone circuit or cable through which a T-carrier runs. See T-1.

TSR

Terminate and Stay Resident. A term for loading a software program in a DOS computer in which the program loads into RAM and is always ready for running at the touch of a combination of keys.

TTI

Transmit Terminal Identification. The telephone number and words on top of a received fax document, identifying its point of origin.

This information does not originate with the telephone company, but with the sender, who programs it into the fax.

TVM

See Time-Varying Media.

TVM Object

An encapsulation of an atomic piece of time-varying media. This encapsulation may be the data itself or a reference to the data.

Twisted Pair

Two insulated copper wires twisted around each other to reduce induction (thus interference) from one wire to the other. The twists, or lays, are varied in length to reduce the potential for signal interference between pairs. Several sets of twisted pair wires may be enclosed in a single cable. In cables greater than 25 pairs, the twisted pairs are grouped and bound together in a common cable sheath. Twisted pair cable is the most common type of transmission media. It is the normal cabling from a central office to your home or office, or from your PBX to your office phone. Twisted pair wiring comes in various thicknesses. As a general rule, the thicker the cable is, the better the quality of the conversation. And the longer cable can be and still get acceptable conversation quality. However, the thicker it is, the more it costs.

Two-Dimensional Coding

A data compression scheme that uses the previous scan line as a reference when scanning a subsequent line. Because an image has a high degree of correlation vertically as well as horizontally, two-dimensional coding schemes work only with variable increments between one line and the next, permitting higher data compression.

Two-Wire Circuit

A transmission circuit composed of two wires -- signal and ground -- used to both send and receive information. In contrast, a four wire circuit consists of two pairs. One pair is used to send. One pair is used to receive. All trunk circuits -- long distance circuits -- are four wire.

A four wire circuit costs more but delivers better reception. All local loop circuits -- those coming from a Class 5 central office to the subscriber's phone system -- are two wire, unless you ask for a four-wire circuit and pay a little more.

UDP

User Datagram Protocol. A TCP/IP protocol describing how messages reach application programs within a destination computer. This protocol is normally bundled with IP-layer software. UDP is a transport layer, connectionless mode protocol, providing a (potentially unreliable, unsequenced, and/or duplicated) datagram mode of communication for delivery of packets to a remote or local user.

UDP/IP

User Datagram Protocol/Internet Protocol. See UDP.

Ultra Enterprise Server

Sun's line of new servers for network computing. Ultra Enterprise 1 and Ultra Enterprise 2 servers share the architecture, components, and packaging of Sun's Ultra 1 and Ultra 2 desktop workstations. The largest (Ultra Enterprise 6000) can handle as many as sixteen slots for processor and I/O boards, 30GB of memory, 10TB of storage and 30 CPUs. Uses UltraSPARC processors.

UltraSPARC Processor

Sun's superscalar 64-bit processor used in large-scale symmetric multiprocessing. Has two integer arithmetic and logic units, fast floating-point, graphics unit and both on-chip and secondary cache memory. Uses 128 bit wide external bus interface. Executes up to four instructions per clock cycle and permits out-of order completion, enabling very fast integer and floating point computations. 167MHz and above.

Umux

Universal Data Stream Multiplexor. Part of the SunXTL Teleservices architecture. Provides a uniform means for applications to access and share data channels associated with a telephone call. Umux is a streams pseudo-device driver used to connect data channels to applications and to control access to data. The server and the provider work together to deliver the stream to the application with Umux.

Unified Messaging

A telecomputing application that treats incoming and outgoing electronic mail, voice messages, and fax messages together. A mixed-media messaging control mechanism.

Unimodem

Universal Modem Driver. A standard part of TAPI and Win 95. Unimodem is implemented as a Telephony Service Provider Interface DLL and as a Virtual Communication Device Driver. It provides a layer of services for data and fax modems and voice. With it, users and app developers don't have to learn or maintain AT modem commands to dial, answer and configure modems. Rather, Unimodem does these tasks automatically by using mini-drivers written by modem vendors. Other service providers (like those supporting things such as an ISDN adapter, a proprietary PBX phone or an AT-command modem) can also be used with TAPI. Port drivers are specifically responsible for communicating with I/O ports, which are accessed through the VCOMM driver. Port drivers provide a layered approach to device communications.

Unimodem V

The newest release of Unimodem, the Windows universal modem driver/telephony service provider for data/fax modems. It supports both AT+V and AT#V voice modem features. Unimodem V includes the most commonly requested features to support data/fax/voice modems, including wave playback and recording to and from the phone line, wave playback and recording to and from the handset, and support for speakerphones, caller ID, distinctive ringing, and call forwarding. Unimodem V also supports G3 fax media.

Prior to this release, TAPI could not address fax processing directly, as these were handled by MAPI. Now TAPI apps can control the processing of fax transactions directly. The Unimodem V software consists of the following parts: 1. The Unimodem V Telephony Service Provider (TSP) and VxD. The TSP handles program requests, such as dialing and answering, which are passed down from TAPI. The TSP hides modem-specific details of how telephony operations, such as dialing, are handled. The VxD is called by the TSP to send command strings to the modem. It is called by the VCOMM VxD to change modem settings and send/receive data to/from the modem; 2. Operator Agent, a program that identifies whether an incoming call is from a person, a fax machine, or a data modem. It then forwards the call to the appropriate program, such as the answering machine or fax program. If Operator Agent can't automatically determine the type of call, it asks the caller to identify it. If the caller cannot identify it, Operator Agent forwards the call to a program specified by the user; 3. A wave driver for serial port modems that supports the following formats: IMA ADPCM at 4800 kHz, 7200 kHz, or 8000 kHz; and Rockwell ADPCM; 4. A wave wrapper for use with modems with a separate audio hardware interface, where synchronization is needed between the modem and audio via AT commands. To install Unimodem V, you must have a PC running Win 95.

Unicode

Unicode is a 16-bit, fixed-width character encoding standard that encompasses virtually all of the characters commonly used on computers today-this includes most written languages, plus publishing characters, mathematical and technical symbols, and punctuation marks. Unicode support makes it easier and faster for developers to create products and localize them different languages.

UNIX

An immensely powerful and complex operating system for computers. UNIX is developed, marketed, championed and trademarked by AT&T. UNIX is very powerful and very complex to use. Because of its power it needs a computer with a large amount of RAM memory.

It has been extensively used in universities, and is only now emerging onto the business scene. UNIX allows a computer to handle multiple users and multiple programs simultaneously. It works on many different computers. This means you can often take applications software which runs on UNIX and often move it -- with little changing -- to a bigger, different computer. Or to a smaller, different computer. This process of moving programs to other computers is known as "porting." UNIX was developed in 1969 by Ken Thompson of AT&T Bell Laboratories.

URL

Uniform Resource Locator. The address of a page on the Internet. Our URL is ComputerTelephony.com. By employing URLs, you can locate pages on the Web.

V.17

ITU recommendation for simplex modulation technique for use in extended Group 3 facsimile applications only. Provides 7,200, 9,600, 12,000, and 14,400 bps trellis-coded modulation.

V.21

ITU recommendation for 300 bps duplex modems for use on the switched telephone network. V.21 modulation is used in a half-duplex mode for Group 3 fax negotiation and control procedures.

V.22

ITU recommendation for 1,200 bps duplex modems for use on the switched telephone network and on leased lines.

V.22bis

ITU recommendation for 2,400 bps duplex modems for use on the switched telephone network. V.22bis also provides for 1,200 bps operation for V.22 compatibility.

V.29

ITU recommendation for 9,600 bps modem for use on point- to-point leased circuits. This is the modulation technique used in Group 3 fax for image transfer at 7,200 and 9,600 bps. V.29 uses a carrier frequency of 1,700 Hz which is varied in both phase and amplitude. V.29 can be full duplex on four-wire leased circuits, or half duplex on two-wire and dial-up circuits.

V.32

ITU recommendation for 9,600 bps two-wire full duplex modem operating on regular dial-up lines or two-wire leased lines. V.32 also provides fallback operation at 4,800 bps.

V.32bis

ITU recommendation for full-duplex transmission on two-wire leased and dial-up lines at 4,800, 7,200, 9,600, 12,000, and 14,400 bps. Provides backward compatibility with V.32. It includes a rapid change renegotiation feature for quick and smooth rate changes when line conditions change.

V.33

ITU recommendation for 14.4 kbps and 1.2 kbps modem for use on four-wire leased lines.

V.42

ITU recommendation, primarily concerned with error correction and compression for modems. The protocol is designed to detect errors in transmission and recover with a retransmission.

Vari-A-Bill

A new 900 service of AT&T whereby the call's price varies depending on certain events -- the caller punching out some tones on his phone, or a service technician coming on line, etc.

VCOMM

Win32 (virtual) communications device driver. It protect-modes services and lets Windows apps and drivers use ports and modems.

To conserve system resources, comm drivers are loaded into memory only when in use by an app. VCOMM also uses new Plug and Play services in Windows 95 to help configure and install comm devices. The Win32 communications APIs provide an interface for using modems and comm devices in a device-independent fashion. Applications call the Win32 APIs to configure modems and perform data I/O through them. Through TAPI and Unimodem, apps can control modems or other telephony devices with VCOMM. Unimodem routes TAPI/Unimodem AT+V commands through VCOMM. In addition, data from the Win 32 Comm API and Audio from the Multimedia Wave API are routed through VCOMM. The virtual device driver in turn communicates through port drivers and serial virtual device drivers to the hardware.

Visual Call Control

A type of telephony application that allows users to manage their telecommunications from their computer screen. For example, with such an application a fax could be sent to another person by dragging an icon of the fax across the screen and dropping it on the name of the recipient. The telephony application would work behind the scenes to look up the persons telephone number, dial it, establish a communications link, and transmit the fax. Similarly, a conference call could be arranged by clicking and dragging the names of four individuals onto a conference call icon.

Visual J++

Visual J++ is Microsoft's Visual Java development environment. It uses the same user interface found in Developer Studio and Visual C++. With it, you can build Java applications and applets. Visual J++ has an Integrated Development Environment (IDE) and a collection of integrated components to create, test, tune and deploy Java code on multiple platforms. Visual J++ is certified as Java Compatible and allows users to build cross-platform applications, with specific optimization for the most common environments, such as Windows95 and Windows NT, without affecting portability. Microsoft has embraced the Java language and is integrating it with ActiveX. The latter provides a way to add functions not natively supported by Java (i.e. playing a video file).

Microsoft has a license agreement with Sun to create the reference implementation of Java for Windows and has freely licensed this technology back to Sun. Visual J++ built applications will work with any major browser, including Internet Explorer 3.0 and Netscape Navigator.

Visual Basic

Microsoft Visual Basic is a Microsoft programming system for Windows. You can create everything from simple programs to advanced, enterprise-wide client/server applications. It is a visual development tool supporting all Windows platforms--including Windows 95 and Windows NT. The foundation of this programming system is OLE, Microsoft's open object model. Together, OLE and VBX controls offer the world's largest and fastest-growing object library of pre-built components you can buy, use, and reuse in your programs. The Visual basic development environment uses a flexible programming language, and data-access capabilities. It uses pre-built OLE components and includes an integrated database engine and data-aware controls for developing database programs.

VOCODER

A family of voice-coding technologies for speech compression that are based on an underlying model of human speech. For example, many vocoders recognize voiced and unvoiced phonemes and code them differently. Voxware's MetaVoice technology is a new type of vocoder.

Voice Board

Also called a voice card or speech card. A Voice Board is an IBM PC- or AT-compatible expansion card which can perform voice processing functions. A voice board has several important characteristics: It has a computer bus connection. It has a telephone line interface.

It typically has a voice bus connection. And it supports one of several operating systems, e.g. MS-DOS, UNIX. At a minimum, a voice board will usually include support for going on and off-hook (answering, initiating and terminating a call); notification of call termination (hang-up detection); sending flash hook; and dialing digits (touch-tone and rotary). See VRU.

Voice Circuit

The typical analog telephone channel coming into your house or office. It has a bandwidth between 300 Hz and 3000 Hz. This is not sufficient for high fidelity voice transmission. You'd probably need at least 10,000 Hz. But it's sufficient to recognize and understand the person on the other end.

Voice Commands

High-level interface portion of the Microsoft Speech API for speech recognition. It is designed to provide command and control speech recognition for applications. With this interface, a user gives the computer simple commands, such as "Open the file," and can answer simple yes/no questions. The voice-command design mimics a Windows menu in behavior, providing a "menu" of commands that users can speak.

To use voice commands, a developer designs a Voice menu that corresponds to a window or state within the application. Most programs will have one Voice menu for the main window and one for every dialog box. Contained within every Voice menu is a list of voice commands users can say. When they say one, the app is notified which command was spoken. "Open a file" and "Send mail to ail name" are typical voice commands. Each voice command has information in addition to the spoken command, such as a description string and a command ID.

Voice Data

Encoded audio data.

Voice Digitization

The conversion of an analog voice signal into binary (digital) bits for storage or transmission.

Voice Frequency

VF. An audio frequency in the range essential for transmission of speech. Typically from about 300 Hz to 3000 Hz.

Voice Mail

A specialized form of Interactive Voice Response. Includes Voice Messaging and Automated Attendant systems. Denotes customer premise or central office systems that are aimed at automating and enhancing message taking and ATTENDANT console styled services. See VOICE MAIL SYSTEM.

Voice Mail System

A device to record, store and retrieve voice messages. There are two types of voice mail devices -- those which are "stand alone" and those which profess some integration with the user's phone system. A stand alone voice mail is not dissimilar to a collection of single person answering machines, with several added features; you can instruct the machines (voice mail boxes) to forward messages amongst themselves; you can organize to allocate your friends and business acquaintances their own mail boxes so they can dial, leave messages, pick up messages from you, pass messages to you, etc.

You can also edit messages, add comments and deliver messages to a mailbox at a pre-arranged time. In addition, messages can be tagged "urgent" or "non-urgent" or stored for future listening. The range of voice mail options varies among manufacturers.

An integrated voice mail system includes these features. First, it will tell you if you have any messages. It does this by lighting a light on your phone, and/or putting a message on your phone's alpha-numeric display. Second, if your phone rings for a certain number of rings (which number you set), the phone will transfer your caller automatically to your voice mail box, which will answer the phone, deliver a little "I am away" message and then receive and record the caller's message.

Voice Message Service

A service typically over dial up phone lines which provides the ability for a phone user to access a voice mail system and leave a message for a particular phone user. See VOICE MAIL SYSTEM.

Voice Messaging

Recording, storing, playing back and distributing phone messages. New York Telephone has an interesting way of looking at voice messaging. NYTel sees it as four distinct areas: 1. Voice Mail, where messages can be retrieved and played back at any time from a user's "voice mailbox"; 2. Call Answering, which routes calls made to a busy/no answer extension into a voice mailbox; 3. Call Processing, which lets callers route themselves among destinations via their touch-tone phones; and 4. Information Mailbox, which stores general recorded information for callers to hear.

Voice Modem Extensions or Windows 95

Microsoft's standards-based architecture for integrating voice modems into communications apps. The architecture allows developers to use TAPI and Win 32 to create sophisticated voice/fax/data apps. The standards used in these extensions include: 1. The ANSI IS-101 "Facsimile Digital Interfaces – Voice Control Interim Standard for Asynchronous DCE (aka AT+V); 2. TAPI; 3. Unimodem V; 3. Multimedia Wave API; 4. Win 95 Virtual Communication Driver (VCOMM); 5. Win 32 Comm API. Voice modems handle traditional fax and data transmission and also have the ability to digitize the human voice for real-time transmission, recording and playback.

Voice Processing

1) Term which applies to all forms of technology used for the digitization, decoding, storage, or manipulation of the human voice. 2) Form of Computer Telephony which uses interactive voice response elements and/or call processing elements in order to interconnect computer-based information to telephone callers. See also INTERACTIVE VOICE RESPONSE.

Voice Recognition

The ability of a machine (obviously a computer) to understand human speech.

Voice Response Unit

VRU. Another term for INTERACTIVE VOICE RESPONSE. Think of a Voice Response Unit as a voice computer. Where a computer has a keyboard for entering information, an IVR uses remote touch-tone telephones.

Where a computer has a screen for showing the results, an IVR uses a digitized synthesized voice to "read" the screen to the distant caller. An IVR can do whatever a computer can, from looking up train timetables to moving calls around an automatic call distributor (ACD).

The only limitation on an IVR is that you can't present as many alternatives on a phone as you can on a screen. The caller's brain simply won't remember more than a few. With IVR, you have to present the menus in smaller chunks.

Voice Store And Forward

Voice mail. A PBX service that allows voice messages to be stored digitally in secondary storage and retrieved remotely by dialing access and identification codes. See VOICE MAIL SYSTEM.

Voice Switched

A device which responds to voice. When the device hears a voice, it turns on and transmits it, while muting the receive side. The most common voice-switched device is the common desk speakerphone.

With voice switching, it's easy to hog a circuit. Just keep making a noise. Watch out for voice hogging. If you're calling someone and waiting for them by listening in on your speakerphone, mute your speakerphone. This way you'll hear them when they answer.

Voice Text

High-level interface portion of the Microsoft Speech API for text-to-speech. Using C++, applications call "Register" so the voice text knows the name of the application and to what audio device the speech will be played, since some applications will be telephone aware. Also, an application can provide a notification sink so that it's alerted when speaking starts or stops, but this isn't necessary. The next step is to send out text to be spoken. Finally, when the application is finished using voice text it releases the object. There are two objects involved in doing text-to-speech: Voice-Text Object.

This is the object that handles speech and the only one that the application deals with. Notification Sink. The notification sink code is optional and is supplied by the application. The voice-text object calls methods in this object when audio starts or stops playing, or with mouth-animation cues.

VRU

See Voice Response Unit.

VS&F

Voice Store and Forward. Voice is digitally encoded, sent to large storage devices and later forwarded to the recipient. See VOICE MAIL.

Wabi

Sun software for running Microsoft Windows applications on Solaris. Runs on Intel and SPARC.

WabiServer

Allows multiple and simultaneous users to run Wabi on Intel and SPARC.

WAN

Wide Area Network. A network using common carrier-provided lines that cover an extended geographical area. WANs are data networks typically extending a LAN outside the building, over telephone common carrier lines to link to other LANs in remote buildings or other geographic areas.

WATS

Wide Area Telecommunications Service. WATS is basically discounted toll service. It is provided by all long distance and local phone companies. AT&T started WATS but forgot to trademark the name, so now every supplier uses it as a generic name. There are two types of WATS services -- in and out WATS, i.e. those WATS lines that allow you to dial out and those which allow you to receive incoming calls (the typical 800 line service).

You subscribe to in and out-WATS services separately. In the old days you needed separate in and out lines to handle the in and out WATS services. But these days you can choose to have in and out WATS on the same line. This is not particularly brilliant traffic engineering, since you can't receive an incoming 800 call if you're making an outgoing call.

WAV
Wave Audio, the Win32 function for handling sound.

Wave Wrapper
Microsoft's standard DLL for modems using a separate audio port for voice communication (as opposed to just the serial port for modem and voice). This is used with Unimodem to synchronize audio data with standard modem functions. The Voice application in this case calls multimedia system functions, which in turn calls the Wave Wrapper DLL (wavewrap.drv).

The Wrapper tells Unimodem to send AT+V commands to the modem to place it into voice transfer mode. The Wrapper makes subsequent calls to the multimedia system DLL to play the audio from the device. In this case, the device vendor supplies a modem (hardware) wave device interface and does not communicate directly with the serial port.

Web Browser
Client software used to view information on Web servers. Can display graphics. Netscape is the most popular, accounting for an alleged 75% of all browsers out there.

Web Server
A server dedicated to storing data (such as Web pages in HTML format) and distributing it on the World Wide Web. There were an estimated 6,000 Web servers at the beginning of 1994. Today there are around 200,000

WHOIS
A command on some systems that reveals the actual user's name, based on that person's username.

Win32 API

Win32 Application Programming Interface (API). Microsoft's uniform API for all of Microsoft's 32-bit platforms. Win32 APIs include Win32 functions, messages, and structures. Win32 apps are able to do Win 95 operating system functions in the areas of window management, graphics device interface (GDI), system services, and extension libraries.

The idea is to allow the development of applications that run on all platforms while still using the unique features and capabilities of any given platform. Some Win32 APIs carry out their tasks only on the more powerful platforms. For example, security functions are available only in the Windows NT operating system. Most other differences are system limitations, such as restrictions on the range of values or the number of items a given function can manage.

Win 32 COM API

Microsoft's tool for the creation of Win 95 communications applications. The API is used to send ASCII or binary data under Telephony application control. It works alongside TAPI and the Multimedia Wave API and sits on top of Unimodem. The COM API does not handle call control, but rather streaming of data.

Windows Open Services Architecture (WOSA)

WOSA is a standard set of interfaces to connect a variety of applications with back-end devices and services, such as messaging, telephony, databases, security, etc. MAPI, TAPI and SAPI are a part of WOSA. Microsoft's scheme is to hide programming complexities from both users and application developers. With WOSA, Microsoft hopes to provide seamless access across a variety of systems. To do this, WOSA defines a common set of APIs. These allow developers to write applications and back-end services for distributed computing environments.

Windows Telephony API

TAPI. Part of the Windows Open System Architecture, TAPI lets developers create telephony applications.

TAPI is a group of Win 32 APIs for managing phone calls. TAPI is not used for sending binary data. Transmission of data under Windows is done with the Win 32 COM API or the Multimedia WAVE API. TAPI provides a standard way for communications apps to control telephony functions for data, fax and voice calls. It manages all signaling between a computer and phone network, including basic functions such as dialing, answering and hanging up a call. It also includes supplementary call-handling things such as hold, transfer, conference and call park that are often found in PBXs, Centrex, ISDN and other phone systems.

In general, TAPI services arbitrate requests from apps to share comm ports and devices. Win32-based apps can use TAPI to make outgoing calls while others are waiting for incoming calls. TAPI is tightly integrated with Windows NT and Windows 95. TAPI 2.0 is now included in the Win NT Server 4.0 and Win NT Workstation 4.0. TAPI is an open industry standard. It is switch fabric independent so TAPI-compatible applications can be run on a wide variety of PC and telephony hardware and support a variety of network services. Developer can telephone-enable virtually any general purpose application with TAPI Assisted Telephony. TAPI 2.0 supports Unicode, so it is easier to make apps work globally. TAPI also integrates more than just call control, with complementary support from other WOSA elements. TAPI can be used to run several telephony applications simultaneously on a client or server PC. The combination of TAPI, the Windows platform, additional communications-related Windows APIs, and ActiveX Controls provides the ideal platform for developing and using telephony applications.

Wink

A momentary interruption in SF (single frequency) tone indicating the distant central office is ready to receive the digits you just dialed.

Wink Operation

A timed, off-hook signal, normally of 140 ms, which indicates the availability of an incoming register for receiving digital information from the calling office.

Wink Release

On most modern central offices when the person or device at the other end hangs up, your local central office will send you a single frequency tone. That tone is called wink release. Such a tone can be used to alert a data device that the device at the other end has hung up. (Remember it can't tell by just listening -- like you and me.) When a data device hears a wink release, it usually takes it as a signal to hang up also.

Wink Start

Short duration off-hook signal. See Wink Operation.

WinSock

Windows Sockets or WinSock is API for Microsoft Windows and TCP/IP. It's how WWW browsers, FTP clients, email packages, and word processors talk to the Internet. WinSock was born at Interop in 1991. Led by Martin Hall, the group defined a common way for MS Windows applications to talk to TCP/IP.

Thousands of applications have been written to WinSock. WinSock has played a instrumental role in the growth and use of the Internet, groupware, video-conferencing, email, and other communications technologies for MS Windows. The WinSock specifications are produced by the WinSock Group - a loose coalition of vendors made up of developers and technologists from around the world who work together to define, refine and extend this popular interface definition.

WOSA

See Windows Open Services Architecture.

WWW

World Wide Web. A vast, worldwide hypermedia system consisting of a network of networks that allows you to browse through information on hundreds of thousands of databases. WWW is the multimedia-enabled layer of the Internet.

X.5

ITU recommendation on digital data communication networks and services, dealing with interfaces.

X.25

Possibly one of the most important international standards ever recommended by the ITU. From the beginning, it has provided a common reference point by which mainframe computers, word processors, minicomputers, VDUs, microcomputers, and varied equipment from different manufacturers can operate together over a packet switched network.

X.25 defines the interface between a public data network and a packet-mode user device. Typically used to connect WANs, packet switching breaks network data into smaller packets and sends them from point to point through interconnected switches. X stands for packet switched network.

X.38

ITU recommendation for the access of Group 3 facsimile equipment to the Facsimile Packet Assembly/Disassembly (FPAD) facility in public data networks situated in the same country.

X.39

ITU recommendation for the exchange of control information and user data between a Facsimile Packet Assembly/Disassembly (FAPD) facility and a packet mode data terminal equipment (DTE) or another pad, for international internetworking.

X.400

ITU recommendations for the transmission of electronic text and graphic mail between unlike computers, terminals, and computer networks. Briefly, X.400 describes how mail messages are encoded, and X.25 sets up how they are transmitted.

X.500

The ITU standard defining directory services, most commonly for E-mail systems. It provides for common naming and addressing in the networks of large-scale enterprises.

xBASE

The collective term for all database management systems using dBASE-compatible (.DBF format) data files, with a variety of enhancements and differing index formats (e.g., dBASE III+, dBASE IV, FoxBASE, FoxPro, Clipper).

XENIX

Microsoft trade name for a 16-bit microcomputer operating system derived from AT&T Bell Labs' UNIX.

XFN. X/Open Federated Naming

Used in Sun's Enterprise Server. Provides enterprise directory name Service for simplified access to and federation among multiple naming services. Works with Distributed Computing Environment (DCE), ONC, and Internet Domain Name Service (DNS).

XFN/NIS+

Part of Sun's ENOS Networking Solutions. It's a secure repository of network information.

XGL

Part of Sun's Imaging and Graphics Solutions. Solaris' graphics library for 2-D/3-D applications development.

XIL

Part of Sun's Imaging and Graphics Solutions is an open Imaging API used for image enhancement, scaling, compression, color conversion, and display.

XtlCall Object

Part of the SunXTL Teleservices architecture. C++ objects created by developers with the SunXTL API. Each XtlCall object corresponds to a telephone call. An XtlCall object has command methods to query the current state of a call or request a change in state (to put a call on hold). An XtlCall object also has callback methods for the asynchronous notification of state changes.

Xtltool

A GUI tool supplied to browse and edit the provider configuration database in a SunXTL environment.

Yellow Alarm

A T-1 alarm signal sent back toward the source of a failed transmit circuit in a DS-1 2-way transmission path. A yellow sends 0's (zeros) in bit two of all time slots. See also T-1.

Yellow Pages

A directory of telephone numbers classified by type of business. It was printed on yellow paper throughout most of the twentieth century until it was obsoleted in the late 1990s by dial-up yellow page directories operated by voice processing systems and in the early 21st century by electronic directories delivered on disposable laser disks. As a concession to history, the laser disks are now painted bright yellow. Actually, yellow pages remain one of the phone companies' most lucrative sources of revenues. Advertising rates are not cheap. There is now competition. There are many "Yellow Pages" directories, since AT&T never trademarked the term "Yellow Pages." Some "yellow page" directories are better value than others.

And some are more legitimate than others. Some actually never get printed or, if they are printed, are not printed in great quantity and are not distributed as widely as their sales literature implies.

Many businesses have been suckered into paying money for listings and advertisements in directories that never appeared. This fictitious directory scam also has happened with "telex" and "fax" directories. This "scam" is fraud by mail and is heavily stomped upon by the US Postal Service. As a result, many fake directories (especially the telex ones) are "published" abroad.

Z

Zero Code Suppression

The insertion of a "one" bit to prevent the transmission of eight or more consecutive "zero" bits. Used primarily with digital T1 and related telephone-company facilities which require a minimum "ones density" keep the individual subchannels of a multiplexed, high-speed facility active. Several different schemes are currently employed to accomplish this. Proposals for a standard are being evaluated by the ITU. See also ZERO SUPPRESSION.

ZeroFill

A traditional fax device is mechanical. It must reset its printer and advance the page as it prints each scan line it receives. If the receiving machine's printing capability is slower than the transmitting machine's data sending capability, the transmitting machine adds fill bits (also called zero fill) to pad out the span of send time, giving the slower remote machine the additional time it needs to reset prior to receiving the next scan line.

Zip Tone

Short burst of dial tone to an ACD agent headset indicating a call is being connected to the agent console.

Appendix C

Killer App Makers and Voice Processing Experts

The companies and listed below were invaluable resources in putting this book together. Please call them and share your ideas. These are the brightest minds in the galaxy.

Active Voice Corporation
2901 Third Avenue
Seattle, WA 98121
206-441-4700
206-441-4784 (fax)
www.activevoice.com

Aculab Ltd.
Aculab House
Old Road
Leighton, Buzzard
Bedfordshire LU7 7RG, UK
44-1525-371393
44-1525-381284 (fax)
www.aculab.com

AdVoTec, Ltd.
615 North Oakhurst Court
Huron, OH 44839
419-433-7566
419-433-8418 (FAX)
www.advotec.com

Alcom Corporation
1616 N. Shoreline Blvd.
Mountain View, CA 94043
415-694-7000
415-694-7070 (fax)
www.alcom.com

Alterna Technologies Group, Inc.
5920 Macleod Tr. S., # 700
Calgary, Alberta T2H 0K2
403-253-5531
403-253-5580 (fax)
www.alterna.com

AltiGen Communications
47801 Fremont Blvd.
Fremont, CA 94538
510-252-9712
510-252-9738 (fax)
www.altigen.com

Amcom Software, Inc.
5555 W 78th St.
Edina, MN 55439
612-829-7445
800-852-8935 (fax)

AnswerSoft, Inc.
3460 Lotus Dr. Suite 123
Plano, TX 75075
214-612-5100
214-612-5198 (fax)
www.answersoft.com

Apex Voice Communications, Inc.
15250 Ventura Blvd. 3rd Floor
Sherman Oaks, CA 91403
818 379 8400
818 379 8410 (fax)
www.apexvoice.com

Applied Language Technologies
215 First Street
Cambridge, MA 02142
617-225-0012
617-225-0322 (fax)
www.altech.com

Applied Voice Technology, Inc.
11410 NE 122nd Way Box 97025
Kirkland, WA 98083
206-820-6000
206-820-4040 (fax)
www.appliedvoice.com

Artisoft, Inc.
Stylus Product Group
Five Cambridge Center
Cambridge, MA 02142
617-354-0600
617-354-7744 (fax)
www.artisoft.com/visualvoice

Aurora Systems, Inc.
33 Nagog Park Suite 200
Acton, MA 01720
508-263-4141
508-635-9756 (Fax)

Bosch/TTSI
Technical Telephone Systems, Inc.
18 Worlds Fair Drive
Somerset, NJ 08873
908-563-6600
908-356-1366 (fax)

Boston Technology
100 Quannapowitt Parkway
Wakefield, MA 01880
617-246-9000
617-246-4510 (fax)
www.bostontechnology.com

CallStream Communications, Inc.
871-9 Equestrian Court
Oakville, Ontario L6H6L7
905-847-5362
905-847-3421 (fax)
www.callstream.com

CallWare Technologies, Inc.
2323 Foothill Dr.
Salt Lake City, UT 84109
801-486-9922
801-486-8294 (fax)
www.callware.com

Cardiff Software, Inc.
6351 Yarrow Dr #E
Carlsbad, CA 92009
619-931-4500
619-931-4550 (fax)

CCOM Information Systems
120 Wood Avenue South
Iselin, NJ 08830
908-603-7750
908-603-7751 (fax)

Centigram Communications Corporation
91 E Tasman Dr.
San Jose, CA 95134
408-428-3837
www.centigram.com

Comdial Telecommunications Corporation
1180 Seminole Trail
P.O. Box 7266
Charlottesville, VA 22906-7266
804-978-2200
804-978-2293 (fax)
www.comdial.com

Commetrex Corporation
One Meca WAy
Norcross, GA 30093
770-564-5522
770-564-5649 (fax)

Compro Technologies, Inc.
848B West Bay Ave., Suite 2
Barnegat, NJ 08005-2126
609-698-2700
609-698-0858 (fax)
www.comprotech.com

CompuSolve Data Systems, Inc.
421-13 Route 59 Suite 309
Monsey, NY 10952
914-426-6410
914-356-9013 (fax)
www.add-ons.com

Computer Talk Technology
Richmond Hill, Ontario
905-882-5000
www.computertalk.com

Concepts In Communications
336 Fourth Ave. Times Bldg.
Pittsburgh, PA 15222
412-391-7862
412-339-1424

Copia International, Ltd.
1342 Avalon Court
Wheaton, IL 60187
708-682-8898
www.copia.com

Creative Advanced Technologies, Inc.
500 Throckmorton
Fort Worth, TX 76102
817-332-5899
817-332-2156 (fax)

DAC Systems
60 Todd Road
Shelton, CT 06484
203-924-7000
203-944-1618 (fax)
nickson@ibm.net

DATATAL AB
Jarnvagsgatan 4
Klintehamn, S-620 20
Sweden
46-498-240-405
46-498-240-051 (fax)
www.datatal.se

Dialogic Corporation
1515 Route 10
Parsippany, NJ 07054
201-993-3000
201-631-9631 (fax)
www.dialogic.com

Dictaphone Corporation
3191 Broadbridge Ave.
Stratford, CT 06497-2559
203-381-7000
203-381-7102 (fax)
www.dictaphone.com

Dragon Systems, Inc.
320 Nevada Street
Newton, MA 02160
617-965-5200
617-572-0372 (fax)
www.dragonsys.com

Edify Corporation
2840 San Tomas Expressway
Santa Clara, CA 95051
408-982-2000
408-982-0777 (fax)
www.edify.com

Enhanced Systems, Inc.
6961 Peachtree Industrial Blvd.
Norcross, GA 30092
770-662-1504
770 242 1630 (fax)
www.esisys.com

Enterprise Integration Group, Inc.
1320 El Capitan Dr. Suite 202
Danville, CA 94526
510-328-1300
510-328-1313 (fax)

Excel, Inc.
255 Independence Dr.
Hyannis, MA 02601
508-862-3000
508-862-3030 (fax)
www.xl.com

Executone
478 Wheelers Farms Road
Milford, CT 06460
203-876-7600
203-877-6551 (fax)
www.executone.com

Expert Systems, Inc.
1301 Hightower Trail, # 201
Atlanta, GA 30350
770-642-7575
770-587-5547 (fax)
www.easey.com

Frank Solutions, Inc.
9250 East Costilla Ave. Suite 100
Englewood, CO 80112
303-792-5500
303-799-3621 (fax)

G&H Consulting
603 Shenipsit Lake Road
Tolland, CT 06084
860-870-7517

Genesys
Genesys Telecommunications Laboratories, Inc.
1155 Market St. 11th Floor
San Fransicso, CA 94103
415-437-1100
415-437-1260 (fax)
www.genesyslab.com

GlobeStar Systems
1815 Ironstone Manor #7
Pickering, Ontario
Canada L1W 3W9
905-839-0893

Hammer Technologies, Inc.
A Teradyne Company
226 Lowell St.
Wilmington, MA 01887
508-694-9959
508-988-0148 (fax)
www.teradyne.com/hammer/

Ibex Technologies, Inc.
550 Main Street # G
Placerville, CA 95667
916-939-8888
916-621-2004 (fax)
www.ibex.com

Information Products
17 Connecticut South Drive
P.O. Box 558
East Granby, CT 06026-0558
860-653-7822
860-653-0255 (fax)

Info Systems
1110 Finch Avenue West.
Downsview, Ontario M3J 2T2
800-825-5434
416-665-7638
www.talkie.com

InnoMediaLogic, Inc. (IML)
3653 Chambly Road, Longueuil
Quebec, Canada J4L1N9
514-651-8188
514-651-8833 (fax)
www.iml-cti.com

Inter-Tel, Inc.
7300 West Boston Avenue
Chandler, AZ 85226
602-961-9000
602-961-9224 (fax)
www.inter-tel.com

Interactive Communication Systems, Inc. (ICS)
611 North Weber St. Suite 102
Colorado Springs, CO 80903
719-444-0554
719-444-0150 (fax)
www.icstelephony.com

Interactive Intelligence, Inc.
3500 DePauw Blvd., Suite 1060
Indianapolis, IN 46268
317-872-3000
317-637-9071 (fax)
www.inter-intelli.com

Interface Alternative, Inc.
Village Plaza
Interface Alternative
1075 Easton Ave Twr 3 Suite #1
Somerset, NJ 08873-4866
908-545-2600
908-545-7705 (fax)
www.iface.com

Invotec
1480 Terrell Mill Road, Suite 813
Marietta, GA 30067
770-933-9288
770-951-2742 (fax)

iNTELiTRAK Technologies, Inc.
816 Congress Ave. Suite 1100
Austin, TX 78701
512-480-2211
512-473-3680 (fax)
www.intelitrak.com

Intellivoice Communications, Inc.
3490 Piedmont Rd Suite 1025
Atlanta, GA 30305
810-488-0180
810-488-2101 (fax)
www.intellivoice.com

Key Voice Technologies, Inc.
1919 Ivanhoe St.
Sarasota, FL 34231
941-922-3800
941-925-7278 (fax)
www.keyvoice.com

Kiman Corp. (aka Worldly Voices)
1805 Hayes St.
Nashville, TN 37203
615-321-5155
615-321-8804 (fax)
kimancorp@aol.com

Latitude Communications
2121 Tasman Drive
Santa Clara, CA 95054
408-988-7200
408-988-6520 (fax)
www.latitude.com

Lotus Development Corporation
One Rogers Street 6 North
Cambridge, MA 02142-1245
617-693-5700
617-693-5562 (fax)
www.lotus.com

MacroVoice Corporation
6001 Park of Commerce Blvd
Boca Raton, FL 33487
800-662-7689
www.macrovoice.com

Madge Networks
2310 North First St.
San Jose, CA 95131
800-876-2343
408-952-0790 (fax)
www.madge.com

Mastermind Technologies
1005 N Glebe Rd.
Arlington, VA 22201
703-276-9300

MediaSoft Telecom
8600 Decarie Blvd #215
Mount Royal, Quebec H4P2N2
514-731-3838
514-731-3833 (fax)
www.medaisoft.ca

Melita International
6630 Bay Circle
Norcross, GA 30071
770-446-7800
770-409-4445 (fax)
www.melita.com

Microware, Inc.
5313 Arctic Blvd. #201
Anchorage, AK 99518
907-562-7705
907-562-5497 (fax)

Mitel, Inc.
350 Legget Drive
Kanata, ON
Canada K2K 1X3
613-592-2122
613-592-4784 (fax)
www.mitel.com

Moscom Corporation
3750 Monroe Avenue
Pittsford, NY 14534
716-381-6000
716-383-6800 (fax)
www.moscom.com

NetPhone, Inc.
313 Boston Post Road West
Marlboro, MA 01752
508-787-1000
508-787-1030 (fax)
www.netphone.com

NewVoice, Inc.
1893 Preston White Dr., #120
Reston, VA 20191-5449
703-648-0585
703-648-9430 (fax)
www.newvoice.com

NexPath
4701 Patrick Henry Dr. Suite 701
Santa Clara, CA 95054
408-235-8916
408-748-0664 (fax)
www.nexpath.com

Nice Systems, Inc.
150 Broadway, Suite 1200
New York, NY 10038
212-267-3545
212-267-3669 (fax)
www.nice.com

Nortel (Northern Telecom)
2221 Lakeside Blvd.
Richardson, TX 75083
972-684-1000
800-598-6726 (fax)
www.nortel.com

Nuance Communications
333 Ravenswood Ave. Bldg. 110
Menlo Park, CA 94025
415-462-8200
415-462-8201 (fax)
www.nuancecom.com

Nuntius Corporation
8048 Big Bend Blvd. Suite 110
St. Louis, MO 93119
314-968-1009
314-968-3163 (fax)
www.nuntius.com

Octel Communications Corporation
1001 Murphy Ranch Road
Milpitas, CA 95035
408-321-2000
408-321-2100 (fax)
www.octel.com

Omtool
8 Industrial Way
Salem, NH 03079
603-898-8900
603-890-6756 (fax)
www.omtool.com

OPTUS Software
100 Davidson Ave.
Somerset, NJ 08873-9931
908-271-9568
908-271-9572 (fax)
www.facsys.com

Paceart Associates, L.P.
22 Riverview Drive
Wayne, NJ 07470
201-696-1122
201-696-3357 (fax)
www.paceart.com/paceart

Pacific Telephony Design
140 Gardenside Drive, Suite 101
San Francisco, CA 94131
415-647-9911
415-647-9912 (fax)
www.phonezone.com

PAGE TeleCOMPUTING, Inc.
1561 S. Congress Avenue
Delray Beach, FL 33445
407-279-0340
407-279-0187 (fax)
www.page-tel.com

Parity Software Development Corp.
One Harbor Drive
Sausalito, CA 94965
415-332-5656
415-332-5657 (fax)
www.paritysw.com

PC Voice
Paoli Technology Center
Paoli, PA 19301
610-251-9401
610-251-9402 (fax)

Periphonics Corporation
4000 Veterans Memorial Hwy
Bohemia, NY 11716
516-467-0500
516-981-2689 (fax)
www.peri.com

Philips Speech Processing
(Philips Dialogue Systems)
52072 Aachen Kackert Str 10
Germany
011-49-241-8871-251
011-49-241-8871-140 (fax)
www.philips.com

PhoNet Communications, Ltd.
60 Medinat Hayehudim
P.O. Box 12739
Herzlia 46725 ISRAEL
972-9-9502134
972-9-9567432 (fax)

PIKA Technologies, Inc.
155 Terrence Matthews Crescent
Kanata, ON
Canada K2M 2A8
613-591-1555
613-591-1488 (fax)
www.pika.ca

Precision Systems, Inc.
11800 30th Court N.
St. Petersburg, FL 33716
813-572-9300
813-573-9193 (fax)
www.talkpsi.co

Priority Call Management
110 Fordham Road
Wilmington, MA 01887
508-658-4400
508-658-4400 (fax)
www.prioritycall.com

Prodigy Concepts, Inc.
1165 Hopkins Street
Berkeley, CA 94702
510-559-9100
510-559-9120 (fax)
www.prodigyconcepts.com

Pronexus
112 John Cavanaugh Rd. RR2
Carp, Ontario
Canada CD KOA 1LO
613-839-0033
613-839-0035 (fax)
www.pronexus.com

Prospect Software, Inc.
1737 N. First Street, Suite 240
San Jose, CA 95112
408-451-2451
408-451-2454 (fax)
www.prospectsoftware.com

PureSpeech, Inc.
100 Cambridge Park Dr.
Cambridge, MA 02140
617-441-0000
617-441-0001 (fax)
www.speech.com

Rockwell International
1431 Opus Place
Downers Grove, IL 60515
708-960-8000
708-960-8165 (fax)
www.switch.rockwell.com

Scopus Technologies
1900 Powell Street, 7th Floor
Emeryville, CA 94608
510-597-5800
510-597-5964 (fax)
www.scopus.com

Sphere Communications, Inc.
Two Energy Drive
Lake Bluff, IL 40044
847-247-8200
847-247-8290 (fax)
www.spherecom.com

Summa Four, Inc.
25 Sundial Ave.
Manchester, NH 03103-7251
603-625-4050
603-641-6238 (fax)
www.summa4.com

Swiss Railways (SBB)
Schweizerische Bundesbahnen
Direktion Informatik
Bollwerk 10
CH-3030 Bern
Switzerland
41-512-20-2882
41-512-20-4187 (fax)

Stellar Communications, LLC
26500 Agoura Rd. Suite 569
Calabasas, CA 91302
800-711-8963
818-889-4771 (fax)
Stellcom@aol.com

Syntellect Inc.
1000 Holcomb Woods Parkway
Building 410A
Roswell, GA 30076-2585
770-587-0700
770-587-0589 (Fax)
www.syntellect.com

Syscom Services, Inc.
1010 Wayne Avenue
Silver Spring, MD 20910
301-650-9131
301-650-9139 (fax)

T4 Systems, Inc.
3 Innwood Circle
Little Rock, AK 72211
501-227-6637
501-227-6245 (fax)
Frank Carollo, VP Sales and Marketing
www.t4.com

TALX Corporation
1850 Borman Court
St. Louis, MO 63146
314-434-0046
314-434-9205 (fax)
www.talx.com

Technically Speaking
a BrookTrout Company
333 Turnpike Rd.
Southborogh, MA 01772
508-229-7777
508-229-8777 (fax)
www.brooktrout.com

Technology Solutions Company
205 North Mihigan Ave., Suite 1500
Chicago, IL 60601
312-228-4500
312-228-4501 (fax)
www.techsol.com

Tern Systems
27 Grasshopper Lane
Acton, MA 01720
508-266-1966
800-787-4913 (fax)

Treasure Coast Engineering
6615 SouthWest 13th St.
Gainesville, FL 32608
904-335-7670
352-335-0294 (fax)
www.tceinc.com

TTM & Associates, Inc.
271 E. Imperial Highway, Suite 641
Fullerton, CA 92635
714-449-8700
714-449-8705 (fax)
www.ttminc.com

Universal Interactive Systems (UIS)
9444 Haines Canyon Ave.
Tujunga, CA 91042
818-352-7800
818-352-5445 (fax)
www.unisystems.com

Venturian Software
1300 Second Street South
Hopkins, MN 55343
612-931-2432
612-931-2459 (fax)
www.venturian.com

Video Voice & Data, Inc.
292 5th Ave. Room 201
New York, NY 10001
212-714-3531
212-714-3510 (fax)

Voice Processing Plus, Inc.
5745 West Maple Ave Ste 218
West Bloomfield, MI 48322
810-737-9550
810-737-9558 (fax)
www.vpplus.com

Voysys Corporation
48634 Milmont Drive
Fremont, CA 94538
510-252-1100
510-252-1101 (fax)
www.voysys.com

Wygant Scientific, Inc.
813 S W Alder
Portland, OR 97205
800-688-6423
503-242-9700
www.wygant.com

ZEUS Phonstuff
3107-D Medlock Bridge Road
Norcross, GA 30071-1423
770-263-7111
770-263-0049 (fax)
www.zeusphonstuff.com

Index

A

B

C

D

E

F

G

H

I

J

L

M

N

O

P

Q

R

S

T

U

V

W